Perspectives, Science an
for Novel Silicon on Insula

NATO Science Series

A Series presenting the results of activities sponsored by the NATO Science Committee. The Series is published by IOS Press and Kluwer Academic Publishers, in conjunction with the NATO Scientific Affairs Division.

A. Life Sciences	IOS Press
B. Physics	Kluwer Academic Publishers
C. Mathematical and Physical Sciences	Kluwer Academic Publishers
D. Behavioural and Social Sciences	Kluwer Academic Publishers
E. Applied Sciences	Kluwer Academic Publishers
F. Computer and Systems Sciences	IOS Press
1. Disarmament Technologies	Kluwer Academic Publishers
2. Environmental Security	Kluwer Academic Publishers
3. High Technology	Kluwer Academic Publishers
4. Science and Technology Policy	IOS Press
5. Computer Networking	IOS Press

NATO-PCO-DATA BASE

The NATO Science Series continues the series of books published formerly in the NATO ASI Series. An electronic index to the NATO ASI Series provides full bibliographical references (with keywords and/or abstracts) to more than 50000 contributions from internatonal scientists published in all sections of the NATO ASI Series.
Access to the NATO-PCO-DATA BASE is possible via CD-ROM "NATO-PCO-DATA BASE" with user-friendly retrieval software in English, French and German (WTV GmbH and DATAWARE Technologies Inc. 1989).

The CD-ROM of the NATO ASI Series can be ordered from: PCO, Overijse, Belgium

Series 3. High Technology – Vol. 73

Perspectives, Science and Technologies for Novel Silicon on Insulator Devices

edited by

Peter L. F. Hemment
University of Surrey,
Guildford, Surrey,
United Kingdom

V. S. Lysenko
Institute of Semiconductor Physics,
Kyiv, Ukraine

and

A. N. Nazarov
Institute of Semiconductor Physics,
Kyiv, Ukraine

Kluwer Academic Publishers

Dordrecht / Boston / London

Published in cooperation with NATO Scientific Affairs Division

Proceedings of the NATO Advanced Research Workshop on
Perspectives, Science and Technologies for Novel Silicon on Insulator Devices
Kyiv, Ukraine
12-15 October 1998

A C.I.P. Catalogue record for this book is available from the Library of Congress.

ISBN 0-7923-6116-4 (HB)
ISBN 0-7923-6117-2 (PB)

Published by Kluwer Academic Publishers,
P.O. Box 17, 3300 AA Dordrecht, The Netherlands.

Sold and distributed in North, Central and South America
by Kluwer Academic Publishers,
101 Philip Drive, Norwell, MA 02061, U.S.A.

In all other countries, sold and distributed
by Kluwer Academic Publishers,
P.O. Box 322, 3300 AH Dordrecht, The Netherlands.

TABLE OF CONTENTS

Preface   xi
Committee Members   xiii
Invited Speakers   xv
Workshop Photographs   xvii

<u>Paper No</u>   <u>Title and Authors</u>   <u>Page No</u>

SECTION 1: INNOVATIONS IN MATERIALS TECHNOLOGIES

1.1 *Invited*   "SMART-CUT® Technology: Basic Mechanisms and   1
Applications", M Bruel
CEA/LETI-Departement de Microtechnologies,
17 rue des Martyrs, 38054 Grenoble Cedex, France

1.2 *Invited*   "Polish Stop Technology for Silicon on Silicide on   17
Insulator Structures", H S Gamble
School of Electrical and Electronic Engineering,
The Queen's University of Belfast, Belfast,
BT7 1NN, UK

1.3 *Invited*   "Homoepitaxy on Porous Silicon with a Buried Oxide   29
Layer: Full-Wafer Scale SOI", S I Romanov[a,b],
A V Dvurechenskii[a,c], V V Kirienko[a], R Grotzschel[d],
A Gutakovskii[a], L V Sokolov[a], M A Lamin[a]
[a]Institute of Semiconductor Physics, Pr Lavrent'ev 13,
630090 Novosibirsk, Russia
[b]Tomsk State University, Sq Revolution 1,
634050 Tomsk, Russia
[c]Novosibirsk State University, Pirogova 2,
630090 Novosibirsk, Russia
[d]Research Center Rossendorf Incorporated,
PO Box 510119, D-01314 Dresden, Germany

1.4   "Structural and Electrical Properties of Silicon on Isolator   47
Structures Manufactured on FZ- and CZ-Silicon by
SMART-CUT Technology", V P Popov, I V Antonova,
V F Stas, L V Mironova, E P Neustroev, A Gutakovskii,
A A Franzusov, G N Feofanov
Institute of Semiconductor Physics, RAS 630090,
Lavrentieva 13, Novosibirsk, Russia

Paper No	Title and Authors	Page No
1.5	"Development of Linear Sequential Lateral Solidification Technique to Fabricate Quasi-Single-Crystal Super-Thin Si Films for High-Performance Thin Film Transistor Devices", A B Limanov[a], V M Borisov[b], A Yu Vinokhodov[b], A I Demin[b], A I El'Tsov[b], Yu B Kirukhin[b], O B Khristoforov[b] [a]Institute of Crystallography, Russian Academy of Sciences, Leninski Prospect 59, 117333 Moscow, Russia [b]TRINITI, Troitsk, Moscow Region, Russia	55

SECTION 2: ECONOMICS AND INNOVATION APPLICATIONS

Paper No	Title and Authors	Page No
2.1 *Invited*	"Low Temperature Polysilicon Technology: A Low Cost SOI Technology?", F Plais, C Collet, O Huet, P Legagneux, D Pribat, C Reita, C Walaine Thomson CSF, Central Research Laboratories, Domaine de Corbeville, 91404 Orsay Cedex, France	63
2.2 *Invited*	"A Novel Low Cost Process for the Production of Semiconductor Polycrystalline Silicon from Recycled Industrial Waste", B N Mukashev, M F Tamendarov, B Bekelov, S Zh Tokmoldin Institute of Physics and Technology, Almaty 480082, Kazakstan	75
2.3 *Invited*	"Tetrahedrally Bonded Amorphous Carbon for Electronic Applications", W I Milne Department of Engineering, University of Cambridge, Trumpington Street, Cambridge, CB2 1PZ, UK	85
2.4 *Invited*	"Diamond Based Silicon-on-Insulator Materials and Devices", S Bengtsson, M Bergh Solid State Electronics Laboratory, Department of Microelectronics ED, Chalmers University of Technology, S-41296 Goteborg, Sweden	97
2.5 *Invited*	"Low-Temperature Processing of Crystalline Si Films on Glass for Electronic Applications", R B Bergmann, T J Rinke, L Oberbeck, R Dassow Institut fur Physikalische Elektronik, Universitat Stuttgart, Pfaffenwaldring 47, D-70569 Stuttgart, Germany	109

Paper No	Title and Authors	Page No

2.6   "β-SiC on SiO_2 Formed by Ion Implantation and Bonding   121
for Micromechanics Applications", C Serre[a],
A Perez-Rodriguez[a], A Romano-Rodriguez[a], J R Morante[a],
L Fonseca[b], M C Acero[b], J Esteve[b], R Kogler[c], W Skorupa[c]
[a]EME, Department d'Electronica, Unitat Associadada
CNM-CSIC, Universitat de Barcelona, Avda Diagonal
6450647, 08028 Barcelona, Spain
[b]Centre Nacional de Microelectronica CNM-CSIC,
Campus UAB, 08193 Bellaterra, Spain
[c]Forschungzentrum Rossendorf, PF 510119,
D-01314 Dresden, Germany

2.7   "Laser Recrystallised Polysilicon Layers for Sensor   127
Applications: Electrical and Piezoelectric Characterisation",
A A Druzhinin, I I Maryyamova, E M Lavitska,
Y M Pankov, I T Kogut
Lviv Polytechnic State University, Kotlarevsky Street 1,
Lviv 290013, Ukraine

SECTION 3: CHARACTERISATION METHODS FOR SOI

3.1 *Invited*   "Optical Spectroscopy of SOI Materials",   137
A Perez-Rodriguez, C Serre, J R Morante
EME, Department d'Electronica, Unitat Associadada
CNM-CSIC, Universitat de Barcelona, Avda Diagonal
6450647, 08028 Barcelona, Spain

3.2 *Invited*   "Computer Simulation of Oxygen Redistribution in SOI   149
Structures", V G Litovchenko, A A Efremov
Institute of Semiconductor Physics, 45 Prospect Nauki,
252028 Kiev-28, Ukraine

3.3 *Invited*   "Electrical Instabilities in Silicon-on-Insulator Structures   163
and Devices During Voltage and Temperature Stressing",
A N Nazarov, I P Barchuk, V I Kilchytska
Institute of Semiconductor Physics, Prospect Nauki 45,
252028 Kiev-28, Ukraine

3.4   "Hydrogen as a Diagnostic Tool in Analysing SOI   179
Structures", A Boutry-Forveille[a], A Nazarov[b], D Ballutaud[a]
[a]LPSB-CNRS, 1 Place Aristide Briand, 92195 Meudon
Cedex, France
[b]Institute of Semiconductors, Academy of Sciences,
Prospekt Nauki 45, 252650 Kiev, Ukraine

Paper No	Title and Authors	Page No
3.5	"Back Gate Voltage Influence on the LDD SOI NMOSFET Series Resistance Extraction from 150 to 300 K", A S Nicolett[a], J A Martino[b], E Simoen[c], C Claeys[c] [a]Faculdade de Tecnologia de Sao Paulo, Brazil [b]Laboratorio de Sistemas Integraveis, Universidade de Sao Paulo, Brazil [c]IMEC, Leuven, Belgium	187
3.6	"Characterisation of Porous Silicon Layers Containing a Buried Oxide Layer", S I Romanov[a,b], A V Dvurechenskii[a,c], Yu I Yakovlev[a], R Grotzschel[d], U Kreissig[d], V V Kirienko[a], V I Obodnikov[a], A Gutakovskii[a] [a]Institute of Semiconductor Physics, Pr Lavrent'ev 13, 630090 Novosibirsk, Russia [b]Tomsk State University, Sq Revolution 1, 634050, Tomsk, Russia [c]Novosibirsk State University, Pirogova 2, 630090 Novosibirsk, Russia [d]Research Center Rossendorf Incorporated, PO Box 510119, D-01314 Dresden, Germany	195
3.7	"Total-Dose Radiation Response of Multilayer Buried Insulators", A N Rudenko, V S Lysenko, A N Nazarov, I P Barchuk, V I Kilchytska, T E Rudenko, S C Djurenko, Ya N Vovk Institute of Semiconductor Physics, Prospect Nauki 45, 252028 Kyiv-28, Ukraine	205
3.8	"Recombination Current in Fully-Depleted SOI Diodes: Compact Model and Lifetime Extraction", T Ernst[a,d], A Vandooren[b], S Cristoloveanu[a], T E Rudenko[c], J P Colinge[b] [a]LPCS, ENSERG, BP 257, 38016 Grenoble Cedex, France [b]Department of Electrical Engineering, University of California, Davis, CA 95616, USA [c]Institute of Semiconductor Physics, Prospect Nauki 45, 252650 Kiev-28, Ukraine [d]STMicroelectronics, 38920 Crolles, France	213

Paper No	Title and Authors	Page No
3.9	"Investigation of the Structural and Chemical Properties of SOI Materials by Ellipsometry", L A Zabashta[a], I O Zabashta[a], V E Storizhko[a], E G Bortchagovsky[b], B N Romanyuk[b], V P Melnik[b] [a]Institute of Applied Physics, Petropavlovskaja St 58, 244030 Sumy [b]Institute of Semiconductor Physics, Nauka Av 45, 252650 Kiev, Ukraine	217
3.10	"Experimental Investigation and Modelling of Coplanar Transmission Lines on SOI Technologies for RF Applications", J Lescot[a], O Rozeau[a,b,c], J Jomaah[b], J Boussey[b], F Ndagijimana[a] [a]Laboratoire d'Electromagnetisme, MicroOndes et Optoelectronique [b]Laboratoire de Physique des Composants a Semiconducteurs, LEMO-LPCS, BP 257, 38016 Grenoble Cedex 1, France [c]ST Microelectronics, 38920 Crolles, France	225

SECTION 4: PERSPECTIVES FOR SOI STRUCTURES AND DEVICES

Paper No	Title and Authors	Page No
4.1 *Invited*	"Perspectives of Silicon-on-Insulator Technologies for Cryogenic Electronics", C Claeys[a,b], E Simoen[a] [a]IMEC, Kapeldreef 75, B-3001 Leuven, Belgium [b]KU Leuven, ESAT-INSYS, Kard Mercierlaan 94, B-3001 Leuven, Belgium	233
4.2 *Invited*	"SOI CMOS for High-Temperature Applications" J P Colinge Department of Electrical and Computer Engineering, University of California, Davis, CA 95616, USA	249
4.3 *Invited*	"Quantum Effect Devices on SOI Substrates with an Ultrathin Silicon Layer", Y Omura High-Technology Research Center and Faculty of Engineering, Kansai University, 3-3-35 Yamate-cho, Suita, Osaka 564-80, Japan	257
4.4 *Invited*	"Wafer Bonding for Micro-ElectroMechanical Systems (MEMS)", C A Colinge Department of Electrical and Electronic Engineering, California State University, 6000 J Street, Sacramento, CA 95819-6019, USA	269

Paper No	Title and Authors	Page No
4.5	"A Comprehensive Analysis of the High-Temperature Off-State and Subthreshold Characteristics of SOI MOSFETs", T E Rudenko, V S Lysenko, V I Kilchytska, A N Rudenko Institute of Semiconductor Physics, Prospect Nauki 45, 252650 Kiev-28, Ukraine	281
4.6	"Influence of Silicon Film Parameters on C-V Characteristics of Partially Depleted SOI MOSFETs", D Tomaszewski[a], A Jakubowski[b], J Gibki[b], T Debski[a], M Korwin-Pawlowski[c] [a]Institute of Electron Technology, Al Lotnikow 32/46, 02-668 Warsaw, Poland [b]Institute of Microelectronics and Optoelectronics, Warsaw Technical University, ul Koszykowa 75, 00-662 Warsaw, Poland [c]General Semiconductor Ireland, Macroom Co, Cork, Ireland	295
4.7	"Effect of Shallow Oxide Traps on the Low-Temperature Operation of SOI Transistors", V S Lysenko, I P Tyagulski, I N Osiyuk, Y V Gomeniuk Institute of Semiconductor Physics, National Academy of Sciences of Ukraine, Prospect Nauki 45, 252650 Kiev, Ukraine	307
4.8	"Nanoscale Wave-Ordered Structures on SOI", V K Smirnov[a], A B Danilin[b] [a]Institute of Microelectronics RAS, 3 Krasnoborskaya Street, 150051 Yaroslavl, Russia [b]Centre for Analysis of Substances, 1/4 Sretensky Blvd, 103405 Moscow, Russia	315
4.9	"Thin Partial SOI Power Devices for High Voltage Integrated Circuits", F Udrea[a], H T Lim[a], D Garner[a], A Popescu[a], W Milne[a], P L F Hemment[b] [a]Department of Engineering, University of Cambridge, Cambridge, CB2 1PZ, UK [b]School of Electronic Engineering, Information Technology and Mathematics, University of Surrey, Guildford, Surrey, GU2 5XH, UK	321

Keyword Index — 329

Author Index — 335

PREFACE

This proceedings volume contains the contributions of the speakers who attended the NATO Advanced Research Workshop on "Perspectives, Science and Technologies for Novel Silicon on Insulator Devices" held at the Sanatorium Pushcha Ozerna, Kyiv, Ukraine from 12th to 15th October 1998.

This meeting was the second NATO Silicon on Insulator (SOI) Workshop to be held in the Ukraine where the first meeting (Gurzuf, Crimea, 1st to 4th November 1994) focussed upon the physical and technical problems to be addressed in order to exploit the advantages of incorporating SOI materials in device and sensor technologies. On this occasion emphasis was placed upon firstly, promoting the use of SOI substrates for a range of novel device and circuit applications and secondly, addressing the economic issues of incorporating SOI processing technologies and device technologies within the framework of the resources available within the laboratories and factories of the Newly Independent States (NIS).

The primary goal of both workshops has been the breaking of the barriers that inhibit closer collaboration between scientists and engineers in the NATO countries and the NIS. Indeed, it was a pleasure for attendees at the first meeting to renew acquaintances and for the first time attendees to make new contacts and enjoy the warm hospitality offered by our hosts in Kyiv. An outcome was the forging of new links and concrete proposals for future collaborations.

The SOI technologies discussed in this meeting included SOS, ZMR, porous silicon, bonded wafer, SIMOX and 'SmartCut'/UNIBOND. The device and circuit applications ranged from conventional CMOS to power transistors, MEMS and Large Area Display Systems and included Cryogenic Microelectronics and Quantum devices. The meeting consisted of four sessions, namely

- Innovations in Materials Technologies
- Economics and Innovative Applications
- Characterization Methods for SOI
- Perspectives for SOI Structures and Devices

The Opening Plenary Session included papers which linked technologies with commercial opportunities where the Guest Speaker was Professor A P Shpack, Principal Scientific Secretary of the National Academy of Sciences of Ukraine. In total there were thirty-eight oral presentations and twelve poster papers, which includes seventeen Invited Papers.

The meeting was very well received and the attendees wish to express their gratitude to the NATO International Scientific Exchange Programme, Ministry of Science and Technology of Ukraine and the National Academy of Sciences of Ukraine whose financial support made the meeting possible. Thanks are extended to the management of the Sanatorium Pushcha Ozerna for their attention to the comfort of the attendees. The organizers offer their thanks to V I Kilchytska for ensuring that the local technical and social arrangements ran so smoothly and to I Barchuk, G Rudko and M Vovk for Clerical and Technical Assistance and to K E M Arthur for assistance in preparing this proceedings volume.

Peter L F Hemment   Alexei N Nazarov
University of Surrey   Vladimir S Lysenko
                              Institute of Semiconductor Physics, Kyiv

COMMITTEE MEMBERS

Workshop Chairmen:

P L F Hemment
School of Electronic Engineering,
Information Technology and Mathematics
University of Surrey
Guildford
Surrey
GU2 5XH
UK

V S Lysenko
Institute of Semiconductor Physics
National Academy of Science of Ukraine
252028 Kyiv-28
Ukraine

International Organising Committee:

A N Nazarov, Ukraine
V G Malinin, Russia
J P Colinge, USA
V S Lysenko, Ukraine
P L F Hemment, UK

Sponsors:

NATO International Scientific Exchange Programme
Ministry of Science and Technology of Ukraine
National Academy of Science of Ukraine

Workshop Secretariat:

Dr A N Nazarov
Institute of Semiconductor Physics
NAS of Ukraine
Prospect Nauki 45
252028 Kyiv-28
Ukraine

Dr V I Kilchytska
Institute of Semiconductor Physics
NAS of Ukraine
Prospect Nauki 45
252028 Kyiv-28
Ukraine

INVITED SPEAKERS

S Bengtsson
Solid State Electronics Laboratory, Chalmers University of Technology, Sweden

R B Bergmann
Institut für Physikalische Elektronik, Universitat Stuttgart, Stuttgart, Germany

V P Bondarenko
University of Informatics and Radioelectronics, Minsk, Belarus

M Bruel
CEA/LETI-Departement de Microtechnologies, Grenoble, France

C Claeys
IMEC, Leuven, Belgium

J P Colinge
University of California, Davis, USA

C A Colinge-Desmond
California State University, Sacramento, USA

H S Gamble
The Queen's University of Belfast, Belfast, UK

J S Im
School of Engineering and Applied Science, Columbia University, New York, USA

V G Litovchenko
Institute of Semiconductor Physics, Kiev, Ukraine

P P Maltsev
Applied Problems, Russian Academy of Science, Moscow, Russia

W I Milne
Department of Engineering, University of Cambridge, Cambridge, UK

B N Mukashev
Institute of Physics and Technology, Almaty, Kazakstan

A N Nazarov
Institute of Semiconductor Physics, Kyiv, Ukraine

Y Omura
Faculty of Engineering, Kansai University, Osaka, Japan

F Plais
Thomson CSF, Central Research Laboratories, Orsay, France

S I Romanov
Institute of Semiconductor Physics, Novosibirsk, Russia

V Dobrovolski, V Zotov, I Tyagulski, H Gamble, I Osiyuk, D Tomaszewski, V Gomenjuk, E Evtuh, M Bruel, J-P Colinge, D Ballutaud.

G Ninidze, B Mukashev, J Im, O Zabashta, A Druzhinin, W Milne, J Lescot, I Antonova, C Patel, L Ishchuk, C Desmond-Colinge.

V Bondarenko, S Romanov, A Okunev, F Plais, S Bengtsson, V Kilchytska, T Rudenko, E Terukov.

H Lim, C Claeys, R Bergmann, Y Omura, A Nazarov, P Hemment.

V Smirnov, P Smertenko, V Lysenko, B Mukashev, Y Gomenjuk, V Mordkovich, P Maltsev, B Katchmarshyi, I Osiyuk, I Barchuk.

D Ballutaud, S Bengtsson, V Bondarenko, S Romanov, Y Omura, A Nazarov, V Zotov, V Turchanikov, A Druzhinin.

D Tomaszewski, W Milne, I Antonova, I Tyagulski, A Okunev, O Zabashta, H Gamble, E Tovmach.

C Claeys, H Lim, R Bergmann, E Terukov, J Im, V Borisov, L Ishchuk.

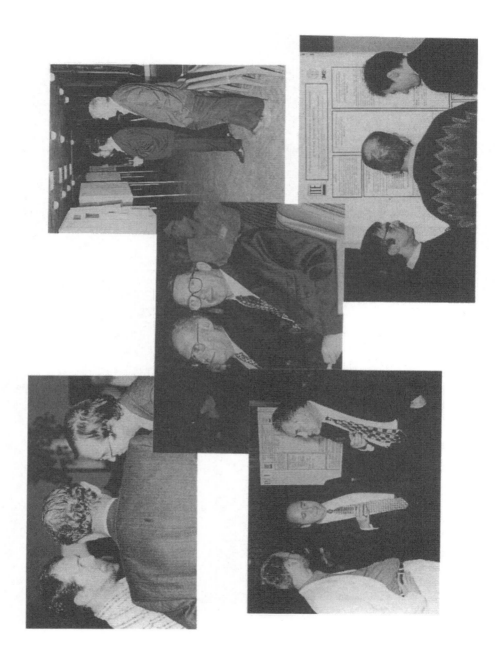

SMART-CUT® TECHNOLOGY: BASIC MECHANISMS AND APPLICATIONS

Michel BRUEL

CEA/LETI- Département de Microtechnologies

17 rue des Martyrs

38054 GRENOBLE CEDEX

FRANCE

ABSTRACT :

The Smart-Cut® process is presented here, a generic process enabling practically any type of monocrystalline layer to be achieved on any type of support. The Smart-Cut® process is based on proton implantation and wafer bonding. Proton implantation enables delamination of a thin layer from a thick substrate to be achieved whereas the wafer bonding technique enables different multilayer structures to be achieved by transferring the delaminated layer onto a second substrate. The basic mechanisms involved in the splitting effect induced after hydrogen implantation in silicon are assessed through transmission electron microscopy (T.E.M) observations, Infra-Red absorption spectroscopy in the Multiple Internal Reflection (MIR) configuration and by measurement of the duration ($T_{splitting}$) of the heat treatment necessary for splitting to occur. Study of the splitting time dependence with annealing temperature shows two temperature ranges associated to two different activation energies indicating two mechanisms. Yet in both cases the kinetics are shown to be related to hydrogen diffusion.

It is shown that this industrially economic process is particularly well suited to achieving very high-quality SOI material. Other examples of industrially developed applications of the process are also given.

1

P.L.F. Hemment et al. (eds.),
Perspectives, Science and Technologies for Novel Silicon on Insulator Devices, 1–15.
© *2000 Kluwer Academic Publishers. Printed in the Netherlands.*

INTRODUCTION

Silicon material is essentially the basis of the electronic component industry which today represents a world market of some 150 billion dollars. In this market, the relative demand for complex integrated circuits (microprocessors and memories) operating at very low voltage with very low power consumption is ever increasing and could in the next few years reach a growth rate of 40 %. This demand is motivated by increasing requirements for mobile phones and portable electronics in general, and by the necessity to reduce the power densities consumed by more and more complex circuits to limit overheating problems. Among the approaches enabling integrated circuits operating at very low voltage with very low power consumption to be achieved, the use of Silicon-on-Insulator wafers is one of the most promising. Using a Silicon-on-Insulator (S.O.I.) wafer instead of a conventional wafer to achieve integrated circuits provides the following advantages : reduction of stray capacitances, elimination of latch-up phenomena, better operation at very low voltage and simplification of the manufacturing process. All these advantages lead to better operation at low voltage, and lower power consumption, while preserving or improving the dynamic performances.

In a Silicon on Insulator (S.O.I) structure a thin layer of monocrystalline silicon rests on a dielectric layer, generally amorphous, itself situated at the surface of a silicon wafer. As such a structure cannot be achieved by conventional methods, different ways had to be imagined [1] which to varying extents provided a solution to this problem. Yet none of these ways seemed to answer all the requirements for a SOI material technology suitable for mass market applications, i.e. high quality material, high fabrication yield, low production cost, large volume production. So, the driving force which led to invention and development of the SMART-CUT process was the need for a new process for fabrication of S.O.I. wafers providing an answer to these requirements.

GENESIS OF THE PROCESS

The basic physics phenomenon which led the author [2] to invent the process generally known under the name of Smart-Cut® is the blistering phenomenon. Blistering, and also flaking and exfoliation, are visible macroscopic effects which have been known for a long time [3] and are induced by high-dose implantations of inert gas or hydrogen ions in materials.

These macroscopic effects are in fact the cooperative result of the microscopic effects induced in-depth by penetration of particles. The microscopic effects of hydrogen or rare gas implantation such as the creation of microcavities, microblisters [4] or microbubbles [5] (close to the penetration depth R_p corresponding to the maximum concentration) have been known for a long time. These microcavities enhance propagation of inter-cavity fractures where their density (depending on statistical fluctuations) reaches a percolation threshold. This leads to formation of a local cluster where all the microcavities are joined by a fractured zone, resulting in a blister at the surface. The driving force of this mechanism is the gas pressure in the microcavities and the stresses in the layer. From physical analysis and observation of the blistering phenomenon, the author had the intuition that by preventing blisters from forming, the increased pressure of the gas phase would result in a homogeneous force exerted over the whole surface of the layer and could therefore make it possible to split the whole surface layer. The additional idea was also that doing this would enable any plastic deformation of the layer to be avoided and would result in a defect-free layer because the light ions implanted through the layer create relatively few defects.

The idea which then quite naturally springs to mind is to join the surface to an undeformable solid : the stiffener. As the presence of the stiffener at the surface prevents the implantation process, implantation had to be carried out before fitting the stiffener, at a dose such that the surface is not deformed, and a way then had to be found of activating the physical mechanism able to increase the size and density of the microcavities so as to achieve their total splitting over the whole surface. It was found that the blistering phenomenon, which is in general induced directly by high-dose implantation (typically around 1.5 x 10^{17} ions cm^{-2}) around the ambient temperature, could also be obtained after implantation of a medium dose of about 5 x 10^{16} Cm^{-2} by thermal activation around a few hundred degrees C. Figure (1) shows the undamaged surface of a silicon wafer after implantation at a dose of 5.5 x 10^{16} H+cm^{-2} and the appearance of blisters and their evolution in the course of post-implantation heat treatment.

The basic principle of the process was thus laid down :

- Implantation of hydrogen ions or inert gas ions into a wafer

- Strengthening the wafer surface with a stiffener (bonding of a handle wafer at the surface or deposition of a thick layer on the surface).

- Heat treatment inducing splitting.

- (Optional) Fine polishing of the surface to get rid of the splitting-induced microroughness.

Since the process was disclosed [6], studies have been carried out at LETI and in an increasing number of laboratories on the basic physics phenomena involved and the principle of the process has been applied to other semiconductors [7,8].

The process development efforts were considerably boosted at the beginning of 1993 with the decision to set up a common LETI-SOITEC development program to get industrialization of this process under way by 1995 for fabrication of SOI wafers [9] under the Unibond trademark.

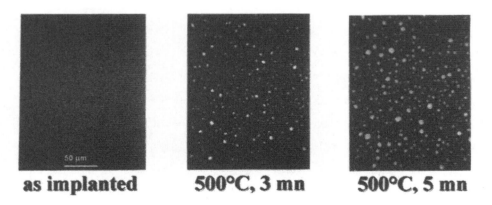

as implanted 500°C, 3 mn 500°C, 5 mn

Figure 1 : Blistering of the silicon surface induced by hydrogen implantation and annealing

BASIC PHYSICS MECHANISMS

The basic mechanisms are assessed by transmission electron microscopy (T.E.M) observations which enable the crystallographic defects to be studied, Infra-Red absorption spectroscopy in the Multiple Internal Reflection (MIR) which gives information on the hydrogen chemical bonds inside the silicon matrix and by measurement of the duration ($T_{splitting}$) of the heat treatment necessary for splitting to occur which provides information on the kinetics of the splitting phenomenon.

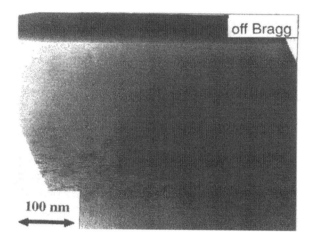

Figure 2 Transmission Electron Microscopy observations of in-depth induced microcavities

T.E.M. shows that hydrogen implantation induces formation of a defective region in which small microcavities or platelets are embedded. The dimensions of these platelets are about 100Å and 1 or 2 atomic rows thick. Such a typical population is shown in figure 2. They appear mostly aligned parallel to the surface in a (100) plane. Their sizes are typically around 10 nm. High resolution TEM confirms that they all have the same thickness of about 1 nm (fig. 2). Only a few of these cavities lie on the (111) planes and mostly in the deeper region of the sample. After implantation, the nucleation stage of the (100) cavities is not completed and a 5-minute annealing at 500°C is required to enter the coarsening regime of the cavities [10][11]. This means that, depending on the implantation conditions, the nucleation stage can be finished or still be in progress after the implantation step.

Beyond the nucleation stage, increasing the annealing time leads the microcavities to grow in size but reduce in density as usually observed during the coarsening stage of precipitates. It is to be noted that N_b the number of atoms trapped in the cavities, remains quite constant during annealing, during the nucleation stage as well as during coarsening.

This is interpreted in terms of Ostwald ripening mechanisms. All the H atoms lost by one cavity are captured by another. In fact, such a coarsening process involves H diffusion between the cavities in a population so that the larger cavities grow at the expense of the smaller ones. It is noteworthy that (111) cavities do not grow during annealing and do not play a significant role in splitting. The MIR observations [11] are consistent with the TEM observations. Before annealing, hydrogen is bonded to silicon mainly via Si-H bonds. However Si-H bonds on (111) surfaces and $Si-H_2$ bonds on (100) surfaces are present ; they correspond to the presence of (100) and (111) platelet-like defects in the implanted region which are observed by TEM.

The post-annealing heat treatment induces changes of configuration and leads to enhancement of $Si-H_2$ and $Si-H_3$ bonds which is consistant with the observed growth of cavity-like structures. In addition an increase of the band contribution related to the (100) internal surface and a decrease of the contribution of the Si-H (111)-related vibrational feature are observed on annealing.

$T_{splitting}$ can be easily and accurately measured. The duration of the splitting phenomenon itself is extremely short (probably about a few milliseconds) and generates a small noise which can be detected. The measured $T_{splitting}$ range is very large (a few seconds to many hours) and depends on temperature but also on implantation conditions (dose and energy), material specificity (doping level) and bonding parameters. Remarkably, whatever the conditions, two temperature ranges with two different activation energies can be identified from an Arrhenius-type plot of $1/T_{splitting}$ (Fig.3). Whatever the conditions, in a high temperature range an activation energy of about 0.5 eV is obtained. This value is comparable to the activation energy for atomic hydrogen diffusion in low-doped silicon i.e. $E_d=0.48$ eV, thus evidencing that in the high temperature regime splitting is only controlled by atomic hydrogen diffusion.

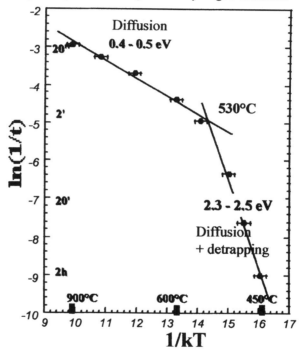

Figure 3 : Annealing time (t) necessary for splitting as a function of reciprocal annealing temperature (T)

In the low temperature range (<500°C), the activation energy is larger and appears to be the sum of two energies involving two combined mechanisms : hydrogen diffusion (0.48 eV) and dissociation of Si-H bonds whose related energy figure is greatly dependent on experimental conditions.

With this cleaning, the hydrophilicity and microroughness of the two bonded wafers lead to a room temperature bonding energy in the 0.1-0.2 J/m^2 range. During the low temperature annealing necessary for splitting, the bonding energy increases to about 0.5 J/m^2. With this cleaning procedure transferred layers are obtained free of macroscopic defects.

APPLICATIONS

Smart-Cut advantages :

The advantages of the Smart-Cut process derive from the physics of light particle implantation into materials. Use of light ions means low stopping power, mainly electronic stopping which induces :

- few atomic displacements. : few defects created

- low ion straggle : very precise cutting

- ease of getting deep penetration (rate of about 8nm / keV in Silicon)

- the delaminated layer thickness depends only on the ion energy

- very good layer thickness homogeneity due to the fact that ion energy dispersion is quite low

- ease of thickness choice by tuning the implantation energy

- implantation doses comparable to those of standard microelectronic technologies.

Smart-Cut : a generic process

The Smart-Cut process can be used in a very general manner : by its principle this process enables transfer of any monocrystalline film on any type of substrate. The following conditions are required :

- ability of the implanted material to generate blisters due to the effect of implantation of protons or inert gas ions. This would appear to be a fairly general phenomenon.

For "standard" conditions (69 keV, 5.5 x $10^{16}H^{+}cm^{-2}$, wafer p-doped 5 x 10^{14} cm^{-3}), the low temperature activation energy is about 2.5 eV (Fig. 3) which is the sum of 1.8 ±0.2 eV, the energy necessary to break the Si-H bonds [12], plus the diffusion energy of 0.48 eV. The transition temperature between low and high regimes is about 550°C which is comparable to the temperature where all the Si-H bonds are broken [13], thus confirming the proposed mechanism.

CLEANING AND BONDING

After medium dose implantation and before annealing, while many microcavities and platelets are already present at the depth of the maximum hydrogen concentration, no deformation or blisters are seen on the surface of the implanted wafer. Hence, this surface can be

bonded onto another wafer. This "handle" wafer acts both as a stiffener and as a support for the transferred layer. To obtain the splitting effect on the full wafer without any macroscopic defects, surfaces must be free of particle contamination before bonding, otherwise voids or un-transferred areas exist in the transferred layer. The sensitivity of the process to particle-induced defects is greatly dependent on the bonding energy at the splitting temperature ; to minimize this sensitivity a high-quality bonding at low temperature (<500°C) is necessary. Particles with a size over 0.3 μm can lead to void generation for a bonding energy of 0.5 J/m^2 [14].in the case of unoptimized cleaning conditions ; in this case a strict correlation between particle and voids can be obtained. In the case of low bonding energy, trapped particles are responsible for small defects at the bonding interface.

These defects are too small to reduce the stiffening effect sufficiently to avoid splitting. These small defects tend to inflate during annealing as soon as splitting occurs resulting in plastic deformation of the SOI surface. This kind of defect is easily detectable just after splitting. If larger particles are present, the interface bonding defect is large enough to induce an effective local lack of stiffening effect, thus preventing splitting from occurring in this area and leading to untransferred areas [14]. To avoid these effects a specific cleaning was developed based on modified RCA to eliminate particles >0.3 μm [14].

- possibility of joining the surface of the implanted material to a stiffener (substrate, handle wafer, thick layer) with sufficient bonding energy to prevent debonding during the splitting heat treatment

- existence of a sufficiently low splitting temperature range to prevent debonding or fractures induced by the stresses due to differential expansion in the case of different materials.

(It is important to note that the stiffener must remain affixed to the implanted surface during the heat treatment inducing splitting, otherwise the delamination phenomenon does not take place and all we get is a blistering phenomenon on the implanted surface [11]).

Fabrication of S.O.I. substrates

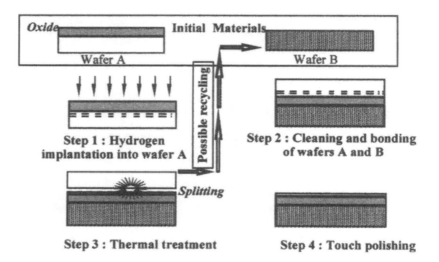

Figure 4 : Schematic showing the principle of the Smart-Cut process

To achieve SOI wafers, the Smart-Cut process basically comprises four main steps (Fig. 4) :

- Step 1: Hydrogen implantation into a wafer A (dose in the 3.5×10^{16} to 1×10^{17} H$^+$ cm^{-2} range). In this case, wafer A is silicon and is capped with thermally grown SiO$_2$.

- Step 2: Hydrophilic bonding at room temperature of wafer A to a wafer B (wafer B is either bare or capped). Wafer B plays a key role in the Smart-Cut process as a stiffener and as a handle wafer under the buried oxide in the SOI structure.

- Step 3: Two-phase heat treatment of the two bonded wafers. During the first phase (400-600°C), the implanted wafer A splits into two parts giving rise to a SOI structure and the remainder of wafer A. Once splitting has occurred, high temperature annealing is performed at 1100°C for 2 hours to remove any silanol groups from the bonding interface. The bonding energy is then as high as the rupture energy of bulk silicon (about 2.5 J/m^2) and the bonded interface is comparable to a Si-thermally grown SiO2 interface (Fig. 5).

- Step 4: Final touch polishing - after splitting, the layer of the SOI structure exhibits micro-roughness which makes polishing of the surface necessary for the elimination of the disturbed region. It is noteworthy that after splitting wafer A, which has just been peeled off from a very thin layer (less than or around one micrometer thick),[9] can be recycled and become wafer B of the next pair of wafers being processed. On average, a single bulk wafer is needed to achieve one S.O.I. wafer.

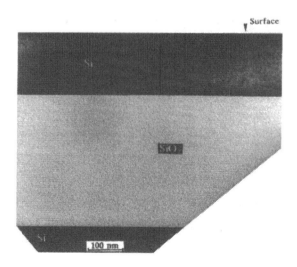

Figure 5 :TEM Cross section of a SOI Structure obtained through the Smart-Cut process .

In addition to the general advantages presented above , this process presents attractive benefits compared to the other processes used to produce SOI : all the technological steps can be performed on standard microelectronics equipment. Unlike BESOI technologies [15], only one bulk wafer is required to achieve one SOI wafer by means of the Smart-Cut process.

This process is compatible with any wafer size and is successfully applied today in production to obtain 100-150 and 200 mm SOI wafers, and recently 300 mm SOI wafers.

Other applications

At LETI interesting results related to single-crystal silicon on glass or quartz for display applications have been obtained. The process has also been applied successfully to silicon carbide [7] for making silicon carbide on insulator with a 200 nm thick SiC monocrystalline layer on Silicon Carbide and on Silicon wafers and more recently thin layers of monocrystalline GaAs on 100mm Silicon wafers were obtained [8]. The process also applies for transferring structured and patterned layers [16] or even electronic devices. In this case a structure is obtained with a monocrystalline silicon layer situated above an interconnection level. One application could be fabrication of the storage capacitors of a DRAM /S.O.I. below the memory cell. Other authors have also verified that blisters can be formed in germanium and diamond [17]. These semiconductors should therefore support the « smart-cut » process for the fabrication of thin monocrystalline layers.

CONCLUSION

In the genesis of the Smart-Cut® process, observation of the blistering phenomenon plays an important role. From a phenomenon which is commonly considered as being a parasitic effect, a generic process has been developed which is highly suitable for achieving large-size thin single-crystal layers by delamination from a bulk substrate. Its importance has been demonstrated for the production of SOI wafers and also to fabricate thin films of semiconductors (SiC, GaAs, Ge,...) on silicon or glass. It has also been shown that the process is applicable for the transfer of structured and processed layers.

ACKNOWLEGMENTS

The author is grateful to all his colleagues at LETI and to the staff of SOITEC who in one way or another contributed and still continue to contribute to the development of this process.

REFERENCES

1. Colinge , J.P. *Silicon on Insulator Technology - Materials to VLSI*, 2nd Edition, , Kluwer Academic Publishers, Boston/Dordrecht/London

2. Bruel , M. *Nucl. Instr. and Meth. In Phys. Res. B* 108 (1996) p.313-319

3. Kaminsky, M. *IEEE Trans. Nucl. Sci.* NS-18 (1971) p.208

4. Chu , W.K , Kastl , R.H., Lever, R.F. ,. Mader , S.and Masters , B.J. Radiation damage of 50-250 keV Hydrogen Ions in Silicon, in *Ion Implantation in Semiconductors* 1976, Editors : F. Chernow, J.A. Borders, D.K. Brice, Plenum Press New York and London

5. Evans J.H. *Journal of Nuclear Materials* 68 (1977) p.129-140

6. Bruel , M. *Electron. Lett.* 31 (1995) No. 14, p.1201-1202

7. Di Cioccio, L. , Letiec , Y. , Letertre , F., Jaussaud , C. and Bruel , M. *Electron. Lett.* 32 (1996) No. 12, p.1144-1145

8. Jalaguier, E., Aspar , B., Pocas, S ,. Michaud, J.F., Zussy, M., Papon A.M. and. M.Bruel, M *Electron. Lett.* 34, 4 (1998) p. 408-409

9. Auberton-Hervé, A.J., Lamure, J.M., Barge, T., Bruel, M., Aspar B. and Pelloie, J.L. *Semicond. Int.* 11 (1995) p. 97.

10. Aspar, B., Bruel, M., Moriceau, H., Maleville, C., Poumeyrol, T., Papon, A.M., Claverie, A., Benassayag, G., Auberton-Hervé, A.J., Barge T., *Microelectronic Engineering* 36 (1997) p.233-240.

11. Aspar, B., Lagahe, C., Moriceau, H., Soubie, A., Bruel, M., Auberton-Hervé, A.J., Barge, T., Maleville, C., *Proceedings (to be issued) MRS Spring Meeting 1998 Symposium Defects and Impurities in Semiconductors.*

12. Hubert, K.P. Hezberg, G. *Molecular spectra and molecular structural constants of diatomic molecules*, Van Nostrans, New York (1979).

13. Pearton, S.J., Corbett, J.W., J.T. Borenstein, *Physica B* 170 (1991) p.85.

14. Maleville, C., Aspar, B., Poumeyrol, T., Moriceau, H., Bruel, M., Auberton-Herve, A.J., Barge T.and Metral, F. *Proc. 7th Int. Symp. on SOI Technology and Devices*, eds. P.L.F. Hemment et al., vol. 96-3, The Electrochem. Soc. Series, Pennington (1996) p.34.

15. Maszara, W.P. *Proc. Fourth Int. Symp. on Silicon on Insulator Technology and Devices*, ed. D. N. Schmidt, vol. 90-6, The Electrochem. Soc. Series, Pennington (1990) p.199.

16. Aspar, .B., Bruel M., Zussy, M. , Cartier, A.M. *Electron. Lett.* 32 (1996) No. 21, p.1985.

17. Tong, Q.Y., Scholz, R. ,Goesele, U., Lee, T.H., Huang, L.J., Chao,Y.L. and Tan, T.Y. *Appl. Phys. Lett* 72 (1), 1998, p.49-51

The forward I-V characteristics of the diodes (Fig. 4) are near ideal with an exponential behaviour over nearly 9 orders of magnitude.

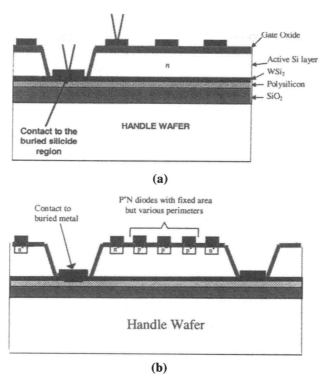

(a)

(b)

Figure 3. (a) MOS Capacitor structure, (b) pn junction diode.

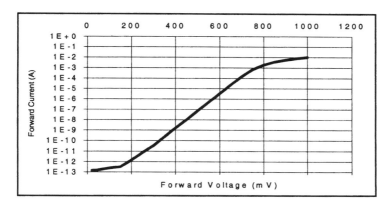

Figure 4. Forward I-V characteristic of a typical diode.

A typical forward ideality factor is 1.005 and the series resistance is less than 20 ohms. Under reverse bias the diodes exhibit a sharp breadkown at 6.8 V when the leakage current is approximately 5 nA. The leakage current under reverse bias was plotted as a function of diode perimeter to enable decoupling of the edge and bulk generated current components. The bulk generation lifetime was calculated to be of the order of 200-300 μs. In the production of these diodes, 5 masks were used and the substrate was thermally cycled to 1000°C during several of the processing steps. From the measured carrier lifetimes and junction leakage currents, there is no evidence of device degradation through stress induced defects or by tungsten migration during the manufacturing process.

High speed bipolar transistors on S^2OI substrates using a single polysilicon process were recently reported by Goody et al.(8). The S^2OI substrates were produced essentially as described above with an arsenic implant used prior to bonding to facilitate the ohmic contact of tungsten silicide to silicon. A wafer with a measured S^2OI thickness of 1.5 μm was selected so that it was not necessary to grow an epitaxial layer to define accurately the intrinsic collector width. Instead the base was implanted directly into the surface of the 4 ohm-cm S^2OI layer. This meant that the variation in S^2OI layer thickness was directly reflected as a variation in intrinsic collector width. The transistor was completed by depositing and patterning an oxide layer followed by deposition of a polysilicon implanted emitter.

An improvement in leakage and yield over devices on standard SOI layers was found. This may be due to the tungsten silicide acting as a gettering site (9) or to modifications made to the trench and field oxides. Circuit functionality was demonstrated by 41 stage CML ring oscillators. As f_t is a function of the intrinsic collector width, which in the absence of an epitaxial layer, is a function of the S^2OI layer thickness, it will in future be necessary to improve the thickness control of the S^2OI layer for tight control of device performance.

4. Polish Stop Technology

Several techniques have been developed to produce accurate thickness control of SOI layers. These include Smart-cut (10), SiGe etch stop layer (11), porous silicon(12), selective plasma thinning (PACE)(13) and polish stop techniques (14). The etch stop and porous silicon techniques require the growth of high quality epitaxial layers, the PACE technique requires expensive equipment and the smart cut approach needs developing for producing S^2OI layers. To enable test structures to be produced on S^2OI substrates for research purposes we have adopted a polish stop technique. The approach is to first produce standard 3 μm S^2OI substrates. This is achieved by implanting the active silicon, depositing layers of CVD tungsten and polycrystalline silicon and bonding to an oxidised handle wafer. The active wafer is ground and polished back to approximately 3 μm. This is a high yield process since there are few pre-bond process steps and no patterning.

To provide the polish stops, a masking oxide is grown and RIE is employed to form a trench grid pattern. The trench pattern is then etched through the silicon and tungsten silicide layer using HBr chemistry. This process etches both the silicon and the WSi_2 without creating an undercut (4). The grid pattern consisted of 10 μm wide trenches on a 110 μm pitch to give 100 μm silicon squares (15). A silicon nitride layer, (of the desired final silicon layer thickness) 0.5 μm thick is deposited by LPCVD at a temperature of 800°C. This silicon nitride layer is patterned with grid lines of width 6 μm which are aligned along the centre of the etched trenches as shown in Fig.5. A 0.1 μm thick layer of polycrystalline silicon is deposited to give an indication of when the nitride polish stops have been reached and to allow removal of the masking oxide.

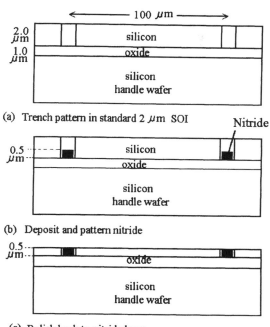

(a) Trench pattern in standard 2 μm SOI

(b) Deposit and pattern nitride

(c) Polish back to nitride layer

Figure 5. Polish stop process.

A CMP process is employed to polish back the polycrystalline silicon to expose the masking oxide. This is removed in a wet etch and the CMP process continued. When the polysilicon on top of the nitride is reached, colour changes will be observed in the polysilicon until the polish reaches the nitride itself. A combination of a chemical/mechanical polish and a mechanical only polish is used. This requires two slurries with different pH values, Nalco 2350 with a pH level of 10.5 is a chemical/mechanical slurry while Nalco 2355 with a pH level of 10 is mainly a mechanical slurry. The polishing rate of silicon can vary from 0.7 μm/min to 1 μm/min depending on polish conditions such as carrier speed, platen speed and pressure. Compared to the removal rate for silicon, the removal rate for silicon nitride is a factor of 10 slower. Therefore it is to be expected that the silicon polishing will cease when the silicon nitride grid is reached. Thus the thickness of the SOI layer is controlled by the thickness of the deposited nitride layer.

'Over polishing' of the silicon in the islands between the nitride polish stops can degrade the layer uniformity which would be unacceptable for deep submicron SOI. The

24

degradation takes the form of depressions or dips which will be largely due to chemical etching if a high pH chemical-mechanical polish is used. Using the lower pH slurry, mechanical only polishing takes place and the dips in the silicon islands are not unacceptably large. Thus, on polishing back to the nitride stop, the lower pH slurry is used. A final touch polish in the higher pH slurry is employed to improve the surface quality of the final SOI layer.

A Tencor Alpha Step profile of the 2 µm standard SOI substrate after trenching is shown in Fig.6a . The distance from the top of the nitride polish stops to the silicon surface is 2.5 µm. After the two step polishing process the resulting SOI layers have a thickness of 0.5 µm and a uniformity of ± 10%. The surface profile given in Fig.6b, shows the silicon level with the nitride polish stops at the edges of the silicon islands with a dip of 0.06 µm at the centre of the islands. The thickness at the centre of the silicon islands varied from 0.4 µm to 0.45 µm across the 100 mm diameter SOI substrate.

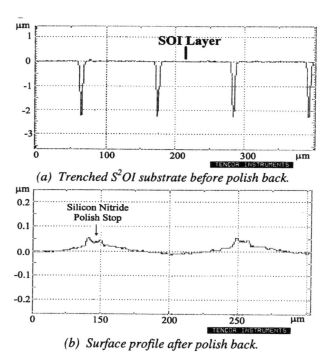

(a) Trenched S^2OI substrate before polish back.

(b) Surface profile after polish back.

Figure 6. Surface profiles.

The post bonding polish stop technique thus can be successfully used for producing sub-micron SOI/S^2OI with a thickness variation of less than ± 0.05 µm. This value also depends on the uniformity of the deposited nitride layer. While large trench widths of 10 µm have been used in this feasibility study, nitride polish stops 0.5 µm wide have been shown to be effective as polish stops. The minimum trench width that can be employed is

3 times the thickness of the desired S^2OI layer. It has also been shown that trenches spaced 400 μm apart can give silicon islands with less than 0.1 μm non-uniformity (16). With optimisation of the polishing process the silicon uniformity over even larger areas should be satisfactory.

5. Other Applications

5.1 INFRARED PHOTODETECTORS

Carline et al. (17) used the buried tungsten silicide in an S^2OI structure as a mirror to improve the performance of a long wave infrared SiGe/Si flat detector. For focal plane arrays (FPAs) imaging in the 8 - 12 μm wavelength atmosphere hybridised n-type GaAs / AlGaAs quantum well infrared plotodetectors deliver the best performance. However uniform arrays are difficult to make. In the new structure an S^2OI substrate with a 0.3 μm thick buried tungsten silicide layer was used as the starting substrate. A 16 period p - $Si_{0.86}Ga_{0.14}$ / Si resonant cavity QWIP was grown epitaxially on the S^2OI substrate at 650°C. The 16 alloy quantum wells were 5nm thick and contained 1.5×10^{12} cm^{-2} boron atoms and the undoped silicon barrier layers were 55 nm thick. Using wet etching mesa structures were formed as shown in Fig.7. The authors found that the normal incidence responsivity of the SiGe/Si QWIPs incorporating a buried silicide reflector were comparable to those of III-V QWIPs being used in long wavelength imaging cameras. Thus the new QWIPs on S^2OI substrates offer the possibility of full integration of FPAs for the 8-12 μm radiation.

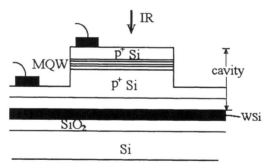

Figure 7. Quantum well infrared photodetectors (17)

5.2 COMPLEMENTARY BIPOLAR PROCESS

Tungsten silicide layers have diffusion coefficients for boron and phosphorus several orders of magnitude greater than that for silicon. This has caused problems in polysilicon gate CMOS processes where the n- and p- channel devices had differently doped gates. However, it is possible to use the very fast dopant diffusion in buried tungsten silicide layers to simplify the production of complementary bipolar circuits while improving their performance.

A tungsten silicide layer is deposited on the surface of the active wafer which is then bonded to an oxidised handle wafer. The handle wafer is ground and polished back to 3 μm ± 0.5 μm. Sub-micron SOI layers of thickness 0.7 μm, say, are then produced by a polish stop technique for test structures or by the PACE process for characterisation of commercially equivalent substrates. The isolation trenches are etched down through the active silicon and the underlying tungsten silicide layer. After refilling the trenches with dielectric the collector sinker contacts are appropriately diffused as shown in Fig.8. The dopant diffuses rapidly along the buried tungsten silicide collector contacts and into the over lying silicon to produce shallow heavily doped layers for ohmic contacts to the intrinsic collectors. The base regions of the n-p-n and p-n-p transistors are then selectively implanted. The complementary transistors are completed by fabricating the polysilicon n^+ and p^+ emitters in the normal way.

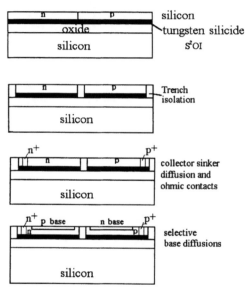

Figure 8 Sinker diffused complementary bipolar SOI process.

The advantages of the above process are; [i] the ohmic contact and sinker diffusions are carried out in one step thus eliminating two photolithographic and two process steps, [ii] the SOI layer is thinner and so trench etch and refill times are reduced and [iii] the thinner SOI layer gives lower parasitic collector-substrate capacitance. Since the high temperature bond anneal is carried out before the sinker/buried collector diffusion the dopant up-diffusion from the buried collectors is much less. The up-diffusion for the n-p-n and p-n-p transistors are individually controllable. Thus the base widths for the p-n-p transistors can be made smaller than those of the n-p-n transistors by carrying out the p-n-p boron sinker diffusion/drive-in before the phosphorus sinker diffusion for the n-p-n transistors. Further adjustments to the performance of the two transistor types can be obtained by controlling their emitter diffusion processes. The performance characteristics of the two transistor types can therefore be optimised for compatibility.

A further refinement of the above process would be to open trenches down to the silicide and refill them with tungsten silicide. The buried collector diffusions could then be carried out further reducing the collector series resistance and transistor area. This latter process would also be advantageous for smart power circuits.

6. Conclusions

Tungsten silicide has proven to be a satisfactory choice for the buried metallic layer for silicon on silicide on insulator substrates. It is stable up to temperatures exceeding that normally used in advanced bipolar integrated circuit processing. While buried tungsten silicide layers produced by the reaction of sputtered tungsten with the silicon can give excellent results the process is highly sensitive. Uneven reaction can lead to voids in the active silicon and in even worse cases delamination of certain areas can occur. Deposition of the tungsten silicide by LPCVD using tungsten hexafluoride and silane chemistry has proved to be very satisfactory. Commercial reactors are available as tungsten is already used for conducting vias in IC manufacture. The tungsten silicide can be trenched at the same time as the active silicon using HBr chemistry without undercut. This is very convenient and avoids the need to pattern the tungsten silicide before bonding. Test structures employing MOSCs, p-n junctions, bipolar transistors and ring oscillators have shown no evidence of any degradation of the silicon layer above the silicide. Carrier lifetimes and junction leakage currents are of similar magnitude as those found on bulk substrates. The conductivity of the tungsten silicide is in the range 30-40 $\mu\Omega$-cm giving a sheet resistance of less than 2 Ω/sq. for a 0.2 μm thick layer. The conductivity of the tungsten silicide layer in contact with silicon and/or silicon dioxide does not degrade with high temperature processing. The tungsten silicide forms a rectifying contact to silicon with doping less than 10^{17} cm^{-3} and so it is necessary to increase the doping concentration at the contact. This can be achieved by pre-implanting the silicon before tungsten silicide deposition, but this results in significant up-diffusion during the post bond anneal. This problem should be eliminated by doping the tungsten silicide after wafer bonding, for example during diffusion of the dopant in the collector sinker. Employing this technique for complementary bipolar transistor circuits should result in a reduced number of masks and other processing steps and allow tailoring of the n-p-n and p-n-p transistors for better matching. A simple post bond polish stop process has been verified for producing sub-micron SOI and S^{2}OI substrates. The technique can be refined for 2 μm wide trenches on a 400 μm pitch. It is therefore a useful technology verifier for complementary bipolar circuits on S^{2}OI. Buried metal silicides can be used to improve the performance of optical devices by acting as a buried reflector. They will also play an important role in reducing drain/collector series resistance in smart power integrated circuits and may be used as ground planes in high performance ICs.

7. References

1. Goody, S.B. and Saul, P.H. (1995) A Fully Oxide-Isolated Bipolar Transistor Integrated Circuit Process, *Electrochem Soc.* **95-1**, 571-572.
2. Murarka, S.P., Silicides for VLSI Applications, Academic Press, New York 1983.

3. Goh, W.L., Campbell, D.L., Armstrong, B.M. and Gamble, H.S. (1994) Buried Metallic layers with Silicon Director Bonding, 2nd Int. Seminar on Power Semiconductors 94, Prague, 7-14.

4. Wilson, R., Quinn, C., McDonnell, B., Blackstone, S. and Yallop, K., Bonded and Trenched SOI with Buried Silicide Layers, *Electrochem. Soc. Proc.,* **95-7**, 535.

5. Nayar, V., Russell, J., Carline, R.T., Pidduck, A.J., Quinn, C., Nevin, A. and Blackstone, S. (1998) Optical properties of bonded silicon silicide on insulator (S^2OI): A new substrate for electronic and optical devices, *Thin Solid Films,* **313-314**, No. 1-2, 276-280, *Proceedings of the 1997 2nd International Conference on Spectroscopic Ellipsometry,* Charleston, SC, USA, May 12-15.

6. Goh, W.L., Montgomery, J.H., Raza, S.H., Armstrong, B.M. and Gamble, H.S. (1997) Electrical Characterization of Dielectrically Isolated Silicon Substrates Containing Buried Metallic Layers, *IEEE Electron Device Letters,* **18**, 232-234.

7. Goh, W.L., Raza, S.H., Montgomery, J.H., Armstrong, B.M. and Gamble, H.S., The Manufacture and Performance of Diodes Made in Silicon on Metal on Insulator (SMI) Substrates.

8. Goody, S.B., Osborne, P.H., Quinn, C. and Blackstone, S. (1998) High Speed Bipolar on Bonded Buried Silicide SOI (S^2OI) ESSDERC '98.

9. Goh, W.L., Campbell, D.L., Armstrong, B.M. and Gamble, H.S. (1995) Buried Metallic Layers with Silicon Direct Bonding, *Electrochem. Soc.,* **95-7**, 1995, 533.

10. Bruel, M., Aspar, B., Maleville, C., Moriceau, H., Auberton-Hervé, A.J. and Barge, T. (1997) Unibond SOI Wafers Achieved by Smart-Cut Process, in Silicon-on-Insulator Technology and Devices VIII, *Electrochem. Soc.,* 3-9.

11. Godbey, D.J., Twigg, M.E., Hughes, H.L., Palkuti, L.J., Leonov, P. and Wang, J.J., (1990) Fabrication of Bond and Etch Back Silicon on Insulator Using a Strained $Si_{0.7}Ge_{0.3}$ Layer as an Etch Stop, *J. Electrochem. Soc.,* **137**, 3219-3229.

12. Yonehara, T., (1997), Canon Inc Workshop on applications of, and developments in SOI technologies.

13. Murnola, P.B. and Gardopie, G.J., (1995), Recent Advances in Thinning of Bonded SOI Wafers by Plasma Assisted Chemical Etching, *Electrochem. Soc.,* Reno, Nevada, **95-1**, 568.

14. Gay, D.L., Tweedie, M., Armstrong, B.M. and Gamble, H.S. (1997), Polish-Stop Formation for Sub-Micron SOI, 192nd Meeting of the Electrochemical Society, Paris, **97-2**, 2443-2444.

15. Gay, D.L., Baine, P.T., Gamble, H.S., Armstrong, B.M. and Mitchell, S.J.N. (1998), Thin SOI Using Polish Stop Technology, 40th Electronics Materials Conference, Charlottesville, Virginia.

16. Baine, P.T., Gay, D.L., Armstrong, B.M. and Gamble, H.S. (1998), The Realisation of Silicon on Insulator Utilising Trench Before Bond and Polish Stop Technology, *J. Electrochem. Soc.* May 1998.

17. Carline, R.T., Hope, D.A.O., Nayar, V., Robbins, D.J. and Stanaway, M.B. (1997), A Vertical Cavity Longwave Infrared SiGe/Si Photodetector Using a Buried Silicide Mirror, *IEDM 97*, 891-894.

HOMOEPITAXY ON POROUS SILICON WITH A BURIED OXIDE LAYER: FULL -WAFER SCALE SOI

S.I. ROMANOV [1][2], A.V. DVURECHENSKII [1][3], V.V. KIRIENKO[1],
R. GRÖTZSCHEL [4], A. GUTAKOVSKII [1], L.V. SOKOLOV [1],
M. A. LAMIN [1]
[1] *Institute of Semiconductor Physics,*
pr. Lavrent'ev 13, 630090 Novosibirsk, Russia
[2] *Tomsk State University,*
sq.Revolution 1, 634050 Tomsk, Russia
[3] *Novosibirsk State University,*
Pirogova 2, 630090 Novosibirsk, Russia
[4] *Research Center Rossendorf, Incorporated,*
P.O.Box 510119, D-01314, Dresden, Germany

Abstract

A new SOI technology is described which incorporates the technologies of porous silicon and molecular beam epitaxy. The principle of the method is to form a buried oxide layer by means of anodic oxidation of silicon wafer through a surface porous silicon layer previously formed by electrochemical anodic etching. Growth of an epitaxial silicon film completes the production cycle. We believe that this new method based only on a high temperature chemical vapor deposition is quite competitive with other present-day SOI technologies for commercial material production.

1. Introduction

Device quality and low cost are the principal requirements for a commercially viable silicon-on-insulator (SOI) technology. There are several competitive technologies for SOI wafer fabrication where Separation by IMplantation of OXygen (SIMOX) and Wafer Bonding (WB) are, currently, the front running technologies for the preparation of SOI device substrates [1-3].

Recent developments in low dose SIMOX processes, namely, ITOX (Internal Thermal OXidation) and ADVANTOX reported by NTT and Ibis Technology, respectively, are promising methods for lowering the SOI wafer cost whilst maintaining high crystal quality [4].

In WB technologies, the key process is to obtain the thin high quality uniform silicon overlayer. There are two principal trends in this field. One is based on the etch back process using both doping-sensitive chemical etching of epitaxial p^+/p silicon device wafers [5] and structure-sensitive chemical etching of porous silicon layers [6]. This

P.L.F. Hemment et al. (eds.),
Perspectives, Science and Technologies for Novel Silicon on Insulator Devices, 29–46.
© 2000 *Kluwer Academic Publishers. Printed in the Netherlands.*

trend is traditionally named BESOI. However, Canon has developed the method involving porous silicon and patented it as ELTRAN® (Epitaxial Layer TRANsfer) [7]. Another process (Smart Cut) is based on implantation treatment and realized by wafer bonding of a hydrogen implanted silicon wafer followed by top layer splitting initiated by radiation-induced microcracks parallel to the bonding interface. The Smart Cut® process has been developed by LETI [8,9] and used by SOITEC to manufacture UNIBOND® wafers. A new approach termed "Smarter Cut" has been reported [10,11] which employs a lower temperature for silicon layer splitting.

In this paper we describe a new SOI materials technology, SOPS BOL (Silicon-On-Porous Silicon with Buried Oxide Layer), which we believe can be quite competitive with other present-day technologies for the commercial production of device grade substrates. The technology was patented in Russia [12,13].

2. SOPS BOL Technology

2.1. GENERAL DESCRIPTION

Conceptually SOPS BOL technology enables a SOI structure to be formed by epitaxial deposition of silicon on partially oxidized porous silicon. The process flow is shown schematically in Figure 1. There are only three principal steps in this technology. Compared to other SOI methods this is a noticeable reduction in the process complexity as only two simple electrochemical treatments and epitaxial growth are required.

Anodic etching (AE) of silicon wafer is used to form a porous silicon layer (PSL) with different porosity in depth. PSL consists of top and bottom layers. The porosity (void fraction) of the top-PSL is lower than that of the bottom-PSL.

A buried oxide layer (BOL) is formed by anodic oxidation (AO) of the wafer with the surface two-layer PS. BOL is located within the bottom-PSL and contains silicon oxide with different density in depth. There are the compact oxide being free from pores at an interface of the bottom-PSL with bulk silicon and the overlying porous oxide including a large number of cavities (Figure 1).

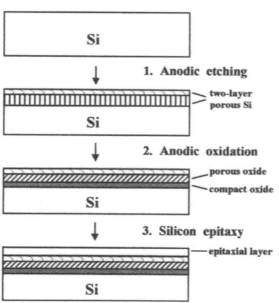

Figure 1. Schematic process route for SOPS BOL technology.

In this study we have used molecular beam epitaxy (MBE) to grow the silicon films but plan to use high temperature CVD in the future. The process is described in detail in the following sections. The details of the samples described in this paper are listed in Table 1. The structures are defined in terms of h_1 – the thickness of the top-PSL; P_1 – the porosity of the top-PSL; h_2 - the thickness of the bottom-PSL; P_2 - the porosity of the bottom-PSL; h_{Si} – the thickness of the epitaxial silicon layer; V_F – the formation potential of the anodic oxide film; T_{ann} – the temperature of annealing.

TABLE 1. Details of the samples used in this work.

Sample	h_1 nm	P_1 %	h_2 µm	P_2 %	h_{Si} nm	V_F V	T_{ann} °C
1/2	200	35	4	45			
2/2	200	35	5	55			
3/2	200	35	8	70			
1.1	50	35	2	65	400	15	
2.3	50	35	2	70	400	25	
1T3	100	40	1	55			
1T4	100	40	1	60			
2T2	100	40	1	65			
11.2	300	40	1	90	10		
0.2	50	35	1.7	60	500		
4T2	300	40	1	70	400	25	900

2.2. ANODIC ETCHING

After RCA-cleaning, the two-layer PS was formed by anodizing 4-inch p^+-type Si (001) wafer with the resistivity of 0.001-0.006 Ω·cm in 20% HF solution (42.4% HF:H$_2$O:C$_3$H$_7$OH ratio 2:1:2) as described in details elsewhere [14]. The top-PSL with the thickness h_1 varing from 50 nm to 300nm and the porosity P_1 = 35%, 40% must meet the requirement of Si MBE. The bottom-PSL with h_2 ranging from 1µm to 8µm

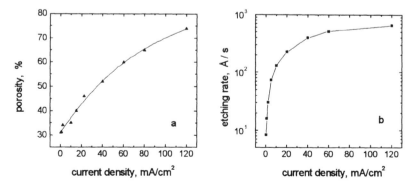

Figure 2. Experimental data showing (a) porosity and (b) etching rate of p^+-Si (001) 0.006 Ω·cm as a function of current density during anodization in a 20% HF solution.

and P_2 = 45–90% will be converted into BOL during subsequent AO. An electrolytic cell has a standard two-electrode configuration with a Pt counter electrode for both PSL and BOL formation.

In Figure 2 we plot the experimental values of the porosity of PS and the silicon etching rate as a function of the forming current density during anodization. The porosity and the layer thickness were measured gravimetrically. By adjusting the anodic current density one can form PSL with different values of porosity from 30% to 80% and higher.

Random and aligned RBS (Rutherford backscattering spectroscopy) spectra from the two-layer PS are shown in Figure 3. For all samples the top-PSL was formed under the AE regime to obtain h_1 = 200nm and P_1 = 35% whereas the preparation of the bottom-PSL involved different forming currents to produce samples with different porosities of 45%, 55%, 70% (samples 1/2, 2/2, 3/2 in Table 1). The aligned yields from the surface layer are similar for all three samples and are independent of the bottom layer. A minimum backscattering yield, χ_{min}, calculated from a ratio of the aligned yield to random yield is 10–16%. This result is of importance when growing Si films on a porous substrate: the layers are formed independently and are unaffected by one another in the two-layer PS.

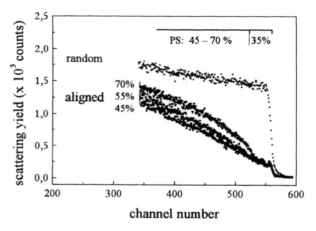

Figure 3. Random and aligned RBS spectra from two-layer PS with fixed porosity (35%) in the top-PSL and different porosities (45%, 55%, and 70%) in the bottom-PSL.

2.3. ANODIC OXIDATION

Nearly all previous characterization studies of the AO of PS have been limited to investigations using photo- and electro-luminescence [15-20]. This is due to that anodic oxidation proves to be an attractive thinning technique and also an effective method for passivation of porous material. Its possible application to SOI fabrication has been shown in two noteworthy papers [21,22].

Bsiesy et al. [23] studied the formation of a BOL in PS on p-type lightly doped silicon and demonstrated the complex nature of the electrochemical oxidation. In the absence

of information concerning AO of p⁺-PSL (that we used in SOPS BOL technology) our
first efforts were focused on a detailed analysis of this process when applied to the two-
layer PS. In so far as the epitaxial growth of silicon on PS BOL surface is concerned we
need to consider (i) what happens to the top-PSL during the initial stage of the
oxidation. It is also important (ii) to examine BOL structure more closely for the
purpose of determining the mechanism of the AO of p⁺- PSL.

To work out this key technological process we have first made a study of the kinetics
of the oxidation of compact (monocrystalline silicon containing no pores) and porous
materials using electronic-grade well-known 1M HCl aqueous solution [24]. This
approach proved to be a success in clarifying the mechanism of the electrochemical
oxidation of p⁺-PSL.

Figure 4 shows the variations in the formation potential V_F of silicon oxide films
during AO of p⁺-Si and p⁺-PSL under galvanostatic conditions (sample 1.1 in Table I).
These are typical oxidation curves for both materials. It is obvious that compact silicon

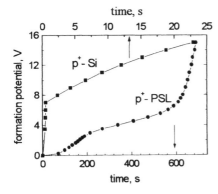
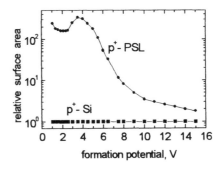

Figure 4. Formation potential V_F of silicon
oxide film grown on p⁺-Si and in p⁺-PSL
as a function of anodic oxidation time at
5 mA/cm² in 1M HCl solution.

Figure 5. Relative surface area s_{eff} (S_{eff}/S_0)
of anodically oxidized p⁺-Si and p⁺-PSL
calculated by means of the
phenomenological equations (1)-(3) from
the kinetics shown in Fig. 4.

and porous silicon exhibit different time dependence of anodic oxidation. For p⁺-PSL
the formation potential initially increases very slowly, much slower than for p⁺-Si, but
after 600 sec, under our experimental conditions, it starts rising very rapidly as in the
case of p⁺-Si (notice that the curves shown in Figure 4 are plotted along different time
axes). We have analysed this behaviour with reference to oxide growth on compact
p⁺-Si. Phenomenological kinetic equations were found as follows:

$$\frac{1}{j}\frac{dV_F}{dt} = 1.21 \cdot j^{0.13} \qquad \text{for } V_F = 0 \text{ - } 3 \text{ V} \qquad (1)$$

$$\frac{1}{j}\frac{dV_F}{dt} = 6.6 \cdot j^{(1-0.14V_F)/4.4} \cdot \exp(-V_F/1.74) \quad \text{for } V_F = 3 \text{ - } 7 \text{ V} \qquad (2)$$

$$\frac{1}{j}\frac{dV_F}{dt} = 0.27 \cdot \exp(-V_F/8.5) \qquad\qquad \text{for } V_F = 7 - 15 \text{ V} \qquad (3)$$

where j represents a current density expressed in mA/cm^2, and V_F (volts) is the formation potential of the anodic oxide film grown under galvanostatic conditions.

We have suggested that the anodic oxidation of p^+-PSL adheres to the same kinetics but is characterized with an effective current density j_{eff} responsible for electrochemical reactions in a porous layer. The value of j_{eff} is substantially below the current density j_0 calculated in terms of an external surface area S_0 of the sample. This effect is clearly due to the very large internal surface area S_{PS} of p^+-PSL [25] that is being oxidized. Using equations (1)-(3) to determine j_{eff} from the data shown in Figure 4 and the simple expression $j_0 \cdot S_0 = j_{eff} \cdot S_{eff}$, where S_{eff} is the effective surface area of the p^+-PSL being oxidized, we calculated a relative surface area $s_{eff} = S_{eff}/S_0 = j_0/j_{eff}$ as a function of V_F. Figure 5 shows this dependence.

The oxidation cycle shown in Figure 5 can be divided into three distinct stages: (1) $V_F < 5V$, (2) $5V < V_F < 8V$, (3) $V_F > 8V$. During the initial stage, (1) $V_F < 5V$, which corresponds to most of the oxidation time, the parameter s_{eff} has values of 160–340. This first phase is followed by the reduction of s_{eff} down to 1.8 at $V_F = 15V$. According to a detailed gas adsorption study by Herino et al. [25] p^+-PS has enormous surface to volume ratio of 200–230 m^2/cm^3. This value essentially does not depend on porosity up to 70%. In sample 1.1 the thickness of the p^+-PSL is about 2 μm (see Table 1). This means that the internal to external surface areas ratio s_{PS} (S_{PS}/S_0) is estimated as 400–460. And so, during the initial stage of AO, the silicon oxide film grows on 40–80 percent of S_{eff} to S_{PS} (s_{eff}/s_{PS}). It is interesting to note that the oxide growth exhibits non-monotonous manner, as seen in Figure 5.

Following this slow 'gentle' phase, (1) $V_F < 5V$, the oxidized area of p^+-PSL begins to shrink rapidly to about 2 percent of S_{eff} to S_{PS} ($s_{eff} = 10$) at $V_F = 7–8V$. This is the intermediate 'cut-off' stage, (2) $5V < V_F < 8V$. We believe that the observed variation of the electrochemical process is associated with breaking of the anodic current flowing through a porous layer and accompanied by the formation of the initial BOL. Thereafter, there is the final stage, (3) $V_F > 8V$, when AO occurs only around this BOL which grows by consuming the p^+ substrate. It follows from observing the continued decrease of the oxidized area down to S_{eff} of roughly 0.4 percent ($s_{eff} = 1.8$) at $V_F = 15V$. This final 'severe' stage of the process will be considered in a further paper by the authors in this Proceedings [26].

The process of AO has previously been described by the authors [23] in qualitative terms but here we demonstrate the new features inherent to AO of porous materials formed on p^+-type silicon. During the initial stage, oxide growth occurs practically over all of the internal surface of the p^+-PSL. The thickness of oxide film is depth dependent, but, more importantly, as evident from Figure 5, the top-PSL takes an active part in AO, when the formation potential is varied through a small range of $\Delta V_F \sim 2$–3V. This value corresponds to the oxide film of thickness 1.0 to 1.5nm on surface pore walls in the top-PSL. This result is of importance when growing Si films on the oxidized porous substrate.

As discussed earlier, BOL is initially formed at the rapid intermediate stage of AO ($5V < V_F < 8V$) and is grown further during the next stage. The BOL is a continuous

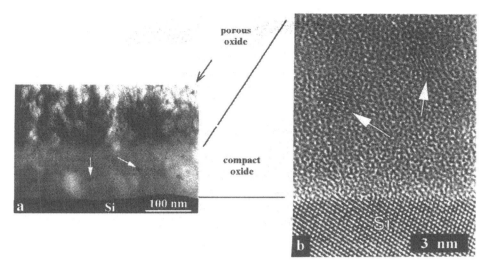

Figure 6. Cross-sectional TEM images of BOL near the bulk silicon substrate: (a) overall view of *porous SiO_x / compact SiO_x / substrate* structure, (b) high-resolution micrograph of *compact SiO_x / substrate* interface. The arrows indicate *Si nanoclusters* in the BOL.

dielectric layer with complex structure, as shown in Figure 6, where the TEM images from sample 2.3 clearly demonstrate two types of anodic oxide. The BOL is seen to be composed of (i) the compact silicon oxide located near the bulk silicon and (ii) the overlying porous silicon oxide. The compact oxide contains no pores but includes Si nanoclusters whereas the porous oxide incorporates both these defects.

Details of this structure are described elsewhere in these Proceedings [26] so here we pay attention to the oxidation of the top-PSL depending upon the porosity P_2 of the bottom-PSL in which the BOL is formed. Figure 7 shows the process in the two-layer PS as P_2 changes. One can see that the onset of BOL formation occurs more rapidly with increasing porosity of the bottom-PSL. This means that the top-PSL involved in AO during the initial stage is oxidized to a smaller degree when the value of P_2 increases (the size of the silicon skeleton decreases respectively). Below we will show how this observed effect impacts upon subsequent silicon epitaxy.

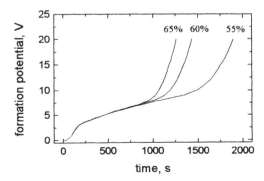

Figure 7. Set of oxidation kinetics curves demonstrating the relationship between the onset of continuous BOL formation and the porosity P_2 of the bottom-PSL. Data are from samples 1T3, 1T4, 2T2 (Table 1).

2.4. EPITAXY

In the present work we used low-temperature MBE to grow the silicon films on both PS and PS BOL substrates under otherwise similar conditions. In such cases it becomes possible to compare directly the results and, thus, elucidate the effect of the BOL on homoepitaxy.

2.4.1. Si MBE on PS

Many workers have studied a variety of low-temperature epitaxial techniques as a means of fabricating island-type SOI structures by thermally oxidizing PS through windows in deposited films. These include MBE [27-30], plasma CVD [31,32], LPCVD [21], rapid thermal CVD [33]. In ELTRAN® technology [6,7,34,35] PS is used to grow an epitaxial film via high-temperature CVD followed by its selective etching. In connection with the main points for consideration in this technology it is also important to note other relevant papers in which specific issues of epitaxial silicon growth on PSL are discussed [36-38].

Taking into account most of the above results we focussed our efforts upon obtaining more detailed information on (i) the internal structure of p^+-PSL, (ii) the relation between the top- and bottom-PSLs, and (iii) the initial stage of homoepitaxial growth.

As soon as AE was over, the wafer was rinsed with deionized water and then slightly oxidized in dry O_2 atmosphere at a temperature of 350°C for 1h. The p^+-PSL wafers were immediately loaded into the 'Katun' MBE system [39]. The substrates were cleaned *in situ* in a low-flux Si beam at 790-820°C. MBE growth of Si films was carried out using an electron-beam evaporation source. The films of thickness h_{Si} varying from 10nm to 500nm were then grown at temperatures $T_g \geq 750$°C. The film quality was monitored *in situ* using RHEED with a 20-keV electron beam during the course of both etching the surface oxide layer and depositing the epitaxial film.

When growing on porous substrate, one issue which must necessarily be resolved is that of bridging the surface pores and achieving a smooth deposited Si layer. We have studied extensively this initial stage of homoepitaxial growth. By use of a surface structure analysis technique, such as *in situ* RHEED, we were able to follow the course of the reactions on the surface of the PSL. The appearance of the Si $(001)2\times1$ structure was noted in the monitored diffraction patterns and just then the thickness of the deposited Si layer was defined as the critical thickness. At this stage the porous surface is believed to be covered with a flat Si layer of 2×1 structure uniformly distributed over the whole sample area. The critical thickness was examined for the one-layer PS in relation to its porosity P and a growth temperature T_g. It has been found that the minimum critical thickness of the epitaxial Si layer can be cut down to 3nm if $T_g \geq 750$°C and $P \leq 40$%.

In [38] the MBE growth of an ultrathin Si film on p^+-PS formed on Si(111) substrate was investigated. The values for T_g and P were determined: $T_g \geq 800$°C and $P < 47$% were required for the Si film thickness to be minimized down to $h_{Si} = 6-9$nm.

To gain greater insight into why the above relationship occurs, we carried out TEM investigations on the internal structure of p^+-PSL over the complete porosity range. Damage of the fine porous structures during preparation of thin specimens for TEM was avoided by protecting the samples with an ultrathin epitaxial Si film. The structure of porous silicon was found to depend very strongly upon the porosity. Two distinct types of porous structure were observed corresponding to (i) low porosity P ≤ 40% and (ii) high porosity P ≥ 65%. The low porosity layers have a *'cheese-like'* microstructure in which individual voids running perpendicular to the wafer surface are completely surrounded by interconnected crystalline material. As regards the high porosity layer, its microstructure is quite different. Given the porosity P ≥ 65%, the material has a *'cellular'* structure which consists of an array of long voids aligned perpendicularly to the wafer surface and separated by silicon rods with a ladder-like construction. Some of the most detailed information on these structures is illustrated in Figure 8 which shows both types of microstructure in two-layer PS. The top-PSL with the *'cheese-like'* structure (sample 11.2 with h_1 = 300nm and P_1 = 40% in Table 1) is apparent by virtue of the high contrast regions (indicated by arrows) due to the individual voids. The voids appear to be faceted and have an average in-plane dimension of 7 - 8 nm. Their distribution is random and a surface density is about $4 \cdot 10^{11}$ cm^{-2}. The bottom-PSL with the *'cellular'* structure (h_2 = 1μm and P_2 = 90%) is imaged as a black contrast network of Si rods, as clearly seen in Figure 8. The average lateral dimension of porous cells

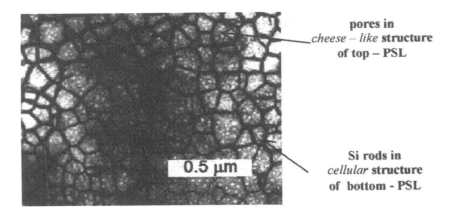

pores in
cheese – like **structure**
of top – PSL

Si rods in
cellular **structure**
of bottom - PSL

Figure 8. Plan-viewed TEM micrograph of the two-layer PS consisting of top- and bottom-PSLs with *'cheese-like'* and *'cellular'* structures, respectively.

decreases with increasing material density from ~ 80nm at P_2 = 90% down to ~ 25nm at P_2 = 65%. At the some time the silicon rods thicken increasing their diameters from ~ 6 - 8 nm to ~ 12 nm by eventual coalescence. As a result the *'cellular'* structure of porous silicon is converted to the *'cheese-like'* structure over the porosity range of 65% → 40%. We conclude that this structural transition correlates with the observed change of MBE growth on porous surface. According to these results the *'cheese-like'* structure of PS is best suited to Si epitaxial growth rather than the *'cellular'* and mixed (40% - 65%) structures. For this reason the PSL used in this technology is fabricated with a graded porosity: the top layer has the porosity of 40% or less and, therefore, is the most suitable surface for silicon epitaxy whereas the bottom layer having over 60% of porosity corresponds to the requirements of the BOL formation.

As a result Si MBE layers grown on these porous silicon substrates had a high crystalline quality controlled by TEM and RBS techniques. A typical cross-sectional TEM micrograph of the epitaxial silicon film on p$^+$-PSL is shown in Figure 9 (sample 0.2 in Table 1). No extended defects are visible in this micrograph. Good crystallinity of

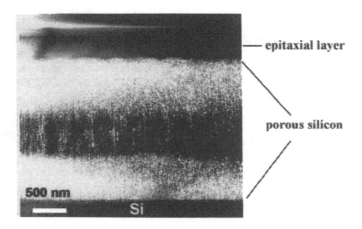

Figure 9. Typical cross-sectional TEM image of SOPS obtained by MBE of Si on PSL.

SOPS structure is also confirmed by RBS analysis for the same sample. Figure 10 shows random and aligned spectra where the surface minimum yield χ_{min} is ≈ 3.2%, a value typical of single crystal. It is noted that this value is a great improvement over the value for the top-PSL having a minimum yield χ_{min} of 10 - 16%.

In such a way we obtained good SOPS structures by means of MBE to give the starting point for the development of Si MBE on PS BOL substrates.

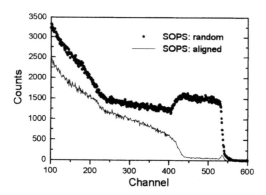

Figure 10. Typical RBS spectra from the SOPS sample reported in Fig. 9.

2.4.2. Si MBE on PS BOL

We first made a series of experiments using PS BOL structure as the substrate for Si epitaxial growth, and using the technological processes described above. The experimental results are in marked contrast with those obtained previously. The epitaxial films turned out to contain a large number of defects such as dislocations and stacking faults as shown in Figure 11 and Figure 12. Experimental values of the defect density ranged up to about $10^{11} cm^{-2}$ that was well compared with the surface density of voids observed on porous substrate, namely, about $4 \cdot 10^{11} cm^{-2}$ reported in Section 2.4.1.

 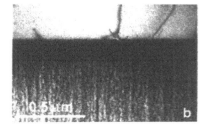

epitaxial layer

top-PSL

porous oxide

Figure 11. Cross-sectional TEM micrographs from SOPS BOL structures obtained by Si MBE: (a) sample 1.1 - h_1 = 50nm, P_2 = 65%, V_F = 15V, defect density (stacking faults and dislocations) > $10^{10} cm^{-2}$; (b) sample 2.3 - h_1 = 50nm, P_2 = 70%, V_F = 25V, dislocation density ~ $10^9 cm^{-2}$.

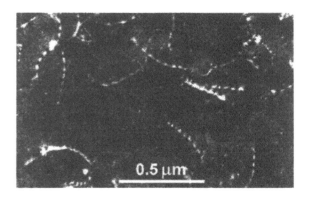

Figure 12. Plan-viewed TEM micrograph from sample 4T2: h_1 = 300nm, P_2 = 70%, V_F = 25V, dislocation density ~ $10^7 cm^{-2}$.

We investigated the structure of the epitaxial films more closely in relation to the thickness h_1, the porosity P_2, and the formation potential V_F. It has been found that the first two parameters, h_1 and P_2, can have a pronounced effect on the defect generation rate whereas the rate is insensitive to V_F. Figure 11 shows that increasing P_2 by 5% (from 65% to 70%) reduces the defect density by more than one order of magnitude. The dislocation density continues to decrease by a factor equal to about 100 as the thickness h_1 of the top-PSL is increased from 50nm to 300nm. This is illustrated in

Figure 11(b) and Figure 12 by cross-sectional and plan-viewed TEM micrographs from SOPS BOL structures (samples 2.3 and 4T2 in Table 1). Here, the extended defects are distinctly seen to be threading dislocations.

The absence of misfit dislocations from the interface between the epilayer and PS BOL suggests that the strain, if it exists, does not reach the critical level for plastic deformation to occur in epilayers [40]. Even though the system was subjected to thermal annealing to induce relaxation (for example, sample 4T2), misfit dislocations were not observed (Figure 12). It was also found that the surface morphology of this epilayer did not undergo significant distortions remaining essentially flat. In addition, it has been demonstrated by Buttard *et al.* [41] using high-resolution x-ray diffraction that the lattice parameter $a_{\parallel}$ of p^{+}-PS parallel to the surface remains practically unchanged ($\Delta a_{\parallel}$ / $a \approx 0 \pm 5 \times 10^{-5}$) during anodically oxidizing. All these data indicate that there is no lattice mismatch in SOPS BOL structures. Therefore we speculate that the source of dislocations in the Si epilayer are the following substrate inhomogeneities.

We think that the heterogeneous nucleation of threading dislocations takes place on the walls of the surface pores which are now covered with anodic oxide. As described in Section 2.4.1, if the porosity of the top-PSL is about or less 40% the porous surface with nonoxidized pores is readily smoothed during growth and threading dislocations are not generated. However, the appearance of anodic silicon oxide in the top-PSL or, more precisely, on the walls of the surface pores leads to the observed deterioration of the Si MBE growth. Furthermore, the defect density is decreased dramatically with increasing the parameter P_2 (Figure 11) which stimulates the rapid BOL formation, as shown in Figure 7. This evidence indicates conclusively that the anodic oxide inside the surface pores is the strongest perturbing factor affecting the Si MBE growth, namely, the smaller silicon oxide is in the top-PSL the higher quality of epitaxial film is achieved. The observed h_1 dependence of the defect density in the epitaxial films shown in Figure 11(b) and Figure 12 may be associated with the same factor, however, there are now no additional data. We believe that anodic oxidation of porous silicon results in (i) a marked decrease in the area of the epitaxy surface by virtue of the Si→SiO$_2$ transition on the pore walls and (ii) initiation of the disordered SiO$_2$ surface area on these pore walls hindering epitaxial growth. As an illustration we refer to the SOPS BOL structure shown in Figure 8. If this structure, as seen, consisting of the defect-free epitaxial silicon film of h_{Si} = 10nm and the two-layer PS (sample 11.2 in Table 1) should be subjected to AO (certainly, without a silicon film) forming an oxide of thickness 2.5nm ($\Delta V_F \sim 5V$) we could obtain the following substrate: (i) the area of epitaxy surface should decrease from an initial 82% to 70%, (ii) the disordered SiO$_2$ surface area should increase from 0% to 27%. The microstucture of the most likely film grown on such a substrate can be seen in Figure 11 and Figure 12. Based upon these considerations we are now working to optimize the processing conditions. It is noted that there is a similar problem in growing epitaxial silicon films on the porous layers by high-temperature CVD [34].

2.5. ELECTRICAL CHARACTERIZATION

A BOL consisting of porous and compact layers of silicon oxide (Figure 6) electrically isolates the epitaxial silicon film from the p^{+} substrate. To estimate its insulation, SOPS BOL structures were patterned using conventional lithography to produce isolated mesas of width 700 μm covered with aluminum contact pads. Current-voltage measurements were made across the SOI structure to determine leakage currents and breakdown voltages before and after thermal annealing at 900°C for 10 min in a nitrogen ambient. The results recorded from wafer 4T2 (Table 1) are shown in Figure 13. The value of the breakdown voltage was initially rather low typically being 15±5V, however, after annealing it increased dramatically up to 120±30V. It is noted that in this case V_F was equal to 25V.

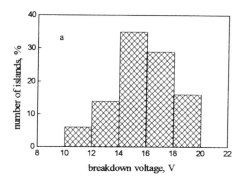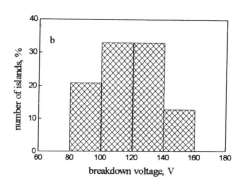

Figure 13. Histograms of breakdown voltage for 700x700 μm² mesa (a) before and (b) after annealing at 900°C. The data were recorded from sample 4T2 (Table 1).

We determined the carrier concentration and Hall mobility in these epitaxial films using the van der Pauw method [42]. A hole mobility of 135 cm²V⁻¹s⁻¹ was determined for a boron doping concentration of $1.3 \cdot 10^{17}$ cm⁻³. The source of the boron was autodoping of epitaxial film during Si MBE on the p⁺ substrate. As indicated earlier (Section 2.4.2), the dislocation density is about 10^7 cm⁻² in the film (see Figure 12) which is assumed to be the reason for the mobility being lower than the theoretical value of 200 cm²V⁻¹s⁻¹ for a perfect single crystal.

3. Advantages of SOPS BOL Technology

We have introduced a MBE-based SOPS BOL technology which is of fundamental scientific importance but has doubtfull value for industrial scale manufacturing. This is due to the fact that the SOPS BOL wafer produced by low-temperature MBE needs be subjected to high-temperature annealing in order for

(i) the porous oxide in BOL to be compacted and to acquire good dielectric qualities;

(ii) the interface of the epitaxial layer with the top-PSL to be modified to produce a planarized interface between the epitaxial layer and the compact oxide with consuming the top-PSL;

(iii) the boron concentration in the silicon film to be reduced to a level suitable for practical devices.

The substrate with BOL allows to make use of any thermal treatment or any epitaxial process in SOI production, such as a high-temperature chemical vapor deposition based on the hydrogen reduction of either $SiCl_4$ or SiH_2Cl_2. CVD method is of great utility in the field of the integrated circuit fabrication process and eliminates the above problem.

When compared to conventional SOI methods, such as SIMOX, and Smart Cut the CVD-based SOPS BOL technology has the following advantages:
1) lower cost and easier manufacture as a result of (i) the marked reduction in process complexity (Figure 1) together with (ii) use of standard semiconductor equipment;
2) high quality of SOI wafers appropriate exclusively to CVD process.
Costs for different Si wafers were estimated from conversations with commercial suppliers, and are shown in Table 2. An important point is that the use of standard equipment in CVD-based SOPS BOL technology is a significant cost saving which reduced the time to install the new technology into the factory. The technology is also compatible with manufacturing equipment being developed for 300 mm wafer processing.

TABLE 2. Comparison of SOI wafer costs between SOPS BOL and SIMOX technologies.

wafer size	4″	5″	6″	8″
	costs in US $			
Si wafer	13	20	30	100
epi-Si CVD	33	50	80	190
SIMOX	125	175	250	>500
SOPS BOL CVD	53	80	130	280

4. Conclusions

SOPS BOL technology is the simplest method for the production of full-wafer scale SOI substrates. Careful investigation of the two key processes, anodic oxidation and Si MBE epitaxy, leads to the following conclusions.

Anodic oxidation of p^+-Si in 1M HCl aqueous solution can be described by the phenomenological kinetic equations in the range of $V_F = 0$–15V. Assuming that the anodic oxidation of p^+-PSL is subjected to the same kinetics we have determined three consecutive stages in the process:

(1) the slow 'gentle' phase up to $V_F \sim 5V$ during which electrochemical oxidation occurs over practically the whole internal surface of PSL;
(2) the 'cut-off' period involving the formation of the initial BOL when $V_F = 5$–8V;
(3) the 'severe' stage when $V_F > 8V$ during which the BOL is grown mainly by sacrificial oxidation of the p^+ substrate.

The top-PSL is excluded from the AO process after the onset of BOL formation. The time taken to form the initial BOL is significantly reduced with increasing porosity of the bottom-PSL resulting in a small amount of silicon oxide in the top-PSL. BOL has a complex structure consisting of porous and compact silicon oxide layers both containing Si nanoclusters.

The structure of p^+-PSL has been shown to depend very strongly upon the porosity. There are two distinct types of porous material which is observed after annealing at 350°C in an oxygen ambient and then at growth temperatures 750–800°C in high vacuum:

(1) a *'cheese-like'* structure which is inherent in low porosity silicon when $P \leq 40\%$ consisting of individual voids running perpendicular to the wafer surface and completely surrounded by interconnected crystalline material;

(2) a *'cellular'* structure which is inherent in high porosity silicon when $P \geq 65\%$ consisting of an array of long voids directional perpendicular to the wafer surface and being separated by silicon rods.

It has been determined that Si MBE growth on the *'cheese-like'* PSL proceeds with the production of a dislocation-free silicon film for growth temperatures $T_g \geq 750°C$. Different porous materials with porosity ranging from 40% and over are not suited to high-quality epitaxial growth.

The epitaxial films grown on PS BOL substrates contain a large number of defects such as threading dislocations and stacking faults. The defect density decreases with increasing P_2 and h_1, but is almost independent of V_F. Heterogeneous nucleation of dislocations is assumed to arise from substrate inhomogeneities, namely, the surface pores when the walls are covered by an anodic oxide film. This assumption is substantiated by our experimental results and is consistent with first order calculations.

The SOPS BOL structure has a breakdown voltage of 15±5V which increases dramatically up to 120±30V after annealing at 900°C. In this case the formation potential of the BOL is equal to 25V. The epitaxial film has a boron doping concentration of $1.3 \cdot 10^{17}$ cm^{-3} and a hole mobility of 135 cm^{2}V^{-1}s^{-1}, or 68 percent of the value in a perfect crystal. The decrease in hole mobility may be presumed to be due to the presence of the observed dislocations in this film.

The advantages of SOPS BOL technology over conventional SOI methods are discussed.

Acknowledgements

The authors would like to thank the State Scientific and Technical Program "Promising Technologies and Devices in Micro-and Nanoelectronics", grant 02.04.1.1.16.□.1, the Siberian Branch of the Russian Academy of Sciences (Grant No.15000-421), the Ministry of High School (Grant No.□□-103-98 via Novosibirsk State University), the Russian Foundation for Basic Research (Grants No.96-02-19301, 97-02-18569, 98-02-17790), and the 'Physics of Solid-State Nanostructures' State Program (Grant No.3-011/4) for part funding of this development.

5. References

1. Colinge, J.-P. (1997) *SILICON-ON-INSULATOR TECHNOLOGY: Materials to VLSI, 2nd Edition,* Kluwer Academic Publishers, Boston.

2. Hemment, P.L.F., Cristoloveanu S., Izumi, K., Houston, T., and Wilson, S. (eds.) (1996) *Silicon-on-Insulator Technology and Devices VII*, Proceedings of the Electrochemical Society, Vol. 96-3, The Electrochemical Society, Inc., Pennington.
3. Colinge, J.P., Lysenko, V.S., and Nazarov, A.N. (eds.) (1995) *Physical and Technical Problems of SOI Structures and Devices*, NATO ASI Series 3: High Technology – Vol.4, Kluwer Academic Publishers, Dordrecht.
4. Alles, M. and Wilson, S. (1997) Thin Film Silicon on Insulator: An Enabling Technology, *Semiconductor International* **20**, No.4, 67-74.
5. Lasky, J.B., Stiffler, S.R., White, F.R., and Abernathey, J.R. (1985) Silicon-on-insulator (SOI) by bonding and etch-back, *Tech. Dig. International Electron Devices Meeting*, Proc. IEDM 85, IEEE, New York, pp.684-687.
6. Sakaguchi, K., Sato, N., Yamagata, K., Atoji, T., Fujiyama, Y., Nakayama, J., and Yonehara, T. (1997) Current Progress in Epitaxial Layer Transfer (ELTRAN®), *IEICE Trans. Electron.* **E80-C**, No.3, 378-387.
7. Yonehara, T. US Patent 5371037 (1991).
8. Bruel, M. US Patent 5374564 (1994).
9. Bruel, M., Aspar, B., and Auberton-Hervé, A.-J. (1997) Smart-Cut: a new Silicon on Insulator Material Technology based on Hydrogen Implantation and Wafer Bonding, *Jpn. J. Appl. Phys.* **36,** 1636-1641.
10. Tong, Q.-Y. and Gösele, U. US Patent application 08/866, 951 (1997).
11. Tong, Q.-Y., Scholz, R., Gösele, U., Lee, T.-H., Huang, L.-J., Chao, O.-L., and Tan, T.Y. (1998) A "smarter-cut" approach to low temperature silicon layer transfer, *Appl. Phys. Lett.* **72**, 49-51.
12. Romanov, S.I. Russian Federation Patent application 97103165 (1997).
13. Romanov, S.I. Russian Federation Patent application 97103424 (1997).
14. Karanovich, A.A., Romanov, S.I., Kirienko, V.V., Myasnikov, A.M., and Obodnikov, V.I. (1995) A secondary ion mass spectrometry study of p^+ porous silicon, *J. Phys. D: Appl. Phys.* **28**, 2345-2348.
15. Bsiesy, A., Vial, J.C., Gaspard, F., Herino, R., Ligeon, M., Muller, F., Romestain, R., Wasiela, A., Halimaoui, A., and Bomchil, G. (1991) Photoluminescence of high porosity and of electrochemically oxidized porous silicon layers, *Surface Sci.* **254**, 195-200.
16. Halimaoui, A., Oules, C., Bomchil, G., Bsiesy, A., Gaspard, F., Herino, R., Ligeon, M., and Muller, F. (1991) Electroluminescence in the visible range during anodic oxidation of porous silicon films, *Appl. Phys. Lett.* **59**, 304-306.
17. Ligeon, M., Muller, F., Herino, R., Gaspard, F., Vial, J.C., Romestain, R., Billat, S., and Bsiesy, A. (1993) Analysis of the electroluminescence observed during the anodic oxidation of porous layers formed on lightly p-doped silicon, *J. Appl. Phys.* **74**, 1265-1271.
18. Halimaoui, A. (1993) Influence of wettability on anodic bias induced electroluminescence in porous silicon, *Appl. Phys. Lett.* **63**, 1264-1266.
19. Vázsonyi, É.B., Koós, M., Jalsovszky, G., and Pócsik, I. (1993) The role of hydrogen in luminescence of electrochemically oxidized porous Si layer, *J. Luminescence*, **57**, 121-124.
20. Rigakis, N., Hilliard, J., Abu Hassan, L., Hetrick, J. M., Andsager, D., and Nayfeh, M.H. (1997) Effect of oxidation treatment on photoluminescence excitation of porous silicon, *J. Appl. Phys.* **81**, 440-444.

21. Bomchil, G., Halimaoui, A., and Herino, R. (1989) Porous silicon: the material and its applications in silicon-on-insulator technologies, *Appl. Surf. Sci.* **41/42**, 604-613.
22. Lee, C.H., Yeh, C.C., Hwang, H.L., and Hsu, K.Y.J. (1996) Characterization of porous silicon-on-insulator films prepared by anodic oxidation, *Thin Solid Films* **276**, 147-150.
23. Bsiesy, A., Gaspard, F., Herino, R., Ligeon, M., Muller, F., and Oberlin J.C. (1991) Anodic Oxidation of Porous Silicon Layers Formed on Lightly p-Doped Substrates, *J. Electrochem. Soc.* **138**, 3450-3456.
24. Allegretto, E.M. and Bardwell, J.A. (1996) Characterization of as-grown and annealed thin SiO_2 films formed in 0.1 M HCl, *J. Vac. Sci. Technol.* **A14**, 2437-2442.
25. Herino, R., Bomchil, G., Barla, K., Bertrand, C., and Ginoux, J.L. (1987) Porosity and Pore Size Distributions of Porous Silicon Layers, *J. Electrochem. Soc.* **134**, 1994-2000.
26. Romanov, S.I., Dvurechenskii, A.V., Yakovlev, Yu.I., Grötzschel, R., Kreissig, U., Kirienko, V.V., Obodnikov, V.I., and Gutakovskii, A. (1999) Characterization of porous silicon layers containing a buried oxide layer, *in this issue the following paper*.
27. Konaka, S., Tabe, M., and Sakai, T. (1982) A new silicon-on-insulator structure using a silicon molecular beam epitaxial growth on porous silicon, *Appl. Phys. Lett.* **41**, 86-88.
28. Lin, T.L., Chen, S.C., Kao, Y.C., Wang, K.L., and Iyer, S. (1986) 100-μm-wide silicon-on-insulator structures by Si molecular beam epitaxy growth on porous silicon, *Appl. Phys. Lett.* **48**, 1793-1795.
29. Lin, T.L. and Wang, K.L., (1986) New silicon-on-insulator technology using a two-step oxidation technique, *Appl. Phys. Lett.* **49**, 1104-1106.
30. Zheng, D.W., Cui, Q., Huang, Y.P., Zhang, X.J., Kwor, R., Li, A.Z., and T. A. Tang, T.A. (1998) A Low Temperature Silicon-on-Insulator Fabrication Process Using Si MBE on Double-Layer Porous Silicon, *J. Electrochem. Soc.* **145**, 1668-1671.
31. Takai, H. and Itoh, T. (1983) Isolation of silicon films grown on porous silicon layer, *J. Electronic Materials* **12**, 973-982.
32. Takai, H. and T. Itoh, T. (1986) Porous silicon layers and its oxide for the silicon-on-insulator structure, *J. Appl. Phys.* **60**, 222-225.
33. Oules, C., Halimaoui, A., Regolini, J.L., Perio, A., and Bomchil, G. (1992) Silicon on Insulator Structures Obtained by Epitaxial Growth of Silicon over Porous Silicon, *J. Electrochem. Soc.* **139**, 3595-3599.
34. Sato, N., Sakaguchi, K., Yamagata, K., Fujiyama, Y., and Yonehara, T. (1995) Epitaxial Growth on Porous Si for New Bond and Etchback Silicon-on-Insulator, *J. Electrochem. Soc.* **142**, 3116-3122.
35. Sato, N., Sakaguchi, K., Yamagata, K., Fujiyama, Y., Nakayama, J., and Yonehara, T. (1996) Advanced Quality in Epitaxial Layer Transfer by Bond and Etch-back of Porous Si, *Jpn. J. Appl. Phys.* **35**, 973-977.
36. Beale, M.I.J., Chew, N.G., Cullis, A.G., Gasson, D.B., Hardeman, R.W., Robbins, D.J., and Young, I.M. (1985) A study of silicon MBE on porous silicon substrates, *J. Vac. Sci. Technol.* **B3**, 732-735.
37. Ito, T., Yasumatsu, T., and Hiraki, A. (1990) Homoepitaxial growth of silicon on anodized porous silicon, *Appl. Surf. Sci.* **44**, 97-102.

38. Yasumatsu, T., Ito, T., Nishizawa, H., and Hiraki, A. (1991) Ultrathin Si films grown epitaxially on porous silicon, *Appl. Surf. Sci.* **48/49**, 414-418.
39. Multi-chamber 'Katun' complex for molecular beam epitaxy, *An advertisement pamphlet (1992) of Institute of Semiconductor Physics, Novosibirsk, Russia.*
40. Fitzgerald, E.A. (1991) Dislocations in strained-layer epitaxy: theory, experiment, and applications, *Materials Science Reports* **7**, 87-142.
41. Buttard, D., Bellet, D., and Dolino, G. (1996) X-ray-diffraction investigation of the anodic oxidation of porous silicon, *J. Appl. Phys.*, **79**, 8060-8070
42. Van der Pauw, L.J. (1958) A method of measuring specific resistivity and Hall effect of discs of arbitrary shape, *Philips Res. Repts.*, **13**, 1-9.

STRUCTURAL AND ELECTRICAL PROPERTIES OF SILICON ON ISOLATOR STRUCTURES MANUFACTURED ON FZ- AND CZ-SILICON BY SMART-CUT TECHNOLOGY

V.P.Popov, I.V.Antonova, V.F.Stas, L.V.Mironova, E.P.Neustroev,
A.Gutakovskii A.A. Franzusov, G.N.Feofanov
Institute of Semiconductor Physics, RAS
630090, Lavrentieva 13, Novosibirsk, Russia

Abstract

The evolution of the structural and electrically active defects during SOI fabrication by the Smart-Cut technology was investigated. The important roles of hydrogen and boron concentrations in the defects observed in the top layer of SOI structures are shown.

1. Introduction

Nowadays the Smart-Cut technology for the manufacture of the silicon-on-isolator (SOI) structures is an attractive way to create SOI with an ultrathin top layer [1,2]. High fluence hydrogen implantation with subsequent annealing allows one to delaminate a thin layer of silicon from a thick substrate and to connect it to an oxidized silicon wafer. The main positive aspects of this technology are a high quality buried oxide, easy control of the top layer thickness and possibility of repeated reuse of the delaminated wafer. The bonding temperature is generally about 350 - 600°C [1]. The following anneal at higher temperature (1100°C) enhances the bond strength at the bonding interface. Understanding of the basic processes enable us to obtain high-quality SOI wafers. The aim of the present work was to investigate the transformation of the electrical and structural defects in these SOI wafers during annealing in the temperature range of 450 - 1100°C.

2. Experimental

Different types of silicon wafers were used as substrates for SOI fabrication. The main parameters of the initial silicon are presented in Table 1. The 100 mm wafers with high quality surface and high flatness (the warpage should be less than 10 μm) were used for bonding.

Implantation of H_2^+ ions was performed through a oxide layer (500 – 3200 A) with the energy of 135 keV and dose (of H atoms) in the range of 4-6.10^{16} cm^{-2}. The ion projected range was equal to 0.65 μm. The dose was chosen to be below a critical dose above which blistering appears after the implantation. This dose is dependent on the

47

P.L.F. Hemment et al. (eds.),
Perspectives, Science and Technologies for Novel Silicon on Insulator Devices, 47–54.
© 2000 *Kluwer Academic Publishers. Printed in the Netherlands.*

48

Table 1 Main parameters of initial silicon wafers

Type of initial Si	Doping level, cm^{-3}	Oxygen concentration, cm^{-3}	Carriers mobility, cm^2/(Vs)	orientation
Fz-Si, n	6.10^{13}	$< 10^{16}$	1300	<111>
Cz-Si-1, p	5.10^{14}	$7\text{-}8.10^{17}$	350	<100>
Cz-Si-2, p	1.10^{19}	$8\text{-}9.10^{17}$	60	<111>

surface coverage, the energy of hydrogen ions, implantation temperature and boron concentration in the irradiated wafer. The thickness of the oxide layer on the handle wafer was equal to 0.2 μm and as a rule this oxide was used as a buried oxide. This oxide was form by dry oxidation of Si surface at 1050°C. Only in some cases the oxide film on wafer implanted by hydrogen (for oxide thickness of 0.32 μm) was utilized as a buried oxide in SOI structure. To obtain the splitting effect over the whole wafer and high-quality bonding with strong bonding energy special cleaning treatments and surface preparations were used. The bonding regime consists of two steps: annealing at 200°C for the wafer connecting and at 450°C for the delaminating. The regimes with different times for the second annealing step (0.5 h and 3 h) were used. Commonly the annealing at 1100°C and polishing for high quality bonding and defect removal finish the SOI fabrication. According to the aim of this study the as-bonded wafers were subjected to isochronal 30min annealing in the temperature range of 450 –1100°C. Polishing is not made for the investigated SOI wafers. The voltage-capacity (CV) and Hall effect measurements, point- contact transistor (also called pseudo-MOS transistor) technique (PCT) [3], transmission electron microscopy (TEM) and Rutherford back-scattering (RBS) were used for studying the SOI properties. Some experiments were carried out on the silicon samples implanted by protons and annealed in the same regimes as bonding one.

3. Results

The bonded wafers (the top layer is placed first) and some of the as-bonding properties are presented in Table 2. The double line "//" dividing the layers corresponds to position of the bonding interface.

Table 2 Bonded wafers and some of the as-bonded properties

Bonding wafers	Symbol	Delamination anneal	As-bonded carrier concentration, cm^{-3}
Cz-Si-1//2000A SiO$_2$/Cz-Si-1	SOI-I	450°C, 0.5h	$(6\text{-}10)10^{16}$
Cz-Si-2/3200A SiO$_2$//Fz-Si	SOI-II	450°C, 0.5h	$10^{17}\text{-}10^{18}$
Cz-Si-1//2000A SiO$_2$/Cz-Si-2	SOI-III	450°C, 3h	$(2\text{-}8)10^{16}$
Fz-Si//2000A SiO$_2$/Cz-Si-3	SOI-IV	450°C, 3h	$(2\text{-}8)10^{16}$

As-bonded wafers have n-type conductivity in the top layer with electron concentration of $2.10^{16} – 1.10^{17}$ cm^{-3}, independent on the conductivity type of the initial silicon crystals utilized for the top layer. Fig.1a,b show the results of CV measurements of wafers SOI-I and SOI-II annealed at different temperatures. One can see that if the bonding time is short (0.5 h) high n-type conductivity is observed up to a maximum annealing temperature of 1100°C. Moreover the higher boron concentration in the initial crystal leads to a higher electron concentration in the SOI top layer (SOI-II). The PCT and Hall measurements also show n-type conductivity for SOI-I and SOI-II with a high electron concentration close to CV data. The carrier mobility for SOI–I annealed at 900°C is equal to 440 cm²/Vs.

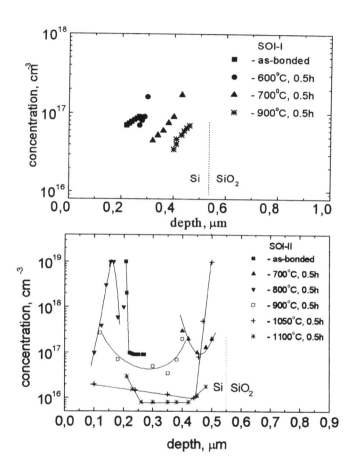

Figure 1. Depth distribution of the electron concentration obtained by CV measurements of wafers SOI-I and SOI-II, annealed at different temperatures

The increase in the bonding time (SOI-III,IV) causes large changes and a complicated situation. The temperature dependence of the carrier concentration for this case is given in Fig 2. N-type of conductivity is observed only in the as-bonded structures or after 500°C annealing. Higher anneal temperature result in p-type conductivity up to 800°C.

50

Fig.3 shows some results obtained from SOI-III and SOI-IV by the PCT technique. Drain current, I_d, includes two components of current: current in the Si/SiO_2 channel and in the silicon film (including surface). Current in the Si/SiO_2 channel is strongly modulated by the gate voltage and its value in the minimum has to be very low (I_{min} <1 nA). Really I_{min} is essentially higher and is determined by the conductivity of the silicon film. The exponential increase in I_d in the vicinity of the minimum allows one to estimate the interface state density D_{it} [3]. The film conductivity σ_f and the interface channel conductivity σ_{sc} can be also obtained from $I_d(V_g)$ measurements.

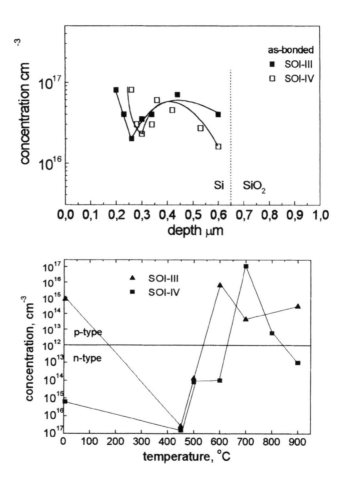

Figure 2. Depth distribution and temperature dependence of the carrier concentration for wafers SOI-III and SOI-IV

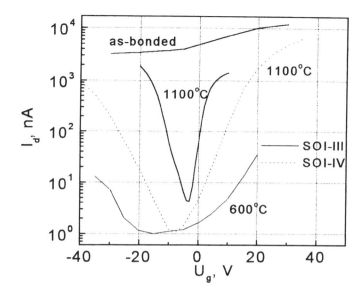

Figure 3 Drain current characteristics versus gate bias measured by the PCT
technique.

Table 3 presents the results of these estimations. Values of σ_f and D_{it} decrease with
increasing anneal temperature from 600°C.

Table.3 Interface state density D_{it}, film conductivity σ_v and interface channel
conductivity σ_{sc}

Wafer	Anneal	D_{it}, cm^{-2}eV^{-1}	σ_f, Ω^{-1}	σ_{sc}, Ω^{-1}
	As-bonding	6.3×10^{13}	1×10^{-5}	1.2×10^{-5}
SOI-III	600°C, 0,5 h	1.6×10^{13}	6.2×10^{-6}	1.2×10^{-9}
	1100°C, 0,5 h	3×10^{12}	2.5×10^{-9}	1×10^{-8}
	As-bonding	9×10^{13}	8.7×10^{-6}	3.7×10^{-6}
SOI-IV	600°C, 0,5 h	2.7×10^{13}	2.5×10^{-9}	2.5×10^{-9}
	1100°C, 0,5 h	1×10^{13}	9×10^{-8}	7.5×10^{-7}

Defects located at some depth centered near the ion projected range and obtained by
TEM for hydrogen implanted samples are presented in Fig.4. The samples annealed at
the temperature of 450°C show clusters of point defects, dislocation loops and
microcavities (bubbles). The increase in anneal temperature up to 650°C annihilates the
point defect clusters. The temperature of 1050°C leads to large bubbles and a dislocation
network. RBS measurements (Fig.5) on SOI annealed at 1050°C also show the presence
of defects located near the film/buried oxide interface.

The properties of the buried oxide are shown in Table 4. It is evident that the radiation defects introduced in the oxide during hydrogen implantation are not annealed up to 1050°C and give rise to high leakage current and low break voltage for SOI-II. The best oxide properties were obtained in the case of SOI-I, SOI-III, SOI-IV when hydrogen implantation was made through a thin oxide (~ 500A) which was removed before bonding with the oxidized wafer.

Table 4. Some properties of the buried oxide for different SOI structures

Wafer	Leakage current (10V), nA	Break voltage, V
SOI-1	<0.1	90
SOI-II	10^3	15
SOI-III	<0.1	100
SOI-IV	<0.1	60

4. Discussion

The bonding temperature is one of the most important temperatures for the removal of hydrogen from the crystal [4]. The main difference between SOI-I,II and SOI-II,IV is the higher hydrogen concentration in the top layer in the first case. SIMS measurements for hydrogen implanted samples show that hydrogen concentration decreases with the increase in the annealing time in the temperature interval of 400 - 700°C but hydrogen still remains in essentially at the same concentration in the SOI after bonding as well as after used annealing [4,5].

TEM also shows the survival of bubbles up to 1050°C. Thus we conclude that the donors which are stable up to 1100°C and observed in SOI-I,II are connected with the high hydrogen concentration in the top layer. Another important point is the boron concentration: the higher boron concentration provides the higher donor concentration. It is known [6] that boron atom at interstitial positions can be a donor but usually it is observed at low temperature. Hydrogen also can create donor centers up to a temperature ~ 520°C [7]. The obtained results allow us to suggest the existence of a high-temperature stable complex of interstitial boron secured by structural defects with the assistance of hydrogen atom(s). Its stability in our opinion is provided by structural defects (dislocation network and/or loops) which as shown in the TEM and RBS data still exist in the structure.

The lower concentration of hydrogen in SOI-III,IV results in other observed defects. The high electron concentration in as-bonded structures annealed at 500°C is most likely caused by the hydrogen enhanced introduction of thermal donors [5] and hydrogen-related defects [8]. Acceptors, which are introduced in the temperature interval of 600-800°C, can be connected with radiation defects in the implanted layer. The annealing of the radiation defects again causes n-type conductivity for both crystals. These donors are most likely the same ones as for the case of SOI-I,II but present in the crystal at a lower concentration.

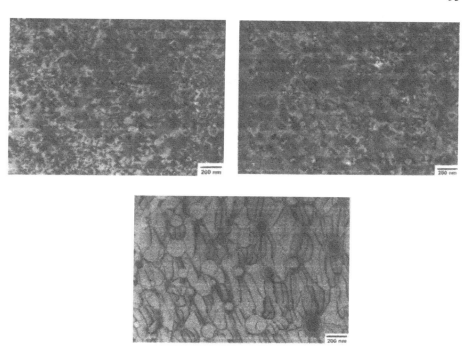

Figure 4. Plan-view TEM for the hydrogen implanted crystals annealed at 450, 650, 1050°C for 0.5 h

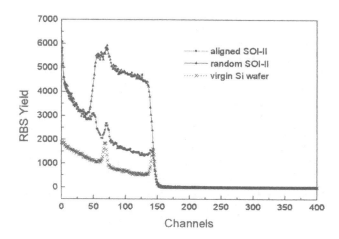

Figure 5. RBS spectra for SOI-II annealed at 1050°C at 05 h.

The high interface state density found in these SOI structures is most likely connected with hydrogen atoms. Investigation of the <111> Si/SiO$_2$ interface heat treated in a hydrogen atmosphere shows that D$_{it}$ is equal to 3.10^{13} cm^{-2} for T > 550°C [9]. This value is about one order of magnitude higher than D$_{it}$ value for the vacuum annealed samples. Authors [9] suggest the increase in D$_{it}$ at temperatures above 550°C caused by activation of the existing interface states is due to diffusion of hydrogen atoms to these states. In the case of SOI structures annealed at 600°C already leads to a decrease in the interface state density which is the results of hydrogen out-diffusion from the top layer.

5. Summary

Post-bonded anneals up to 1050°C does not cause the complete removal of radiation and/or hydrogen-related defects in the top layer. Investigation of the transformation of the electrically active and structural defects shows the strong dependence of the electrical properties on the bonding time. The increase in the bonding time (or time of annealing at 450°C) gives rise to a decrease in the donor concentration in SOI top layer. These results allow us to suggest the existence of a high-temperature stable complex of interstitial boron with structural defects (for example, dislocation networks or loops) with assistance of hydrogen atom(s). Other observed electrically active defects are radiation-related acceptors introduced in SOI with lower hydrogen contamination in the temperature interval of 600-800°C. The high interface state density obtained for SOI structures is most likely connected with hydrogen atoms activating the existing states.

6. References

1. Bruel, M., Aspar, B., Maleville, C., Moriceau, H., Auberton-Herve, A.J., Barge, T. (1997) Unibond SOI wafers achieved by smart-cut process, Electrochem. Soc. Proceedings, 97-23, 3-14.
2. Bruel, M. (1996) Nucl. Inst. Meths. B, 108, 313.
3. Ionescu, A.M, Cristoloveanu, S., Wilson, S.R., Rrusu,A., Chovet A., Seghir, H. (1996) Improved characterization of full-depleted SOI wafers by pseudo-MOS transistor, Nucl. Inst. Meths. B 112, 228-232.
4. Hara, T., Kakizaki, Y., Oshima, S., Kitamura, T. (1998) Annealing effects in the delamination of H implanted silicon, , Electrochem. Soc. Proceedings, 97-23, 33-38.
5. Popov, V.P., Neustroev, E.P., Antonova I.V., Stas, V.F., Obodnikov, V.I. (1998) Donor center formation in hydrogen implanted silicon, European Material Research Society spring meeting, paper AP9.
6. Corbett, J.W., Bourgoin, J.C., Ccheng, L.J., Corelli, J.C., Lee, Y. H, Mooney, P. M, Weigel, C. The Status of Defect Studies in Silicon. - In: Radiation Effects in Semiconductors. Conf. Ser. N 31, Inst. of Phys. London – Bristol, 1977, p. 1 – 11.
7. Stain, H.J., Hahn S. (1995) J.Electrochem. Soc., 142, 1242
8. Tokuda, Y., Ito, A., Ohshima H. (1998) Semicond.Sci.Technol. Study of shallow donor formation in hydrogen-implanted n-type silicon 13, 194-199.
9. Stessmans, A., Afanas'ev V.V, (1998) Appl.Phys.Lett. 72, 2271

DEVELOPMENT OF LINEAR SEQUENTIAL LATERAL SOLIDIFICATION TECHNIQUE TO FABRICATE QUASI-SINGLE-CRYSTAL SUPER-THIN Si FILMS FOR HIGH-PERFORMANCE THIN FILM TRANSISTOR DEVICES

A.B.LIMANOV
Institute of Crystallography, Russian Academy of Sciences, Moscow.
Present address: Columbia University, Henry Krumb School of Mines,
New York, NY. <abl24@columbia.edu>
V.M.BORISOV, A.YU.VINOKHODOV, A.I.DEMIN, A.I.EL'TSOV,
YU.B.KIRUKHIN. O.B.KHRISTOFOROV
TRINITI, Troitsk, Moscow region. <borisov@fly.triniti.troitsk.ru>

1. Introduction

Sequential Lateral Solidification (SLS) [1], and Single-Crystal Sequential Solidification (SCSLS) [2] techniques have been recently proposed by the Columbia University group in order to recrystallize amorphous silicon precursor films into layers with crystalline quality close to single-crystal films. These new methods facilitate superior quality and uniformity of Si films formed on low-temperature glasses compared to the conventional excimer-crystallization techniques. The methods have great significance for Thin Film Transistor (TFT) applications, especially for driver- and system-integrated Liquid-Crystal Displays (LCDs). However, the new techniques, as previously demonstrated [1,2], have certain features that may be further improved. In particular, one notes that: (1) only a small part of the laser energy is actually used for crystallization; (2) if only a single beam is used, then the crystallization rate will be rather low; (3) in addition, previous results have used relatively thick films only.

In this paper we propose a new approach for SLS-processing that is efficient in its use of energy, and which together with a high-power excimer laser optimized for SLS, can provide high crystallization rates even when using a single beam for SLS. Forty-nanometer amorphous-Si films-on-glass have been successfully crystallized by this technique. The microstructure of such thin films have been studied and compared with the microstructure of thick (200 nm) films. Extension of Super Lateral Growth is investigated and optimized under a broad range of SLS-process conditions. Computer modeling is used to study the temperature field, and the velocity of the crystallization front.

2. Linear-SLS-technique

The scheme of crystallization is presented in figure 1 where, for illustrative purposes, the vertical scale is reduced with respect to the horizontal one. We use an excimer laser (1) with a linear-shaped output beam, and a beam delivery system based on cylindrical

P.L.F. Hemment et al. (eds.),
Perspectives, Science and Technologies for Novel Silicon on Insulator Devices, 55–61.
© 2000 *Kluwer Academic Publishers. Printed in the Netherlands.*

optics. The system consists of an optical compressor (2), an expander-homogenizer (3), a bulk-metal mask (4) with a narrow straight slit, and a projection objective (5). Lines LL', KK', and PP' represent the axes of the cylindrical lenses which form these optical components. The laser beam is aligned with the slit in the mask. An linear image (8) of the slit is formed at the sample surface (7) by the objective. In this case, only the narrow central area of the objective is used, which reduces the aberration. It allows the image to be demagnified by twenty times. As a result a high energy density is obtained to melt the Si film in the image plane. This excess energy can be employed to increase the length of the linear image by the addition of an expander installed perpendicular to the objective. Additionally, the optical compressor increases the energy density in order to more prolong the linear image by the expander.

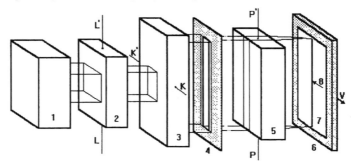

Figure 1. Schematic of the installation for SLS processing with a linear laser beam.

Excimer lasers with energy density of approximately 0.15 J/cm^2 were used. The energy was stable from pulse-to-pulse within 1.5%. Wavelengths of 248 and 308 nm were used. The pulse duration was about 25, 50, and 150 ns (FWHM). The most effective results were obtained using an excimer laser with a linear output aperture of 2x30 mm^2. Due to the small aperture, the laser had a small total energy. As a result of the small energy consumption per pulse, it could operate at repetition rates up to 3 kHz. In such a way, a broad (up to 70-mm-wide) area of the Si film was crystallized by a single pass of the laser beam. The crystallization velocity was 3 mm/s. The width of the melt/illuminated track was limited by the aperture of the optics rather than by the output of the laser.

3. Investigation of the grown films

Corning Glass 7059 was used as a substrate. The substrates were covered with a 300 nm buffer layer of Si_3N_4 deposited by a PECVD process at 300°C.

Amorphous Si films of thickness 200 nm were deposited by LPCVD at 567°C and amorphous Si films of thickness 40 nm were deposited by PECVD at 250°C. Films deposited by the PECVD process were annealed in vacuum at 450°C for 1 hour.

Grown Si films were characterized using an optical microscope, a scanning electron microscope (SEM), and a transmission electron microscope (TEM). Defects were revealed for optical and SEM inspections by selective (Secco) etching [3]. Samples for TEM examination were prepared by peeling off the silicon film under HF acid.

The quality of the films after SLS was found to be independent of the deposition technique (LPCVD or PECVD followed by thermal annealing) and laser wavelength, within our experimental condition. However, a change of pulse duration markedly affected the SLS-processing. A change from 25 to 150 ns improves the film uniformity and increases the growth velocity by a factor of about three due to a longer translation distance per pulse.

Grown films of thickness 40 nm had a smooth surface without any hillocks or protrusions. The grain structure of the films was controlled by competitive growth of the grains during lateral directional solidification. Thus the films consisted of long and narrow grains extended along the growth direction (Fig.2). Only separated dislocations were found within the grains (Fig.2-right) where the width of the grains was about 0.5 μm. The growth direction of the grains deviated within an angle of ± 10° with respect to the direction of the laser beam scanning. Several parallel grains formed extended blocks of width up to 10 μm. Some of the blocks extended up to hundreds of micrometers. All grains within a single block originated from a single grain-seed by sequential partition resulting in boundaries within each block being low-angle boundaries, or

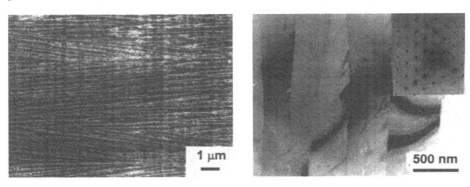

Figure 2. Microstructure of films with thickness 40 nm:
SEM image – left, TEM micrograph (the insert shows a diffraction pattern) – right.

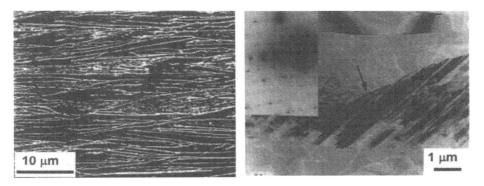

Figure 3. Grown films of thickness 200 nm:
SEM image - left, TEM micrograph of twinning area - right (inset shows diffraction pattern from twins).

subboundaries. This microstructure was confirmed by selected area electron diffraction in the TEM where the insert in figure 2 (left) shows diffraction patterns from two neighboring grains misoriented by about 3°. The diffraction pattern shows that grown films have (110) in-plain orientation, and that the grains grow along a <001> direction.

Grown films of thickness 200 nm had a more random microstructure as shown in fig.3. The average width of the grains was about 1 μm, but the some grains varied in width by as much as a factor of ten. Blocks consisting of twin lamellae were found (Fig.3-right). TEM diffraction showed some deviation of the in-plane orientation of grains from the preferential (110) orientation.

Additional information concerning SLS-grown Si films is given in reference [4].

4. Investigation of Super Lateral Growth

Super Lateral Growth (SLG) is a single stage and key component of the SLS-technique. Its optimization may result in a higher growth velocity and a higher quality of SLS-grown films. SLG is defined as a growth of a thin Si film under a single laser pulse, where the growth occurs under the special conditions that are needed to realize the SLS process; i.e., the Si film is completely melted during the laser pulse along the narrow trace illuminated by the laser.

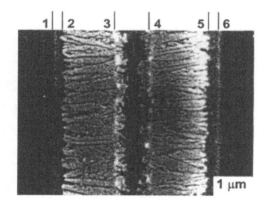

Figure 4. SLG of an amorphous silicon film of thickness 40 nm. Pulse duration was 25 ns.

The typical SEM micrographs showing the microstructure of amorphous silicon (a-Si) films formed by the SLG process are presented in figs. 4 and 5, where the films were exposed to a single shoot of the linear laser beam between boundaries 1 and 6. As a result of the treatment, the film was completely melted between boundaries 2 and 5 and only partially melted between boundaries 1 and 2 and also 5 and 6. Within the areas 1 - 2 and also 5 - 6 the grains grew from an unmelted material near the bottom of the Si film towards the surface resulting in a grain sizes that are similar to the film thickness. It was found that two columns of large grains grew laterally from the partially melted material at boundaries 2 and 5. During lateral growth the temperature of the Si melt along the central part of the treated region dropped significantly below the melting point of silicon resulting in nucleation between boundaries 3 and 4. The growth of grains nucleated within the area 3 – 4 terminated the lateral growth of the large grains along boundaries 3 and 4. The size of the grains within the area 3 – 4 was controlled by the nucleation density in the deep undercooled melt. In the case shown in fig.4 (where the film was thin and the pulse was short), the density of the nucleating sites was higher resulting in a fine-grained film within area 3 – 4. Such fine-grained material was completely removed by Secco etching as well as the remaining precursor a-Si film outside the treated region. The sample shown in figure 5 had a lower nucleation density and, thus larger grains that subsequently were not completely removed by the selective etching within the area 3 - 4. We have studied the "extension" of SLG (i.e., the width of columns 2 - 3 and 4 - 5) in thin (40 nm) as well in thick (200 nm) Si films under different process conditions.

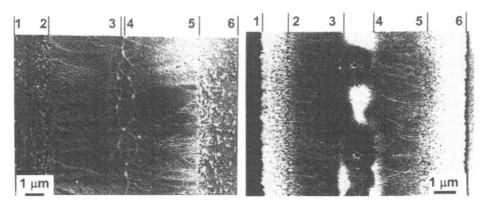

Figure 5. Extension of the crystallized regions following SLG for a pulse duration of 150 ns in Si film of thickness: 40 nm –left, 200 nm – right.

It was found that the largest grains were formed only in the case of maximum energy density of the laser pulse and under perfect focusing of the image of the mask. The energy density was limited by degradation of the treated region (because of agglomeration or scattering by the Si melt). Decreasing of the energy density by about 20% below the degradation threshold did not markedly affect the extension in thick Si films, but decreased the extension in thin Si films by a factor of two. Defocusing of the image resulted in a decrease of the grain size by a factor of four.

The influence of the pulse laser duration on the extension of the lateral growth in SLG structures is shown in fig.6. These results were obtained under ideal imaging conditions and energy densities just below the damage threshold. So, the extension of the SLG regime increased with increasing pulse duration and was greatest for the thinner films. It was found that the extension in thick and thin films is equal for a pulse duration of 50 ns, but was higher in thin films if the pulse duration became greater than 50 ns. This result is shown in fig.5 for a pulse duration of 150 ns. It should be noted that an increase of the pulse duration causes the areas where the film is partially melted (i.e., areas 1 - 2 and 5 - 6) to increase. Probably this is caused by heat dissipation during the longer laser pulse.

60

It should be noted that the acronym "SLG" was first introduced in 1993 by Im et al [5] to describe the treatment of a Si film by a spatially uniform laser beam with an energy density sufficient to effect complete melting although with the survival of same Si seed islands. In our case a special condition is needed to realize the SLS process and we suppose that "Artificial Controlled SLG" (ACSLG) is here a better term.

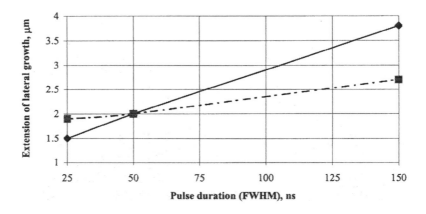

Figure 6. Extension of lateral growth vs. pulse duration for Si films of thickness:
40 nm – solid, and 200 nm dashed line.

5. Investigation of the SLG by modeling of the process

The SLG-process has been studied using a two-dimensional non-equilibrium kinetic model developed by Limanov [6]. In the scope of the model, the temperature at the front controls the growth velocity according to a non-equilibrium kinetic function whilst the temperature is controlled by the heat balance near the front.

Calculations show that there is some difference in the crystallization mechanisms of thin and thick films. The temperature at the crystallization front in thin films constantly drops to 1300 – 1200 K (about 400 K below the melting temperature) during SLG-processing whilst in thick films the temperature saturates near 1600 – 1650 K (between 85 K and 35 K below the melting temperature). As a result, the growth velocity which is controlled by the kinetic function is higher (by 2 - 3 times) in thin films. This effect may explain the experimental observed result that the extension of SLG in thin films can be greater than the extension in thick films.

6. Conclusion

1. New high-efficient approach for SLS-processing has been developed. It can provide crystal growth along broad areas of a Si film with crystallization rates up to 3 mm/s.
2. Stable SLS-growth of Si films of thickness down to 40 nm has been obtained.
3. The microstructure of such thin films has been studied. It was found that thin SLS-grown Si films have a superior microstructure with respect to thicker ones.

4. Extension of Super Lateral Growth has been investigated and optimized under a broad range of SLS-process conditions, especially for a broad range of pulse durations. Extension of SLG up to 4 μm has been obtained for 40 nm a-Si film-on-glass at room temperature by a single laser pulse.

7. Acknowledgments

The authors would like to thank Prof. J.S.Im for helpful discussions.

8. References

1. Sposili R.S. and Im J.S. (1996) Sequential lateral solidification of thin silicon films on SiO$_2$, *Appl. Phys. Lett.* **69**, 2864-2866.
2. Im J.S., Sposili R.S., and Crowder M.A. (1997) Single-crystal Si films for thin-film transistor devices, *Appl. Phys. Lett.* **70**, 3434-3436.
3. Secco d'Aragona. (1972) Dislocation etch for (100) planes in silicon. *J.Electrochem. Soc.*, **119**, 948-951.
4. Limanov A.B., Chubarenko V.A., Borisov V.M., Vinokhodov A.Yu., Demin A.I., Khristoforov O.B., El'tsov A.V., Kirukhin Yu.B. (1999) Investigation in Si films obtained by SLS technique using of 3 kHz excimer laser with stripped output beam. *Russian Microelectronics* **28**, 00-00.
5. Im J.S., Kim H.J., Thompson M.O. (1993) Phase transformation mechanisms involved in excimer laser crystallization of amorphous silicon films. *Appl. Phys. Lett.* **63**, 1969-1971.
6. Limanov A.B. (1997) Simulsation of lateral grain growth at excimer-laser crystallization of amorphous silicon films. *Russian Microelectronics* **26**, 113-118.

LOW TEMPERATURE POLYSILICON TECHNOLOGY:

A low cost SOI technology?

F. Plais, C.Collet, O. Huet, P. Legagneux, D. Pribat, C. Reita, C. Walaine
Thomson CSF, Central Research Laboratories
Domaine de Corbeville
91404 Orsay cedex, FRANCE

1. Introduction

Large Area Electronics (LAE) is becoming a major branch of the electronics industry with 1997 revenues exceeding $ 10b. It is supported by a large industrially-driven research and development effort mainly located in Far-East countries (Japan, Korea, Taiwan). Most of the LAE activity deals with displays along with solar cells or X-ray imagers, and, for all these applications, hydrogenated amorphous silicon (aSi-H) is the material of choice. For about twenty years, polysilicon (polySi) technology has been intensively developed in industrial or public research laboratories but the only successful technology to entre manufacture of large area electronics products was the aSi-H one (and at a much lower scale, the Cd-Se). However, polySi is becoming a challenger for aSi-H as the implementation of low-temperature processes now render this technology (LT-polySi) compatible with large area glass or plastic-type substrates.

Using LT-polySi technology, a large field of applications can be anticipated and, among them, the fabrication of low cost flat panel displays based on active matrix liquid crystal displays (AMLCD[1]) technology has been the "driving force" to introduce it in fabrication lines [2]. At that point poly-Si comes with some advantages over aSi-H such as :

(i) CMOS compatibility (hole to electron carrier mobility ratio of 1 to 3, respectively, as in monocrystalline silicon)

(ii) higher speed of operation (electron mobility 100 times larger than in aSi-H even for the most basic polysilicon process)

(iii) better long-term stability (the crystalline structure of poly-Si is more stable than the metastable structure of aSi-H).

However, some drawbacks such as higher leakage current in the dark or deleterious hot carriers effects have to be carefully addressed just as in monocrystalline silicon technology.

---

[1] The AMLCD market represent 60% of the total 1997 FPD market with 9 G$ revenue and is expected to grow to 80% by year 2002 with 22 G$ revenue [1].

P.L.F. Hemment et al. (eds.),
Perspectives, Science and Technologies for Novel Silicon on Insulator Devices, 63–74.
© 2000 *Kluwer Academic Publishers. Printed in the Netherlands.*

The motivation lying behind LT-polySi technologies for display manufacturing will be explained in section 2, below.

LT-polySi devices and circuits exhibit many more similarities with SOI than with aSi-H. This point will be highlighted in section 3.

Introduction of both technologies (SOI and LT-polySi) have been ultimately limited by material processing issues that will be described in section 4., briefly for SOI (see for example [3]) but in more details for LT-polySi, with some emphasis on Excimer Laser Crystallisation.

In this paper, we will show in section 5 that SOI films can be processed at a temperature as low as 450°C without significant impact on TFT performances. To be compatible with such a low temperature, key processing steps such as low temperature gate oxide deposition, aluminum gate deposition and laser activation have to be implemented in the process flow-chart.

2. The Flat Panel Display environment

Displays are mainly fabricated on large area glass substrates for lower cost although some reflective displays are fabricated on silicon wafers (small displays for projection TV or very small displays for portable applications). Solar cells or imagers also are fabricated on glass substrates. The dimensions of glass substrates in active matrix liquid crystal display (AMLCD) manufacturing plants is currently 550mm x 650mm (generation 3) for the fabrication of six 12.1 inches displays on each substrate. In the LAE business, processed glass substrates form the core of the systems : an AMLCD is physically build on the glass substrate hosting the active matrix ; it is noted that this situation is completely different from the integration of a microprocessor (component) in a computer (system).

For this application, LT-polySi technology is competing with two mature technologies :
• amorphous hydrogenated silicon (aSi-H) technology which has been strenghtened by huge investments in large area substrate production lines and dominates the large size display segment (12"-20") for laptop or desktop computer applications. For this technology, the maximum processing temperature is about 350°C.
• high temperature polysilicon (HT-polySi) technology, which has dominated the small size display segment (<2") for viewfinders or virtual reality. This offers high performances but suffers from the use of fused silica substrates with high cost and limited surfaces. In this technology, the maximum processing temperature is about 950°C.

Figure 1 illustrates different manufacturing solutions for the connection of driver IC's to an AMLCD array. The technique of choice for large volume manufacturing is the tape automated bonding (TAB) one which requires 3 levels of connections. If driving circuitry is fully external (only pixel switch transistors are integrated on the glass plate) the number of connexions is N + M where N (M) is the number of rows (columns). The pitch of connections for column drivers is one third of the image pixel size for a color display. For direct-view displays, with a typical full colour pixel size of 250μm x 250μm (one column every 80 μm), the TAB solution is industrial although it contributes to a decrease in the manufacturing yield. For small displays, such as light valves used in projection systems, with pitch values going down to 50μm or below, this approach is severely limited and alternative techniques such as chip-on-glass (COG) have been introduced with limited success. A more challenging approach consists of integrating the display driving circuits (or a part of) directly on the glass plate using the technology implemented for the transistors of the matrix and adding the processing steps of deposition, lithography and etching which are required for complex circuits fabrication. Depending on their design complexity, the impact on the fabrication yield can render this approach out of interest. However, even though there may be a cost penalty, some advantages of the integrated driver (ID) architecture could justify this approach; among them are :

(i) reduction of module size (important in portable applications or for CRT retrofit)
(ii) better reliability and ruggedness due to a lower number of external connections (very important for avionic or automotive equipments)
(iii) flexibility of design (important for small volume production, as external driver circuits represent an additive cost which depends largely upon volume production).

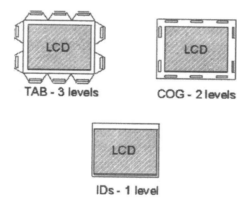

Figure 1 : Different solutions for the integration of driver circuits at the periphery of an LCD array (see text for details).

The first commercial products fabricated using LT-polySi are situated in the mid-size display segment (4"-8") with such products as digital or video cameras, rear projection TV, Personal Digital Assistant, car navigation systems as well as video games or virtual reality. However, due to its high performances and its impact on cost reduction, it is anticipated that LT-polySi will soon compete with HT-polySi or aSi-H in their respective market segments [2]. If one compares the electronic functions implemented using aSi-H and polySi technologies (LT- or HT-), it appears that the former has been mainly used for single device fabrication, pin diodes for photovoltaic

66

applications or switch TFTs for AMLCD control. On the contrary, LT-polySi technology, because it allows the fabrication of self aligned complementary MOS transistors, enable the fabrication of various and sophisticated electronic circuits[2].

LT-polySi technology has been introduced in pilot lines for transmissive type - AMLCD but recent publications indicate that it could be used also for reflective type - AMLCD [4], active matrix organic electroluminescent display (AMOEL) [5] or field emission displays (FED) [6]. As a consequence, LT-polySi technology could become a preferred technology for a large range of applications (display-type or not).

3. Polysilicon thin film transistor and SOI transistor similarities

Looking at the most common polySi TFT structure, namely the top-gate architecture (see figure 2), it is obvious that LT-polySi is very similar to SOI technology.

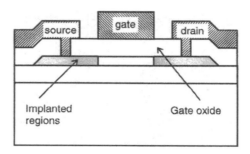

Figure 2 : Cross-sectionnal view of a top-gate poly-Si TFT.

Similarities include the following:

(i) the active layer is insulated from the substrate by an SiO_2 layer;

(ii) it can be very thin (50 nm) and, as a consequence, fully depleted devices can be fabricated;

(iii) isolation between transistors can be achieved by active layer patterning (island technology);

(iv) self aligned regions for source and drain contacts are obtained by ion implantation (gray zone of active layer in fig.2);

(v) LDD are implemented at the channel edge for hot carriers impact reduction (not shown in fig.2);

(vi) the devices are sensitive to defects present at the back (uncontrolled) interface.

---

[2] It must be pointed out that Thomson-LCD is currently using aSi-H TFTs for a partial driver integration on small AMLCDs.

4. Material issues

For many years , material issues have hampered the developpement of SOI technology. As the advantages of the SOI structure become obvious (no well implant, a lower number of masks, reduced parasitic substrate capacitance for higher speed operation) the manufacturing processes for SIMOX or bonded wafer substrates were still generating too many defects for the realisation of high yield device fabrication. In SIMOX substrates for example, electrically active defects have been identified in the silicon film which are associated with oxygen contamination and other defects at the back interface which are associated with the presence of silicon precipitates in the buried oxide. Other SOI technologies have been proposed [3] and abandoned, like Zone Melting Recristallisation or Solid Phase Epitaxy. In recent years, with the availability of large volumes of electronic grade wafers (mostly SIMOX wafers), the volume production of SOI circuits has been pushed and forecasts are very enthusiastic : SIMOX and the newer "smart-cut" substrates of 200mm and 300mm diameters will enter production at the beginning of the new century [7].

The crystalline quality of the polysilicon thin film layers has been continuously improved over the years since the pionnering work of Texas Instruments in the late 70's [8] but unfortunately without reaching an electronic grade level. This was a consequence of synthesising polySi by Solid Phase Crystallisation (SPC) i.e. by annealing for tens of hours at about 600°C, resulting in a high density of crystalline defects not compatible with device and circuit requirements (low threshold voltage values, $V_t < 5V$ and high electron mobility, $\mu > 150$ cm^{2}V^{-1}s^{-1}). For the standard SPC process, typical μ and V_t values on NMOS transistors were about 80 cm^{2}V^{-1}s^{-1} and 7V for a 100nm-thick gate oxide.
It should be mentioned that at the same time, HT-polySi technology was implemented on fused silica substrates for some "niche" applications in projection TV or viewfinders: in that case, as the poly-Si material was obtained by standard SPC at about 600°C, the following high temperature steps (oxidation or implant annealing) were contributing to a dramatic reduction of material defect density.

As for SOI technology with the introduction of SIMOX substrates, polysilicon really became a realistic option for LAE with the opportunity to use industrial excimer laser tools for crystallisation. Excimer lasers have been developed over the past 10 years for industrial applications and will now find applications in fine-rule lithography for next generation IC's. However, this excimer laser crystallisation (ELC) process is still considered as the most critical one in the LT-polySi technological flow-chart. It is outside the scope of this paper to emphasize the physics of laser crystallisation which is well described in ref. [9] and [10]. We will just comment in section 5 on some TFT results in relation to excimer laser crystallisation conditions.
For LT-polySi, the gate insulator is usually deposited rather than grown by thermal oxidation, with a typical thickness of about 100nm to maintain yield at an affordable value even on large area substrates. The maximum processing temperature of LT-polySi technology has been continuously decreased in the past years, from 600°C to 450°C and will reach 350°C in the near future. As a consequence, carrier traps are remaining both

in the polysilicon active layer (density of states, DOS in $cm^{-3}eV^{-1}$) and at the interface with the gate oxide (density of interface traps, D_{it} in $cm^{-2}eV^{-1}$); this impacts on the TFT performance as, for example, on the subthreshold slope S :

$$S = S_0(1 + \frac{q^2 D_{it}}{C_{ox}} + \frac{\sqrt{q^2 \varepsilon_0 \varepsilon_{p-Si} DOS}}{C_{ox}}) \qquad (1)$$

with C_{ox}, ε_0, ε_{p-Si}, and q representing respectively oxide capacitance, vacuum permitivity, polysilicon dielectric constant and electronic charge. In the case of SOI, S is equal to S_o, about 65 mV/dec at T=300K. For LT-polySi, S is in the range 0.2 to 2 V/dec, depending on the crystallisation technique. For a DOS value larger than 10^{18} $cm^{-3}eV^{-1}$ (SPC process), S is not really sensitive to D_{it} in the range 10^{11} to 10^{12} $cm^{-2}eV^{-1}$. For DOS values below 10^{17} $cm^{-3}eV^{-1}$ (laser process), the need for a high-quality SiO_2 deposition system becomes crucial. For this purpose, we have developed over a period of ten years the DECR SiO_2 process which results in excellent interface and bulk electrical properties [11].

5. LT-polySi technology performances

5.1 TFTs RESULTS

The key points of the process include the followings (see figure 2) :

- active layer is synthesised by excimer laser crystallisation (ELC) of an amorphous precursor deposited in a large area PECVD equipment[3]. This technique usually leaves hydrogen contamination in the films at a concentration of around 5 to 10 at%, which is not compatible with direct ELC process. Prior to melting, hydrogen has to be removed by annealing at a temperature of about 450°C in a standard furnace or by multishot/low energy laser irradiations;
- active layer is etched in an ICP reactor using a chlorine-based chemistry; this process exhibits high selectivity to silicon dioxide and tapered edges are easily obtained for a better step coverage, even at low gate oxide thickness (below 50 nm);
- gate oxide is deposited by PECVD in a DECR microwave plasma [11];
- aluminum gate is evaporated or magnetron sputtered (first metal level); in the latter case, which is the most promising technique for large area substrates, much care has to be taken to optimise the deposition parameters, in order to prevent electrical damage to the gate oxide;
- self aligned contact regions are implanted and subsequently activated by low-temperature laser annealing;
- contact holes in passivating oxide are wet-etched;

---

[3] Balzers Process Systems, Palaiseau, France

- source and drain contacts are made with Mo or Ti-Mo layer(s) (second metal level).

In order to assess the overall fabrication sequences of the LT-poly technology, silicon-on-quartz (SOQ) substrates have been processed along with polySi on glass substrates. A one to one proximity aligner is used for lithography, with minimum feature size and mask to mask alignment precision of 3μm. Figure 3 present an overall view of a SOQ substrate[4] (4" diameter) after CMOS processing. The insert shows a magnified view of the circuit and test cells on the plates (each one is 10mm by 10mm in size). In this case, the maximum temperature corresponds to the 450°C forming gas annealing which is necessary to reduce the density of interface states. In the case of LT-polySi substrates this temperature is also used for the removal of hydrogen from the amorphous silicon precursor before laser crystallisation.

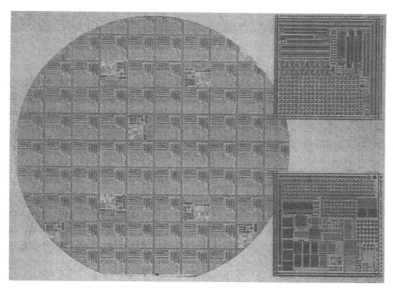

Figure 3 : Global view of 4" SOQ substrate after low-temperature (450°C) CMOS process with circuits and test cells in insert.

Figure 4 shows transfer characteristics of transistors fabricated on SOQ layers. The gate oxyde has been deposited by DECR PECVD to a thickness of 65 nm. Electron and hole field effect mobilities are about 650 and 250 $cm^2V^{-1}s^{-1}$ respectively.

The subthreshold slopes are slightly affected by the use of low temperature deposited oxide with values of 140 mV/dec for NMOS transistors and 65 mV/dec for PMOS. It should be noted that the presence of a small parasitic transistor is evident in in the NMOS characteristics.

---

[4] SOQ substrates have been purchased from SOITEC, France.

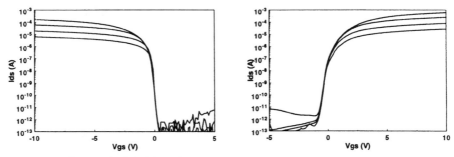

Figure 4 : PMOS and NMOS SOQ TFT (W/L = 10µm/10µm) transfer
characteristics (Vds = 0.1-0.3-1-3 V).

For LT-polySi transistor processing, we use a Lambda Physik excimer laser (λ=308 nm) with a line beam configuration. A multishot technique (10-20 shots per surface area) is implemented and substrates are moved between sequential shots. Impact dimensions on the substrate are 0.7 mm (direction of scanning) by 50 mm (direction perpendicular to scanning). The incident power density uniformity along the major laser axis is defined by the optics, whilst uniformity along the scanning direction can be a problem as it involves independant events. A large number of parameters can impact on uniformity including beam stability, beam edge profile, distance of overlap, shot to shot laser stability and laser fluence. For some particular conditions, we are able to limit the scatter on TFT performances from TFT ot TFT, while keeping reasonable performances. Figure 5 shows the threshold voltages and electron mobility values distributed along the direction perpendicular to scanning. Values correspond to NMOS TFTs with channel width and gate length of 100 and 10 µm, respectively.

The 1σ variation of mobility is below 5%. Using the same lithographic tools and processing equipment, 1σ variation of mobility on SOQ layers is about 2 %. In the later case, scattering of the calculated mobility is due to the combined dispersion of the TFT size and gate oxide thickness.

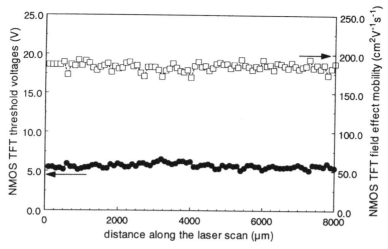

Figure 5 : Threshold voltages and electron field effect mobility of LT-polySi TFTs, distribution along the laser scanning direction (see text for detail).

5.2 A LOW COST SOI TECHNOLOGY ?

With the advent of industrial laser crystallisation tools and an improved understanding of this particular mechanism [9-10], the realisation of DOS values in the low 10^{16} cm^{-3}eV^{-1} range becomes achievable. In that case, Vt and μ values on NMOS transistors are about 2.5V and 300 cm^{2}V^{-1}s^{-1}, respectively, which allow good quality circuit behaviour [12]. Thus, LT-polySi can be considered as a low cost / reduced performances SOI technology. However, the polycristalline structure of the active layer as well as the use of lithography tools compatible with large area glass substrates limit the level of integration of the LT-polySi technology[5]. Minimum gate length and design rules are about 3µm and 1µm, respectively. According to the 1994 LSI SIA roadmap one can see that these values correspond to the ones of LSI circuits fabricated around 1980, with a low level of integration (64K DRAM memory) [13]. Moreover, 3 µm gate length is considered to be the limiting value for LT-polySi, whilst the SIA roadmap still predict a forthcoming reduction of gate length in monocristalline silicon down to 0.1 µm with an associated increase in integration capacity. This difference will never be compensated

---

[5] LT-polySi layers are non-uniform at a microscopic scale, with grain boundaries distributed inside the films. As described in section 4, grain boundary effects can be approached by introducing in the 2D simulator a spatially uniform density of states (DOS); however, as TFT dimensions shrink, gate edge of TFT (region where electrical field is large) can be situated in a grain or at a grain boundary, resulting in high TFT to TFT performance variability. One way to limit the deleterious impact of the random TFT alignment, with respect to grain boundary location, is to keep its minimum dimension at least 10 times larger than the mean grain diameter.

by the use of a substrate area 15 times larger (600mm x 720mm compared to 200mm wafers in 1998 or 1100mm x 960mm compared to 300mm wafers anticipated in 2010).

However, to be exhaustive, it should be mentionned that grain boundary location techniques [14] are currently developped with the idea to fabricate one TFT in a nearly monocristalline grain; if implemented successfully on large area glass substrates, these techniques could allow to reduce the gate length down to the anticipated design rules limits of the lithographic tools, about 0.5 μm on large area glass substrates.

6. Processing tools for LT-polySi technology

Up to now, the process flow-chart of the LT-polySi technology is not standardised as, for example, (i) top and bottom gate TFTs are still under development and (ii) implementation of LDD structures vary from one company to an other. Moreover, the success of a flow-chart relies on a few "magic" tricks which are kept secret by companies working on the topic. As a consequence, LT-polySi fabs are currently captive fabs and only few informations are available about processes.

However, considering that the large area substrate equipments could be easily adapted to accomodate 300 and 400 mm silicon substrates[6], equipement suppliers try to reduce development costs by putting the emphasis on common processing equipments, which will drive LT-polySi technological flow-chart standardisation. These common processing tools or processes are :

ICP etching tools :
High etch rates combined with high selectivities have rendered these reactors very popular in the microelectronics community and the equipment leader of this technology for silicon wafers, has developed a tool for large area glass substrates (up to 620mm x 700mm); ICP etching reactors are used for polySi, SiO_2, SiN or metal (ITO) dry etching.

TEOS SiO_2 PECVD or APCVD tools:
High electrical quality insulators are required for the intermetallic layers of IC's and TFT circuits as well as for the gate insulator of TFTs; the requirements are different with emphasis on (i) step coverage and pinhole density for intermetallic layers and on (ii) interface state density and leakage current for gate oxide layers but equipment suppliers try to address the two different needs with similar equipment.

Implantation systems with substrate motion under a high aspect ratio ion beam :
The big advantages of the self-aligned polysilicon TFT structure have motivated the development of at least three different large area wafer implantation tools where one, based on the constant velocity-motion of a substrate under a linearly shaped ion beam could be used either for silicon wafers or large area glass substrate.

Rapid thermal annealing:

---

[6] Requirements concerning fabrication throughput and processing tool performances are very similar for FPD and IC's fabrication, their main objective is cost reduction.

Activation of ion implanted source-drain regions is usually performed by RTA in standard IC's processing because of the reduced thermal budget; this is of particular interest in the LT-polySi and thus, RTA is used by some manufacturers using e.g. Intevac equipment.

Spin-on-glass (SOG) passivation layers.
These layers are of great interest for intermetallic insulation with a reduced parasitic capacitance[7] because of their reduced relative permitivity. However, they will be first implemented in low cost bottom-gate aSi-H TFT processing in place of the standard silicon nitride passivation layer.

Only a few specific steps are needed for LT-polySi processing, the most important is the laser crystallisation tool[8]. This specific process is still complex and with a high cost but has to be compared to the synthesis of SOI wafers by SIMOX or Smart-Cut techniques. One can simply approach the time required to completely crystallise a large area glass plate by the formula :

$$t = \frac{W_g L_g}{(1-r)wlf} \qquad (2)$$

where Wg and Lg are the glass plate dimensions, w and l are the laser beam dimensions, r is the overlap ratio and f is the laser operating frequency. Considering glass plate dimensions of 600mm x 720mm, laser beam dimensions of 300 mm x 0.3 mm, overlap ratio of 90 % and a laser operating frequency of 300Hz, the treatment time is less than 3 min (20 plates per hour).

7. Conclusions

LT-polySi technology is now being introduced into fabrication lines for small size AMLCDs. The main motivation is to reduce cost by integrating drivers on the glass plate. This technology appears as a major upgrade of the current aSi technology, in particular as carrier field effect mobilities approach those of conventionnal SOI materials. However, considering its limitation in design rules, it is clear that competition with SOI is far away. The main applications for the next few years are expected to be display manufacturing but others like sensors including large area sensors, with add-on electronic functions could become important.

---

[7] These layers will be of particular importance when combined with copper interconnections in the next generation IC's.
[8] Excimer lasers are already used for fine lithography and could be used in the near future for shallow junction activation because of the unique thermal budget of the technique. In that way, the laser is not really specific for polySi technology.

74

Many thanks to F. Petinot (present address: Balzers Process System) for his contribution to this work during the 1995-1998 period and to J. Ballutaud for statistical TFT measurements.

References

[1] J.G. Blake, J.D. Stevens and R. Young (1998) Impact of low temperature polysilicon on the AMLCD market, *Solid State Technnology*, January 1998, 56-62

[2] S. Morozumi (1996) *Information Display*, **12**, No. 11, 18

[3] Physical and Technical Problems of SOI Structures and Devices, Colinge Nazarov (1995) Kluwer Academic Publishers

[4] Y. Iwai, H. Yamaguichi, T. Sekime, H. Kinoshita, S. Fujita, H. Mizuno, T. Hatanaka, T. Ohtani, and T. Ogawa, (1998) A reflective full-color TN-LCD with a single polarizer using low-temperature poly-Si TFTs, *Proceedings of the SID 1998* 225-228

[5] R.M.A. Dawson, Z. Shen, D.A. Furst, S. Connor, J. Hsu, M.G. Kane, R.G. Stewart, A. Ipri, C.N. King, P.J. Green, R.T. Flegal, S. Pearson, W.A. Barrow, E. Dickey, K. Ping, CX.W. Tang, S. Van Slyke, F. Chen and J. Shi (1998) Design of an improved pixel for a polysilicon active-matrix organic LED display, *Proceedings of the SID 1998*, 11-14

[6] Y.H. Song, J.H. Lee, S.Y. Kang, J.M. Park and K.I. Cho, (1998) Monolithic integration of poly-Si FEA and TFT for active-matrix FEDs, *Proceedings of the SID 1998* 189-192

[7] M. Auberton-Hervé and M.Lamure (1998) Soitec goes 300mm, *European Semiconductor*, March 1998, 25-26

[8] Polycrystalline silicon for integrated circuit applications, Ted Kamins (1988) Kluwer Academic Publishers

[9] D. Pribat, P. Legagneux, C. Collet, F. Plais, O. Huet and C. Reita (1998) Excimer laser processing for low-temperature TFTs and circuits, International Workshop on Active-Matrix Liquid-Crystal-Displays - digest of technical papers, 9-12

[10] D. Pribat, P. Legagneux, F. Plais, C. Reita, F. Petinot and O. Huet (1997) Low temperature polysilicon materials and devices, *Mat. Res. Soc. Symp. Proc* Vol 3424 127-140

[11] N.Jiang, M.C.Hugon, B.Agius, T.Kretz, F.Plais, D.Pribat, T.Carrière and M.Puech (1992) Device Quality SiO_2 Deposited by Distributed Electron Cyclotron Resonance Plasma Enhanced Chemical Vapor Deposition Without Substrate Heating, *Japanese Journal of Applied Physics*, **31** L1404-L1407

[12] F. Petinot, F. Plais, D. Mencaraglia, P. Legagneux, C. Reita, O. Huet and D. Pribat (1998) Defects in solid phase and laser crystallised polysilicon thin films transistors, *Journal of Non-Crystalline Solids* **227-230**, 1207-1212

[13] H. Iwai (1998) CMOS - year 2010 and beyond ; from technological side, *Proceedings of the IEEE 1998 Custom Integrated Circuits Conference*, 141-148

[14] M.A. Crowder, P.G. Carey, P.M. Smith, R.S. Sposili, H.S. Cho and J.S. Im (1998) Low-Temperature Singl-Crystal Si TFT's Fabricated on Si Films Processed via Sequential Lateral Solidification, *IEEE Electron Devices Letters*, 19(8), 306-308

A NOVEL LOW COST PROCESS FOR THE PRODUCTION OF SEMICONDUCTOR POLYCRYSTALLINE SILICON FROM RECYCLED INDUSTRIAL WASTE

B N MUKASHEV, M F TAMENDAROV, B BEKETOV and S ZH TOKMOLDIN
Institute of Physics & Technology of the Ministry of Science - Academy of Sciences of the Republic of Kazakstan, Almaty 480082, KAZAKSTAN
E-mail: mukashev@satsun.sci.kz

Abstract

Alumothermic reduction of silica from the waste products of the phosphorous industry has been developed by the Institute of Physics and Technology, Almaty, and is reported in this paper. The slag composition consists of four main components, namely SiO_2, CaO, Al_2O_3 and MgO as well as a few percent of P_2O_5, CaF_2 and other impurities. The presence of fluorine provides favorable conditions for the process because it causes a decrease in the viscosity and the melting temperature of the slag to a temperature in the range 1200 - 1300°C. Also, the process has the benefit of energy saving due to the exothermic nature of the reaction. The yield of the silicide-contained alloy in this process is about 80% by weight of a metal reductant. It is found that there are two possible routes for the purification of semiconductor silicon from the silicide-contained alloy these being chemical refining and silane synthesis by a commonly used pyrolysis process.

1. Introduction

Currently, the refining of silicon used in the manufacture of electronic and photovoltaic devices is based upon silane chemistry [1]. Much of the silicon used for photovoltaics (solar-grade silicon) is derived from material rejected by the silicon microelectronics industry [2]. However, during recent years the rejected material has become scarce and more expensive due to the recent major expansion of the semiconductor industry. As a consequence, the price for solar-grade silicon is rapidly increasing from US$ 10/kg in 1995 to today's figure, just four years later, of about US$ 25/kg. Because of the continuing expansion of silicon based photovoltaic devices, currently at a rate of more than 10% per annum, the photovoltaic crystalline silicon industry will eventually require a dedicated supply of polysilicon feedstock. In addition the overall demand for silicon will increase from 5,000 tons to about 23,000 tons in the year 2000. Therefore developing new sources of silicon for photovoltaics and microelectronics are of strategic importance. This paper presents a novel, environmentally benign technology for solar-grade silicon and silane production by recycling the waste from the phosphorous industry.

75

P.L.F. Hemment et al. (eds.),
Perspectives, Science and Technologies for Novel Silicon on Insulator Devices, 75–84.
© 2000 *Kluwer Academic Publishers. Printed in the Netherlands.*

2. Process Background

At present there are a number of technologies for semiconductor silicon production. These technologies are based on silane chemistry and other gaseous phase processes (see Table 1) [3]. Gaseous compounds of silicon are obtained usually through the carbothermic reduction of silica to metallurgical grade silicon followed by chlorination and hydration. The most popular silicone chloride based technology is the Union Carbide process (see Figure 1) which supports the manufacture of high purity solar- and electronic-grade silicon. Unfortunately this technology needs, at the initial stage of processing, high purity synthetic silica and charcoal for the carbothermic reduction process.

TABLE 1. Current technologies for the production of semiconductor silicon [3].

Manufacturer	Process	Reaction
Dow Corning	Advanced process of reduction in the arc furnace	$SiO_2 + C$
Hemlock Semiconductors	Hydration of silicon tetrachloride and pyrolysis of silicon dichloride	$SiCl_4 \rightarrow SiHCl_3 \rightarrow SiH_2Cl_2 \rightarrow Si$
Texas Instruments	Plasma-chemical reduction	$SiO_2 + C$
Motorola	Chemical vapour deposition	$Si_{tech} + SiF_4$
Union Carbide	Hydration of silicon tetrachloride and pyrolysis of silane	$SiCl_4 \rightarrow SiHCl_3 \rightarrow SiH_4 \rightarrow Si$
Stanford Res Inst	Vapour-solid interaction	$SiF_4 + Na$
Westinghouse	Plasma-chemical reduction	$SiCl_4 + Na$
Battel Col Lab	Vapour-phase interaction	$SiCl_4 + Zn$
Aerochem	Plasma-chemical reduction	$SiCl_4 + H_{at}$
Shumacher	Hydration of silicon tetrabromide and pyrolysis of silicon tribromide	$SiBr_4 \rightarrow SiHBr_3 \rightarrow Si$

The novel process developed and experimentally tested at the Institute of Physics and Technology (IP&T) involves the extraction of silicon by alumothermic processes from waste material from the phosphorus industry. The underlying goal is to utilise a new low-cost raw material - phosphorous industry slag - for the extraction of a new intermediate silicide-contained alloy (SCA, a new type of metallurgical silicon). Our experiments have shown that a simple treatment of the alloy by acids facilitates the production of refined silicon powder or the synthesis of silane. The process are covered by preliminary patents in Kazakstan: "A method of silicon obtaining" No 4627 (1997) and "A method of silane obtaining" No 4628 (1997).

The technological steps in this process for the production of semiconductor silicon are shown in Figure 2. It has been found that the effectiveness of the chemical refining and silane synthesis depends on the initial input feed and values of the parameters controlling the alumothermic reduction process.

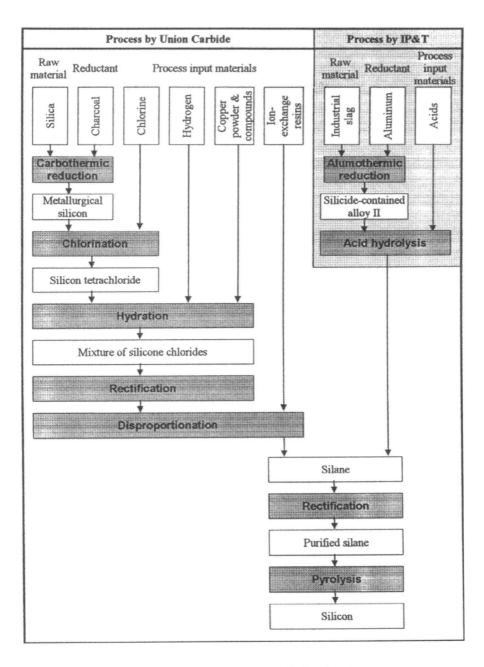

Figure 1. Comparison of the Union Carbide silane based process
with the IP&T process incorporating a standard technology.

78

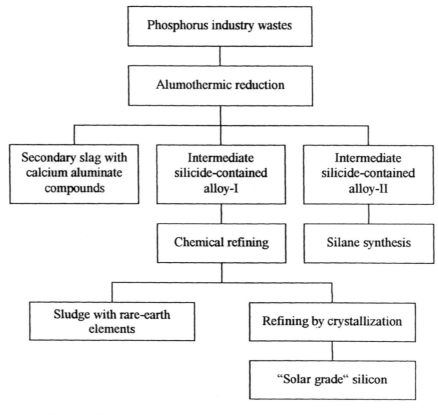

Figure 2. Main technological stages for silicon production from phosphorus waste.

The basic process steps are described below.

2.1 RAW MATERIAL

The phosphorous slag is material waste from the phosphorous industry. It is a chemically stable multi-component system of compounds and impurities with a typical composition listed in Table 2.

TABLE 2. The chemical composition
of the phosphorous slag.

Component	Content, %
SiO_2	45.0 - 47.0
CaO	43.0 - 45.0
MgO	2.2 - 2.5
Fe_2O_3	0.35 - 0.45
Al_2O_3	3.2 - 3.6
Na_2O	0.59 - 0.62
P_2O_5	0.8 - 1.3
CaF_2	4.0 - 4.5
K_2O	0.20 - 0.23
TiO_2	0.19 - 0.22
MnO	0.20
Sr and other rare metals	0.15

2.2 ADVANTAGES OF ALUMOTHERMIC REDUCTION OF PHOSPHOROUS SLAG

The presence of calcium and fluorine compounds decreases the melting temperature (up to 1200 - 1300°C) and viscosity of the slag. The main reaction in this process is

$$3SiO_2 + 4Al \rightarrow 3Si + 2 Al_2O_3 \tag{1}$$

which is initiated at 1200°C when the temperature rises due to the exothermic nature of the reaction. This allows a considerably saving of energy compared to regular carbothermic processing.

A further advantage in having the presence of calcium compounds is that they not only make possible a reduction of the processing temperature but also act as a catalyst for the chemical reactions [4] leading to an increase in the silicon content of the silicide-contained alloy.

A valuable by-product of the alumothermic reduction process is a secondary slag containing calcium-aluminate compounds suitable for production of high quality cements.

2.3 SILICIDE-CONTAINED ALLOY AND SECONDARY SLAG

The chemical composition and phase distribution of the silicide containing alloy and secondary slag, determined by chemical and X-ray analyses, are shown in Table 3 and Table 4.

TABLE 3. Composition of the silicide-contained
alloy and secondary slag.

Chemical analyses			
Alloy		Slag	
Element	Content, weight %	Element	Content, weight %
Si	80	Si	6.9
Ca	8.4	Ca	19
Al	3.5	Al	32.5

TABLE 4. Phase composition of the silicide-contained
alloy and secondary slag.

X-ray analyses of phases			
Alloy		Slag	
Phase	Content, weight %	Phase	Content, weight %
Si	80	$Ca_2Al_2SiO_7$	40 - 45
$CaAl_2Si_{2\#}$	13	$CaAl_4O_7$	40 - 45
$CaSi_2$	7	Other phases	10 - 20

Thus the main components of the silicide-contained alloy are Si, Ca and Al with a percentage composition of silicon (80%), calcium (8.4%) and aluminum (~3.5%). The secondary slag consists of calcium-aluminate compounds. This data is in a good agreement with theoretical predictions carried out using a thermodynamic model.

2.4 THERMODYNAMIC MODELLING OF THE ALUMOTHERMIC REDUCTION PROCESS

Thermodynamic modelling of the process has been carried out using the universal code ASTRA-4, which is suitable for high-temperature processes in multi-component heterogeneous systems [5]. The code is based on the principle of maximum entropy in isolated equilibrated thermodynamic systems and has it's own database of the thermodynamic properties of 3200 substances.

The interaction of aluminum with phosphorous slag was simulated in the temperature range 1000 to 3000 K with temperature increments of 100 K in an inert argon atmosphere at normal pressure. Simulation of the melt phase of both the silicide-contained alloy and secondary slag was performed in the framework of an ideal model of reacting components where about 120 condensed and gaseous substances and elements were taken into account.

The calculations predicted that over a wide range of temperatures, up to 2000 K, the alloy contains Si, Ca and Al. Increasing the aluminum fraction in the input feed by 10% above the "stoichiometric" composition gives the best yield of silicon into the silicide-contained alloy but further increase of the aluminum fraction leads to a considerable fall in the silicon yield (Figure 3). The content of calcium and phosphorous in the alloy also strongly depends upon the aluminum content (Figures 4 and 5). Figure 3 and Figure 5 show that the temperature range 2050 to 2250 K is optimal for alumothermic processing when the composition by weight is about 60% of Si, 30% of Ca, 3% of Al, 1.7 ± 0.2% of Fe, 1.7 ± 0.9% of P.

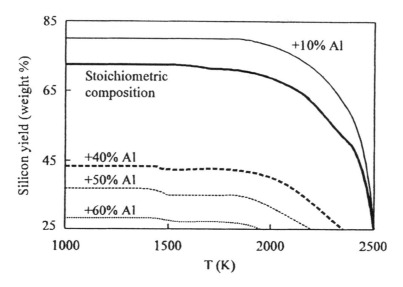

Figure 3. Silicon yields from the SCA for different amounts of aluminum in the input feed.

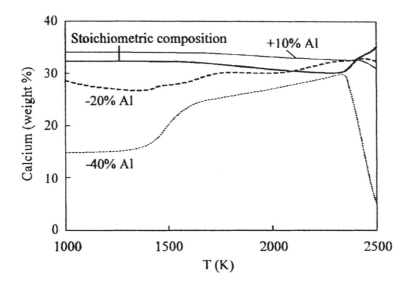

Figure 4. Temperature dependence of the calcium content in the SCA
for different aluminum contents in the input feed.

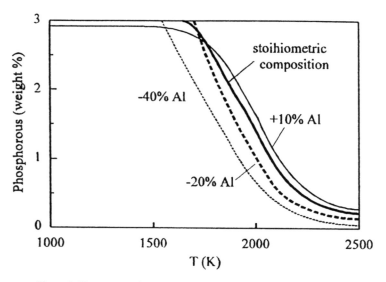

Figure 5. Temperature dependence ot the phosphorous content of the SCA
for different aluminum contents in the input feed.

The secondary slag mainly consists of calcium-aluminate compounds, the total
amount of which exceeds 90% by weight (Figure 6).

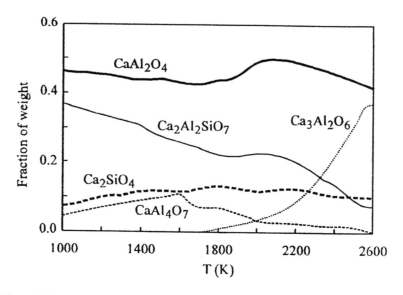

Figure 6. Temperature dependence of the fraction of calcium compounds in the secondary slag.

3. Chemical Refining

Currently, there are three advanced processes under development for the production of solar-grade silicon. These processes are based on (i) the reduction of silicon chlorides and silanes, (ii) the carbothermic reduction of silica using high purity raw materials and (iii) refining the melt of metallurgical silicon by active silica slags and reactive gases [1,3].

For effective chemical refining, in our experiments, the silicide containing alloy was mechanically crushed to grains of 1 to 3 mm diameter. During further chemical treatment using water diluted acids the grains were further reduced in size to less than 100 microns. Subsequently, the powder is washed with water and refined in acid solutions. As a result of such operations we have obtained silicon powder with a purity of better than 99,99%. Table 5 shows the concentration of impurities determined using inductively coupled plasma optical emission spectrometry.

TABLE 5. Concentrations of impurities in chemically refined silicon powder.

Impurity	Concentration, ppm
Mn	14
Cu	7
Cr	1.5
Fe	21
Zn	7
Ba	4
Co	0.06
Ti	15
V	0.24
Ni	18
Al	4
B	20
P	2

4. Crystal Grows

Large-block polycrystalline silicon ingots with a resistivity of about 0,05 - 0,1 Ohm·cm and a dislocation density of 10^3 - 10^4 cm^{-2} have been grown from the chemically refined silicon powder by using the Czochralski method. Such material can be successfully used in photovoltaics. Preliminary experiments have shown that it is possible to decrease the concentration of boron to a level of 1 - 2 ppm by liquid extraction using calcium fluoride in the molten state. This additional refining is made possible by reactive gas blowing of the molten charge.

84

5. Silane Production

A very interesting feature of the silicide alloy (denoted here and in Figure 1 as "alloy-II") is the possibility of using it to synthesise high-purity monosilane by a simple chlorine free reaction using acids at temperatures from 20 to 90°C (see Figure 2). This process based upon silane chemistry is proposed as a simple low cost technology, which is an alternative to the Union Carbide process based on silicon chlorides (see Figure 1).

The rate of the silane synthesis and hence the output of silane depends upon the alloy-II chemical composition, which may be easily controlled by varying the parameters of the alumothermic reduction process. Experiments have shown that a yield of silane exceeding 50% can be achieved , which is significantly higher than the yield of 25 - 30% in the process of silane reduction from Mg_2Si [6].

6. Conclusions

A novel low cost, non-chlorine, environmentally benign technology for the production of photovoltaic grade polycrystalline silicon has been described. The process is based upon the alumothermic reduction of silica in waste products from the phosphorus industry. Preliminary experiments have shown that impurities, including boron, can be removed by chemical refining which provides opportunities for the production of large quantities of feedstock for the production of semiconductor grade silicon. However, further detailed studies of the process are necessary to optimise the processing parameters.

References

[1] Nashelskii, A.Ya. and Pulner, E.O. (1996) Today's technologies for solar-grade silicon, *Vysokochistye veshestva* 1, 102-111.
[2] Gee, J.M., Ho, P., Van Den Avyle, J. and Stepanek, J. (1998) Some thermochemical calculations on the purification of silicon melts, Proc 8th Workshop on Crystalline-Silicon Solar Cell Materials and Processes, Copper Mountain Resort, 5 p.
[3] Grankov, I.V. and Ivanov, L.S. (1985) Intensification of the polycrystalline silicon obtaining process, *Tsvetnye metally* 6, 60-64.
[4] Patent No 152551, Norway, announced at 07.02.83, publication N 830391 at 16.10.85.
[5] Vatolin, N.F., Moiseev, G.K. and Trusov, B.G. (1994) Thermodynamic modelling of high-temperature inorganic compounds, *Metallurgiya*, Moskva.
[6] Zhigach, A.F. and Stasinevich, D.S. (1969) Chemistry of hydrates, *Khimiya*, Leningrad.

TETRAHEDRALLY BONDED AMORPHOUS CARBON FOR ELECTRONIC APPLICATIONS

W.I.MILNE
Department of Engineering
Cambridge University
Trumpington Street
Cambridge CB2 1PZ
U.K.

Abstract:

The preparation, optical and electrical properties of highly sp^3 bonded amorphous carbon films produced using a Filtered Cathodic Vacuum Arc system are described. The sp^3 bonding fraction is highly dependent on incident ion energy reaching a maximum of approximately 85% at a carbon ion energy of 90-100 eV. The 'as grown' material is a p-type semiconductor and has a room temperature resistivity of order 10^8 Ωcm. N -type doping of the films can also be achieved by the incorporation of nitrogen during the deposition process. The optical bandgap also varies as a function of incident ion energy reaching a maximum value of typically 2.3-2.5 eV. The films are highly disordered and have broader optical absorption edges than those observed in a-Si:H and a higher density of paramagnetic states - of order 10^{19} - 10^{20} cm^{-3}. Preliminary results on photoconductivity as a function of temperature, light intensity and photon energy are also presented. Both tetrahedrally bonded amorphous carbon (ta-C) and its hydrogenated form ta-C:H are low mobility solids with $\mu\tau$ products of order 10^{-12} - 10^{-11} cm^2/V at room temperature.

1. Introduction:

There is currently a large interest in the deposition of amorphous carbon thin films containing a sizeable fraction of sp^3 bonds. Such material is usually called Diamond Like Carbon (DLC) and its composition is most conveniently represented on a ternary phase diagram, as shown in figure 1.

It has a number of attractive features for electronic applications including large area deposition at room temperature, high resistivity, possibility of n-doping (and p-doping) and an optical band gap which can vary from 0.8 - 4.5 eV dependent upon method of deposition. Such DLC films can be deposited using several different methods leading to a variety of different DLC types. Techniques such as Plasma Enhanced Chemical Vapour. Deposition (PECVD) [1], and Plasma Beam Source(PBS) [2] produce films containing hydrogen whereas laser ablation [3], Mass Selected Ion Beam (MSIB) [4] and the Filtered Cathodic Vacuum Arc (FCVA) system [5] produce films which are hydrogen free and contain a high percentage of sp^3 bonding- called tetrahedrally bonded

85

P.L.F. Hemment et al. (eds.),
Perspectives, Science and Technologies for Novel Silicon on Insulator Devices, 85–96.
© 2000 *Kluwer Academic Publishers. Printed in the Netherlands.*

amorphous carbon (ta-C). A common feature of all these processes is that the films grow from a beam of medium energy ions. The most complete form of ion filtering is found in

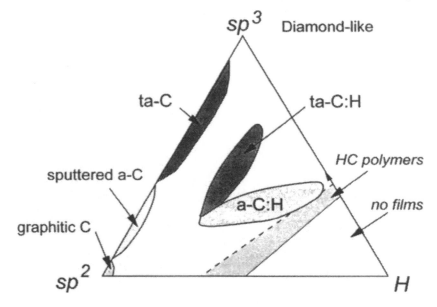

Figure 1 Ternary Phase Diagram of the Various Forms of DLC

the MSIB process but the FCVA is a lower cost source of ta-C suitable for both laboratory use and potential scaling up for industrial application. This paper will therefore describe the FCVA preparation of ta-C films and discuss the optical and electrical properties which make them potentially useful for electronic applications. Discussion of the hydrogenated form of this type of amorphous carbon ta-C:H produced in a Plasma Beam Source will also be included.

2. Experimental Details:

A schematic of a typical FCVA system for deposition of ta-C films is shown in figure 2 where by altering the ion energy of the depositing carbon ions the sp^3 bonding content can be varied up to maximum of approximately 85%. The cathode on which the arc is struck is made of 99.999% pure graphite. The retractable striker is manufactured from a graphite rod of similar purity. The vacuum arc produces a highly ionised plasma stream with energetic ions from which a dense film of amorphous carbon can be grown on substrates held on temperature controlled copper blocks attached to a hexagonal carousel as shown in Figure 2 overleaf.

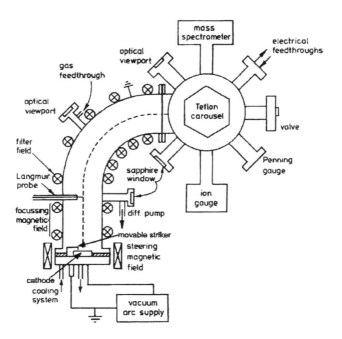

Figure 2 . Schematic of the FCVA Deposition System

Neutral macroparticles are unavoidable by-products from such a system and in order to filter these out the plasma stream is passed through a 90° magnetic filter of strength approximately 50 - 90 mT. The filter does not provide mass or charge selection. The deposition system has been modified to include a scanning coil assembly (not shown in figure) in order to provide uniform deposition over an area of 10 x 10 cm. The effective ion energy is varied by applying a DC bias to the substrate (or a RF bias if non conducting substrates are used). Hydrogenated ta-C films can be deposited in this chamber by injecting Hydrogen into the plasma stream although a much more effective technique of producing such material is by using a Plasma Beam Source [2].

Optical Band Gap measurements were carried out using an Ati-Unicam UV-VIS spectrometer and Photothermal Deflection Spectrometer. Both the Tauc [6] and E_{04} gaps were measured. The sp^3 bonding fraction is estimated from electron energy loss spectroscopy (EELS) using a Philips CM30 TEM fitted with a Gatan 666 EEL spectrometer. For electrical measurements the films were deposited onto Corning 7059 glass and coplanar gap cells of aluminium or chromium were used to derive the conductivity and activation energy. For heterostructure measurements the films were deposited onto p^{++} and n^{++} silicon substrates with resistivities of 0.017 and 0.02 respectively. ESR measurements indicate that ta-C films have high spin densities of order 10^{19} - 10^{20} cm^{-3}.

88

3. Results:

The optical absorption spectrum for a ta-C film deposited at an ion energy of approx. 90 eV is shown in figure 3 . The plot is a combination of data from reflection/transmission using the UV-VIS system and also Photothermal Deflection Spectroscopy.

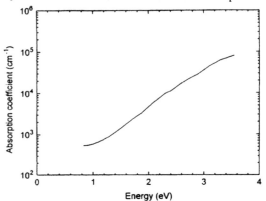

Figure 3 Absorption Coefficient versus Photon Energy

The optical bandgap can either be defined as the energy (E_{04}) at which the absorption coefficient $a = 10^{4}\text{cm}^{-1}$ or as the Tauc gap E_g [6]. The sp^3 bonding fraction, as a function of ion energy is as shown in figure 4 where it can be seen to peak at an ion energy of ~100 eV.

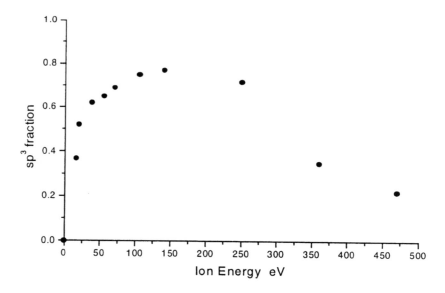

Figure 4   sp3 bonding Fraction as a Function of Ion Energy

Figure 5 plots the variation of optical gap versus sp^2 fraction for FCVA deposited ta-C and shows that it varies in a similar fashion to that of PECVD produced a-C:H and PBS produced ta-C:H. There is only a slight difference in bandgap between the hydrogenated and unhydrogenated carbon films indicating that the gap primarily depends on the sp^2 site density and has only a weak dependence on H content.

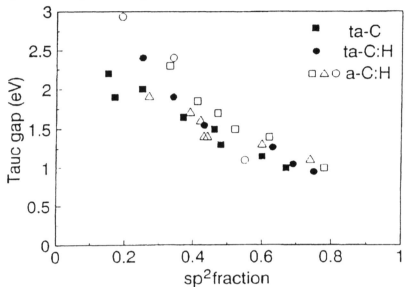

Figure 5 Bandgap as a Function of sp^2 Bonding Fraction

The dependence of the optical gap on the sp^2 fraction may be expected from the electronic structure of a-C since the bandgap is determined by the p states of the sp^2 sites as these states lie closest to the Fermi Level [7]. The gradual change in the bandgap with the sp^2 content suggests that the p bands become narrower and the π - π^* gap widens as the sp^2 fraction decreases. The importance of the π bonding can be deduced from the steady increase in the E_g even when only 20% sp^2 sites remain in the ta-C. The similar variation of the gap in both the a-C:H and the ta-C:H is particularly important at low sp^2 concentrations where the sp^2 sites are embedded in a rigid matrix in ta-C but a rather floppy matrix in the polymeric a-C:H.

From Figure 3 it is obvious that the absorption edge is much broader for ta-C than in other amorphous semiconductors such as a-Si:H indicating a much higher degree of disorder in these films. The lower part of the absorption plot exhibits only an approximately exponential dependence. Nevertheless by drawing a tangent to the log of the absorption coefficient at $\alpha = 2 \times 10^3$ cm^{-1} the Urbach slope can be found to be of order 300 meV which is similar to both a-C:H and ta-C:H but is much larger than the typical 55 meV observed for device quality a-Si:H.

The photoconductivity of ta-C and ta-C:H has been studied as a function of temperature, photon energy and light intensity. At room temperature for optimised material the ratio of light to dark conductivity under approximately AM1 conditions is at best two orders of magnitude for both ta-C and ta-C:H [8]. Slightly nitrogen doped ta-C:H gives the highest ratio of just greater than 200 for σ_{ph}/σ_D. Both ta-C and ta-C:H were found to have low mobilities with a $\mu\tau$ product of order 10^{-12} - 10^{-11} cm^2/V at room temperature.

ESR : Electron Spin Resonance measurements show that ta-C has a high spin density under all deposition conditions. At present the actual values observed depend on the method of deposition and the laboratories in which they were measured. Figure 6 shows the ESR signal as a function of deposition energy for ta-C deposited in the FCVA system by McKenzie et al [9] in Sydney.

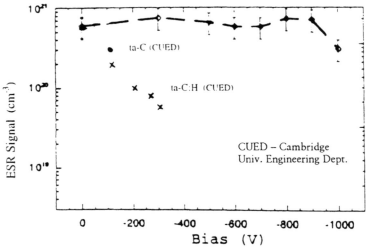

Figure 6 ESR Spin Density as a Function of Ion Energy [9]

We have not made such a comprehensive study as a function of ion energy but the result for our ta-C films grown at the optimum conditions for maximum sp^3 content - 100eV is shown on the diagram alongside some preliminary results on ta-C:H deposited in our laboratory using a Plasma Beam Source system. As can be seen the spin density is of order 10^{20} - 10^{21} cm^{-3} for ta-C and is approximately one order lower for the hydrogenated films. In an effort to reduce the spin density we have annealed the films and although no obvious differences are observed for unhydrogenated films the results for the ta-C:H films, shown in Figure 7, indicate for the first time that annealing can reduce the spin density in such films.

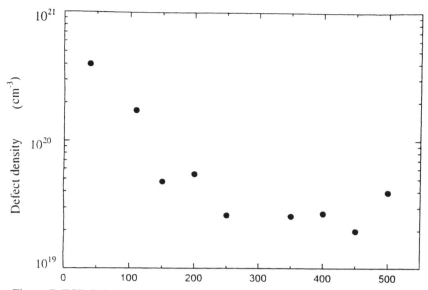

Figure 7 ESR Spin Density for ta-C:H as a Function of Annealing Temp. in °C

The as grown ta-C is found to be p-type with a conductivity activation energy of approximately 0.3 - 0.4 eV. The variation of resistivity with incident ion energy is shown in Figure 8 .

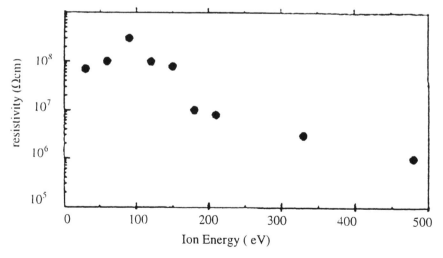

Figure 8 Resistivity as a Function of Ion Energy

The sharp decrease in resistivity at higher energies is attributed to the increase in disorder in the films due to the energetic ion bombardment leading to a higher number of gap states. In order to realise the potential of this material for use in electronic devices however, it is necessary to be able to dope it both n- and p-type. Efforts have been made to make the material n-type using phosphorus as the dopant [10] but it has been shown that the addition of nitrogen to the growing film is a much more effective technique[11]. Typical variations in resistivity, activation energy and optical bandgap as a function of nitrogen incorporation are shown in Figure 9 for films deposited at 100 eV ion energy. Up to nitrogen incorporation of approx 0.5-1.0 At % the Nitrogen acts as a n-type donor and the sp^3 concentration remains constant. Above this nitrogen content the bandgap shrinks and the sp^3 concentration falls indicating that the films are beginning to graphitise.

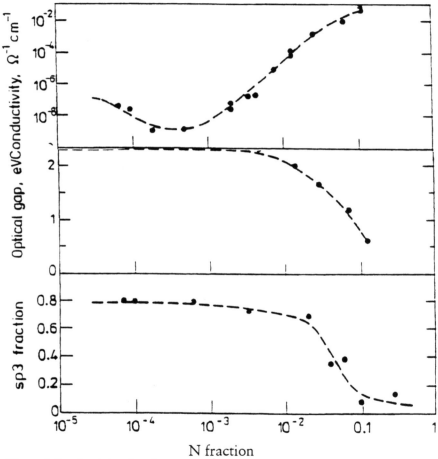

N fraction

Figure 9 Conductivity, Bandgap and sp^3 Content as a Function of N Incorporation

The activation energy is seen to rise from the undoped value of ~ 0.4 eV to 1.1 eV with increasing N incorporation and then it declines on further addition of N to the film. In order to verify that the resistivity changes were not due to the bandgap altering, the optical gap was measured and the variation with N incorporation is also shown in the figure. As can be seen the Eg remains approximately constant for concentrations of

Nitrogen up to 0.5 - 1.0 at %. As more N is added the Eg decreases and a predominantly trigonally bonded C-N alloy is formed. This behaviour is consistent with N acting as an n-type dopant up to N incorporation of order 1.0 at % .

Very little work has been carried out on the p-type doping of ta-C. Ronning et al [12] reported an increase in conductivity due to Boron addition to ta-C produced using MSIB deposition but this was attributed to the formation of localised defects without causing a shift in the Fermi level. The incorporation of Boron in FCVA ta-C was also shown to significantly reduce the stress in such material whilst still maintaining a high sp3 bonding fraction [13]. We have recently carried out a detailed study into the electrical properties of Boronated ta-C [14]. The films were produced by using a solid cathode of graphite impregnated with boron. The room temperature conductivity was found to increase by 5 orders of magnitude with the boron incorporation of order 8%. The conductivity activation energy was found to decrease to 0.15 eV at the same B concentration whilst the bandgap remained constant. Electrical measurements on n-type c-Si/ta-C:B heterostructures showed evidence of increasing rectifying behaviour with increasing Boron incorporation. These preliminary results indicate that it may be possible also to dope ta-C p-type.

4. Devices:

The potential for DLC as a semiconductor device material has been demonstrated by workers at Canon [15] who demonstrated a yellowish green light emission from a simple DLC film containing up to 5% oxygen. The merits of metal/a-C:H/metal MSM sandwich structures for use as the pixel switches in active matrix addressed liquid crystal displays have also been investigated [16] . The high field conduction mechanism has been identified as Poole-Frenkel hopping and the conduction parameters have been determined as a function of the a-C:H bandgap. Also a-C:H MSMs have been shown to have sharper switching characteristics than Silicon Nitride or Ta_2O_5 based MSMs but not as sharp as those made from a a-SiNx:H alloy. An advantage of MSMs made from amorphous carbon is that they can be deposited at room temperature making the use of plastic substrates possible.

The use of ta-C for electronic application is still under investigation. We have successfully demonstrated the fabrication of n-type ta-C/p-type c-Si heterojunctions [17] and this has been recently repeated by a group in Singapore [18]. The blue light sensitivity of ta-C makes its use as a complementary material to a-Si:H for optical applications an attractive possibility but much more work is needed to improve the response.

In spite of the high density of gap states the addition of both B and N to ta-C appears to move the Fermi Level across the bandgap and hence efforts have also been made to manufacture Thin Film Transistors (TFTs) from ta-C [19]. Although a number of carbon based transistors have been previously demonstrated they have either been produced from high temperature CVD diamond or more diverse forms of carbon such as polymeric phenelyne niylene or C_{60} [27]. We have produced both p-channel and more recently n-channel enhancement mode TFTs. Gate control of the drain current has been confirmed but the measured mobility of 10^{-5}-10^{-6} cm^2/Vs is too low for practical applications. In both cases top gate TFT structures were adopted using either PECVD oxide or nitride as the gate insulator.

However the most exciting and potentially most useful application of ta-C for electronic application is as the cathode material in Field Emission Displays (FEDs). Field Emission Displays (FEDs) are an attractive alternative to Active Matrix Addressed Liquid Crystal Displays (AMLCDs) for some flat panel applications. In a FED the electrons are emitted across a vacuum gap from a matrix of microcathodes and excite a screen of phosphor coated pixels. The advantages of FEDs are that they retain the high picture quality associated with the CRT, but are thin and have a much lower power consumption. Current FEDs use Spindt tips of typically Mo or Si as the cathodes. These materials have a high electron affinity and thus need high fields in order to allow electrons to escape. FEDs suffer from reliability and lifetime problems and although these have been overcome to some extent the complicated lithographic processes necessary to produce the tip arrays means that yield is still too low. An alternative is to use thin film materials which naturally emit electrons at low fields due to a low electron affinity. Diamond Like Carbon is a suitable applicant. Figure 10a shows a schematic of a typical triode design Spindt tip FED and Figure 10 b shows a schematic of the alternative DLC based FED.

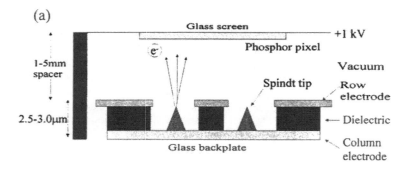

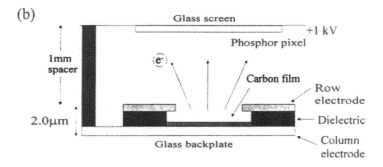

Figure 10 (a) Spindt Tip Based FED and (b) DLC triode Design FED

The emission field is generated by gate electrodes placed close to the cathodes, which can therefore be addressed by low voltages compatible with Si integrated circuits. There have been numerous reports of Field Emission from DLC films- see for example [20-23] and we have carried out a series of experiments to investigate emission from ta-C films. The detailed results are contained in [24-26]. Such films have proven to be suitable for use in FEDs as their emission threshold field can be less than 2-3 V/micron, emission current densities of > 1 mA/cm^2 at fields of 20V/micron are obtainable and with suitable treatment emission site densities greater than 10^4 /cm^2 are achievable. However, much more work is needed to fully understand the emission mechanism and to optimise the material but the future looks very promising.

5. Conclusions

Tetrahedrally bonded amorphous carbon is a new thin film semiconducting material with potential for application in the electronics field. It can be deposited onto Si or glass at room temperature over large areas and is stable up to 700°C. It has an optical bandgap of approximately 2.0- 2.5 eV, is p-type as-grown with a room temperature resistivity of order 10^8 Ωcm. It can be doped n-type by the addition of nitrogen during growth. Although TFTs have been produced using ta-C as the active channel material the estimated mobility of order 10^{-5} cm^2/Vs is too low for practical application in this field. At present the most exciting application for ta-C in large scale electronics is as the cathode emitting material in Field Emission Displays.

6. References

1. Amaratunga G.A.J, Putnis A., Clay K. and Milne W.I., Appl. Phys. Letts., **55**, 634, 1989.
2. Conway N. M.J., Milne W.I., and Robertson J., Diamond and Related Materials, **7**, 477, 1998.
3. Voevodin A.A. et al, J.Appl. Phys., **78**, 4123, 1995.
4. Lifshitz Y., Kasi S.R., Rabelais. J.W., Phys. Rev Letts., **68**, 620, 1989.
5. Fallon P.J., Veerasamy V.S., Davis C.A., Amaratunga G.A.J., Milne W.I, Robertson J. and Koskinen J., Phys Rev B, **48**, 4777, 1993.
6. Tauc J. Grigorovici R. and Vancu A., Phys Stat. Sol., 15, 627, 1966.
7. Robertson J.,O'Reilly and E.P, Phys. Rev. B, **35**, 2946, 1987.
8. Ilie A., Robertson J, Conway N.M.J., Kleinsorge B. and Milne W.I., to be published in J. Appl.Phys November 1998.
9. McKenzie D.R., Rep. Prog. Phys., **59**, 1611, 1996.
10. Amaratunga G.A.J., Veerasamy V.S., Milne W.I., Davis C.A., McKenzie D. and Yuan J., J.Non Crystalline Solids **164-66**, 1119,1993.
11. Veerasamy V.S., Yuan J., Amaratunga G.A.J., Milne W.I., Gilkes K.W., Weiler M. and Brown L.M., Phys Rev B, **48**, 24, 17954, 1993.
12. Ronning C. et al, Diamond and Related Materials, **4**, 666, 1995.

13. Chhowalla M., Davis C.A, Weiler M., Kleinsorge B. and Amaratunga G.A.J., J.Appl. Phys., **79**, 2237, 1996.
14. Kleinsorge B., Ilie A., Chhowalla M., Fukarek W., Milne W.I. and Robertson J., Diamond and Related Materials, **7**, 2-5, 472, 1998
15. Taniguchi Y., Hirabayashi H., Kurihara N. and Ikoma K., European Patent No 93203387.1
16. Egret S., Robertson J., Milne W.I.and Clough F.J., Diamond and Related Materials, **6**, 879, 1997.
17. Veerasamy V.S., Amararunga G.A.J., Milne W.I., MacKenzie H. and Park J.S., IEEE Trans. Elec Devices, **42**, No 3, 1995.
18. Xu Shi, Private Communication.
19. Clough F.J., Milne W.I.., Kleinsorge B., Robertson J, Amaratunga G.A.J. and Roy B.N., Elecs. Letters, 498, 1996.
20. Amaratunga G.A.J. and Silva S.R.P. Appl. Phys. Lett **68**, 2529, 1996
21. Jaskie J. MRS Bull., **21**, 59, 1996.
22. Missert N., Friedmann T.A., Sullivan J.P and Copeland R.G., Appl. Phys. Lett **70**, 1995, 1997.
23. Groning O., Kuttel O.M., Groning P. and Sclapbach L., Appl. Phys. Lett **71**, 2253, 1997.
24. Satyanarayana B.S, Hart. A., Milne W.I. and Robertson J., Appl. Phys. Lett **71**, 1430, 1997.
25. Chung S.J , Moon J.H., Park K.C., Oh M.H., Milne W.I.. and J.Jang, J.Appl Phys **82**, 8, 4047, 1997.
26. Satyanarayana B.S, Hart. A., Milne W.I. and Robertson J, Diamond and Related Materials, **7**, 2-5, 656, 1998.
27. R.C. Haddon, A.S. Perel, R.C. Morris, T.T.M. Palstar, A.F. Hebard and R.M. Fleming, Appl Phys Letters, **67**, 121, 1995

DIAMOND BASED SILICON-ON-INSULATOR MATERIALS AND DEVICES

STEFAN BENGTSSON and MATS BERGH
Solid State Electronics Laboratory
Department of Microelectronics ED,
Chalmers University of Technology
S-412 96 Göteborg, Sweden

Abstract

Silicon-on-insulator materials with buried diamond films were manufactured by a wafer bonding and etch-back procedure. These advanced silicon-on-insulator structures were made to address self-heating effects observed in conventional silicon-on-insulator materials with buried silicon dioxide layers. The compatibility of polycrystalline diamond films to silicon device manufacturing was evaluated by process experiments combined with Raman spectroscopy. A method of encapsulating the diamond film during oxidation procedures in device manufacturing was developed. Resistors, diodes and MOS-transistors were made in the silicon film and were used to characterise the diamond based silicon-on-insulator material. The results show a better heat dissipation capability of the material as compared to conventional silicon-on-insulator materials. Well-working diodes and MOS-transistors were made in the active silicon film. It can be concluded that no fundamental obstacles exist for forming these materials and for transferring a silicon device process onto the material. The main problem during the project came from stress and surface roughness of the diamond films. In the future, when a more mature diamond deposition technology is available, it should be possible to overcome these problems.

1. Introduction

Currently, silicon-on-insulator (SOI) materials [1,2] are expected to receive increasing attention for mainstream CMOS as well as for high frequency and high voltage applications [3]. The advantages of SOI come from dielectric isolation, process simplifications, higher yield and better device performance and reliability [2,3]. On the other hand, important concerns are the availability and cost of the SOI wafers as well as specific peculiarities [2-4] of devices fabricated in SOI materials such as the floating body effect and the low thermal conductivity of the buried insulator. Especially in applications where high power dissipation is expected, it is found that self-heating

P.L.F. Hemment et al. (eds.),
Perspectives, Science and Technologies for Novel Silicon on Insulator Devices, 97–107.
© 2000 *Kluwer Academic Publishers. Printed in the Netherlands.*

effects in the devices limit the applicability of SOI. This may occur for SOI smart power or high voltage applications, where a thick insulating layer has to be chosen in order to inhibit voltage breakdown to the substrate. Other examples of applications can be found in bipolar or BiCMOS devices, where at high frequencies high local current densities, for instance at emitter fingers, may cause device instability if the heat spreading is insufficient. The poor thermal conductivity of the silicon dioxide is in these cases a limiting factor for the heat distribution [4,5]. Heat transfer is limited both vertically to the substrate heat sink and laterally from hot spots. A buried insulator of diamond would have a superior heat spreading capability as compared to silicon dioxide. Diamond is a wide bandgap semiconductor exhibiting high electrical resistivity combined with high thermal conductivity. In Table I properties of silicon, silicon dioxide, poly crystalline diamond and aluminium nitride are compared. Aluminium nitride (AlN) may be considered as an alternative to diamond as an insulator with high thermal conductivity. Both diamond and aluminium nitride based SOI materials have been described in a number of articles and conference contributions [5-11]. In this paper we will describe the work at Chalmers University of Technology and Uppsala University, both in Sweden, in forming and characterising a bond- and etch-back SOI material with a buried film of polycrystalline diamond [5,8,9].

TABLE I. Properties of silicon, silicon dioxide, polycrystalline diamond and aluminium nitride.

Property	Si	SiO$_2$	Diamond	AlN
Bandgap (eV)	1.1	9	5.5	6.2
Dielectric constant	11.9	3.9	5.7	9.1
Dielectric strength (MV/cm)	0.3	10	10	0.15 - 0.4
Resistivity (Ωcm)	- - -	$10^{14\text{-}16}$	$10^{9\text{-}13}$	$10^{10\text{-}13}$
Thermal conductivity (W/Kcm) at RT	1.5	0.015	15[1]	1.5
Thermal expansion coefficient (10^{-6} K^{-1})	2.6	0.5	1.0	4

[1] Single crystal diamond.

2. Formation of Diamond Based Silicon-on-Insulator Materials

The Diamond Silicon-On-Insulator (DSOI) materials were made by a wafer bonding and etch-back procedure shown in figure 1. Polycrystalline diamond films were deposited on 100-mm silicon wafers by hot filament Chemical Vapour Deposition (CVD) [12]. The deposition was made by a commercial firm [13] using state-of-the art techniques. The thickness of the deposited diamond layers were about 3-5 μm. The silicon wafers onto which the diamond films were deposited contained buried etch-stop layers to allow the device layers to be formed by an etch-back procedure after bonding. Both buried silicon dioxide films (SIMOX wafers) and buried epitaxially grown silicon/boron/germanium etch stops were used. The silicon surface was prepared for diamond deposition using different seeding procedures, which involved abrasive treatment by a diamond powder or deposition of diamond growth seeds on top of the silicon surface. For the diamond deposition an ambient of 1-2% of CH_4 in H_2 was used. Other process parameters were a substrate temperature of 650 - 750° C and a pressure in the range of 30 - 50 Torr. The filament was held at a temperature of 2100 - 2200° C and the distance between the filament and the substrate wafer was about 1 cm. Cross section Scanning Electron Microscope (SEM) images of the as deposited diamond layers showed a columnar grain structure. The average diamond grain size was estimated to be less than 1 µm.

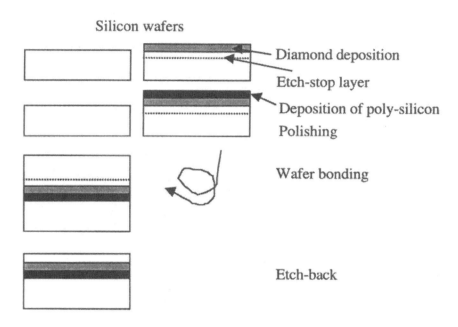

Figure 1. Formation of DSOI materials by wafer bonding and etch-back.

The surface of the deposited polycrystalline diamond films were, by order of magnitude, too rough to allow direct bonding to a silicon wafer. From Atomic Force Microscopy (AFM) analysis on a 10x10 µm surface, an rms roughness of the order of 100 Å was found for the diamond film. This should be compared to rms values of the order of 0.5 Å for polished silicon wafers. In addition, the diamond films also showed a small number of very distinct diamond peaks raising approximately 10 µm above the surrounding film. Wafer bonding is extremely dependent on smooth and clean surfaces [14]. An obvious solution would be to polish the diamond surface to a smooth bondable finish, but this approach would require non-standard manufacturing equipment [15,16] and under all circumstances is difficult. Instead a thick (~15 µm) polycrystalline silicon layer was deposited on top of the diamond film. To protect the diamond from any possible aggressive ambient during the polycrystalline silicon deposition, it was made in two steps. First a protective layer also acting as a seed was deposited using a conventional LPCVD technique at approximately 600° C. Thereafter the thick polycrystalline silicon layer was grown at 1050° C using a high temperature LPCVD technique. The polycrystalline silicon surface was polished using Chemical-Mechanical Polishing (CMP) to prepare the smooth bondable surfaces. This surface was bonded to another silicon wafer followed by lapping, etching and polishing of the silicon wafer onto which the diamond film was deposited to obtain the device layer, as is outlined in figure 1. Figures 2a and 2b show the resulting structure. The main problem of forming the DSOI materials was the very rough surface of the diamond film, and especially the few local distinct peaks (~10 μm of height) of diamond appearing on each wafer. As a result we encountered bonding problems and void generation even when using a polycrystalline silicon layer as the bonding medium. Furthermore, in some early runs of diamond deposition the diamond films were highly stressed, severely warping the silicon wafer. By tuning the seeding procedure as well as the deposition parameters of the diamond CVD process the stress problem was reduced.

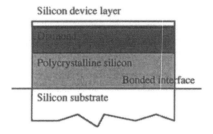

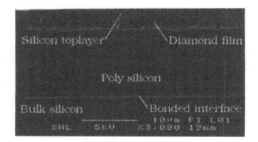

Figure 2a. Schematic structure of the DSOI material.

Figure 2b. Cross-section SEM image of the DSOI structure.

3. Silicon Process Compatibility

Diamond is stable at high temperatures in an inert ambient and resistant against most chemicals, but forms carbon oxide in the presence of oxygen at elevated temperatures. To investigate the need of process modifications in the transfer of an SOI bipolar or MOS process to DSOI, various process experiments were combined with Raman spectroscopy to verify the stability of the diamond film [8]. The diamond films were exposed to standard wet chemical and plasma treatments used in silicon device manufacturing. Raman spectroscopy using a Spectramax from Jobin-Yvon Optics Spectroscopy was used to ensure that the diamond film was not degraded during the processing. Raman spectroscopy is very sensitive to the binding properties of the film and a change from diamond to graphite or any other phase of carbon would be evident in the Raman spectra. Furthermore, the sensitivity in Raman spectroscopy of graphite is about 50 times larger than the sensitivity of diamond. From these experiments we conclude that RCA cleaning, HF and buffered HF etching, etching in KOH, and wet aluminium etching (H_3PO_4:CH_3COOH:HNO_3, 29:5:1 at 40° C) do not adversely affect the diamond film to a measurable extent. Plasma processing was investigated using a Cl_2 plasma as well as plasma stripping of resist in oxygen. The chlorine plasma slowly etched the diamond (30 Å /min). This etch rate gave a selectivity to silicon of about 1:100. Resist stripping in oxygen plasma caused a change in the optical appearance of the diamond surface, but Raman spectroscopy did not reveal any large structural changes. Plausibly the outermost layers of the diamond film are affected by the oxygen plasma treatment. Based upon these results an etching procedure using oxygen plasma in a reactive ion etcher was developed to etch the diamond and allow a patterning of the diamond film. Patterning of the diamond was utilised in the fabrication of device structures (see Section 4).

The influence of annealing steps on the diamond film was investigated. For uncovered polycrystalline diamond films annealing at 1000° C or above, even in an inert atmosphere, caused massive phase conversions in the films. Annealing at 950° C in nitrogen was possible without any measurable change in structure or electrical properties. The electrical analysis was made using current *vs.* voltage measurement after deposition of aluminium gates on top of the diamond film. A method based on encapsulation with silicon nitride was developed [8] to protect the diamond during high temperature processing especially in an oxidising ambient. To evaluate the encapsulation, the diamond film was removed from part of the wafer and the remaining diamond was encapsulated using deposited films of silicon dioxide and silicon nitride. A LOCOS oxidation was made in water vapour and Raman spectroscopy was again used to monitor changes in the diamond film. Figure 3 shows Raman spectra before and after such a sequence of encapsulation and LOCOS oxidation at 1000° C.

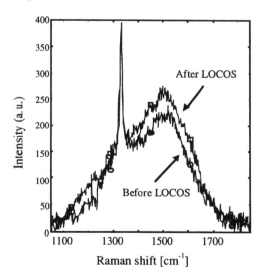

Figure 3. Raman spectra of a diamond film before and after a sequence of encapsulation using a silicon nitride cap and LOCOS oxidation in water vapour at 1000° C.

The diamond peak just above 1300 cm^{-1} can be clearly seen in the Raman spectra both before and after encapsulation and LOCOS oxidation. The graphite peak is located between 1500 and 1600 cm^{-1}. After LOCOS oxidation at 1100° C a small increase in the graphite signal was seen. Encapsulation of the diamond films is, thus, required prior to thermal oxidation. Based on these experiments it was concluded that an SOI device manufacturing process without large modifications could be transferred to DSOI. The exposure of the diamond to oxygen or other oxidising agents at elevated temperature seems to be the most important condition to avoid.

4. Test Devices on DSOI

To evaluate the DSOI material, resistors, diodes and MOS-transistors were manufactured in the silicon films [5,8,9]. Before actual device manufacturing, device islands were formed in the silicon/diamond/silicon structure. The diamond between the silicon islands intended for device manufacturing was removed by reactive ion etching in an oxygen plasma in order to create a trench in the diamond film. This was done to enable a better encapsulation of the diamond for reasons of process compatibility and to avoid ruining large wafer areas if pinholes were present in the device layer. Both the developed encapsulation technique and the plasma etching of the diamond film were thus used in the device manufacturing process. As a first step a low temperature CVD silicon dioxide and LPCVD silicon nitride films were deposited. The films were patterned and used as an etch mask during mesa etching of the silicon top layer in KOH. The exposed diamond was then removed using oxygen plasma reactive ion etching. Holes were opened in the nitride/oxide stack and the devices were formed using standard procedures. Figure 4 schematically shows the procedure for the formation of pn-junctions. Resistors were formed using the same technique as for the formation of pn-junctions.

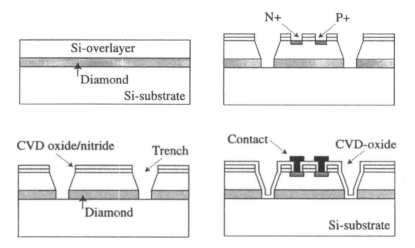

Figure 4. Schematic process flow of pn-junction formation in the DSOI material.

Also, MOS-transistors were formed using the same procedures. After etching to form trenches in the silicon and the diamond films an oxide/nitride stack was deposited to encapsulate the device islands. After opening holes in the oxide/nitride stack, a gate oxide was grown by thermal oxidation, followed by gate formation and source/drain doping by implantation. A CVD-oxide was deposited and the implanted dopants were activated by an anneal in argon at 850 °C. Finally, contact holes were opened and aluminium was evaporated and patterned followed by a low temperature anneal. Figure 5 shows the resulting transistor structure in the DSOI material.

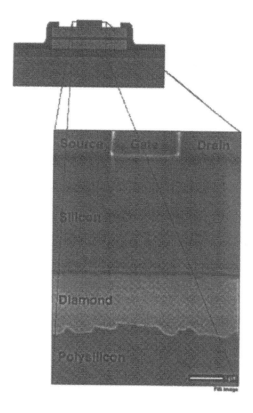

Figure 5. The resulting MOS-transistor structure in the DSOI material.

The major benefit of DSOI comes from the high thermal conductivity of the diamond [5]. The steady state temperature increase *vs.* dissipated power density in the active silicon films was measured using a p^+ resistor with known temperature coefficient. By measuring the resistance *vs.* applied power with the backside of the wafer connected to a heat sink with constant temperature, the heat dissipation properties of different materials were determined. As shown in figure 6 the result for DSOI materials is in-between the results for SOI and bulk silicon.

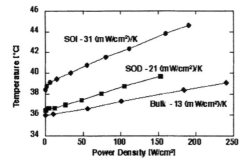

Figure 6. Measured temperature *vs.* applied power density for different materials.

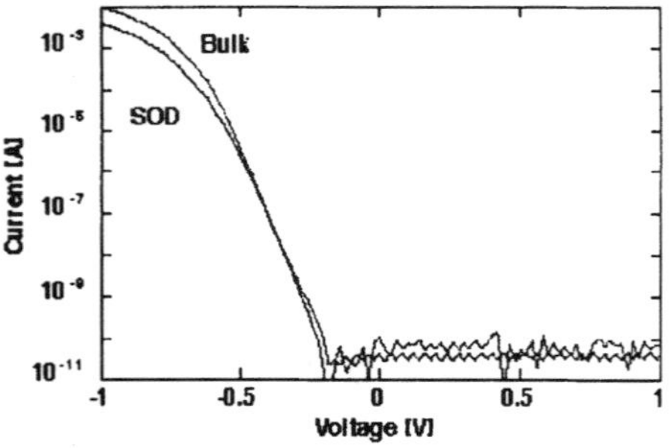

Figure 7. Current *vs.* voltage characteristics for pn-junctions made in bulk silicon and DSOI.

To further evaluate the formed DSOI materials, diodes and MOS-transistors were made in DSOI as well as in bulk silicon. Current *vs.* voltage characteristics for pn-junctions made in DSOI and bulk silicon proved to be very similar as is shown in figure 7. MOS-transistors were manufactured using thermally grown gate oxides. Except for the special steps needed to isolate the silicon islands in the DSOI material the same device manufacturing process was used. Electrical measurements were performed to compare the transistors manufactured in bulk material and in DSOI material. Figures 8 (DSOI) and 10 (bulk silicon) show the drain current *vs.* drain-source voltage. Figures 9 (DSOI) and 11 (bulk silicon) show the drain current *vs.* gate voltage.

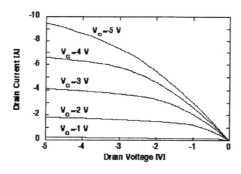

Figure 8. Drain current *vs.* drain-source voltage for a MOSFET fabricated in DSOI material.

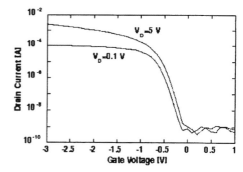

Figure 9. Drain current *vs.* gate voltage for a MOSFET fabricated in DSOI material.

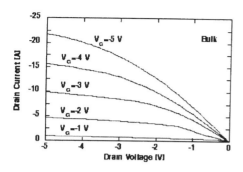

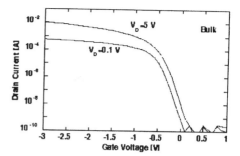

Figure 10. Drain current *vs*. drain-source voltage
for a MOSFET fabricated in bulk silicon.

Figure 11. Drain current *vs*. gate voltage for a
MOSFET fabricated in bulk silicon.

As can be seen in figures 8 to 11 working devices were made both in bulk silicon and in
the DSOI material. Since the silicon films of the DSOI materials were about 2 μm
thick one should expect the transistors to have bulk like characteristics.

5. Discussion

We have shown the possibility of forming SOI materials with a buried diamond film.
The project was made to address self-heating effects in SOI materials and to show the
ability of wafer bonding to form novel structures by integrating otherwise incompatible
materials. Ignoring the self-heating issues a DSOI material is also expected to have
potential for rad-hard applications [17]. Diamond has been shown to be very resistant
against irradiation because of a high charge carrier mobility and high chemical bond
strength [6]. However it seems that the requirements for SOI for niche applications can
be reasonably well solved with current silicon dioxide based SOI materials, although in
the future more demanding applications will probably require a further look into the
possibility of using diamond-based materials or other types of advanced structures.
Although we were able to form the DSOI materials we faced bonding problems mainly
because of the rough diamond surface with a few diamond peaks of the order of 10 μm
high. We needed to use a thick layer of polycrystalline silicon followed by CMP to
reasonably solve the problem. There is no reason why a more mature diamond
deposition technique would not be able to solve the problems related to the peaks.
Therefore the main problems we faced during this project will probably be overcome in
the future.

Diodes and MOS-transistors were successfully manufactured in the diamond based SOI
materials. The fact that the devices behave very similar in bulk silicon and in DSOI
shows that diamond films potentially can be integrated with silicon materials without
degrading the device performance. This is important information since diamond

deposition and especially the seeding procedure before deposition is somewhat brutal and until recently had not been evaluated as a process step in silicon technology.

In addition to buried diamond films also other possibilities exist. In the literature attempts using buried films of aluminium nitride for SOI applications have been reported [10,11]. One may also speculate that it will be possible to use other highly thermally conductive wide bandgap semiconductors for similar applications within silicon technology.

6. Conclusions

We have demonstrated the possibility of manufacturing silicon-on-insulator materials with buried diamond films by a wafer bonding and etch-back procedure. However Tte diamond deposition technology is not yet mature enough to avoid problems with surface roughness and stress. The materials were shown to overcome self-heating effects observed in conventional silicon-on-insulator materials. The silicon device process compatibility of diamond was evaluated by process experiments combined with Raman spectroscopy. A method of protecting the diamond film by means of encapsulation during oxidation procedures in device manufacturing was developed. It can be concluded that no fundamental obstacles exist for forming these materials and for the successful incorporation of the substrates in to a full silicon device process.

7. Acknowledgements

This work was supported by the Swedish National Board for Industrial and Technical Development (NUTEK) within the BELLA consortium. Thanks are due to all co-workers within the project at Chalmers University of Technology and at Uppsala University, to Ericsson Components, Stockholm, for support as well as to Professor Per Svedberg for his enthusiastic co-ordination of the project.

8. References

1. Lasky, J.B. (1986) Wafer bonding for silicon-on-insulator technologies, *Appl. Phys. Lett.* **48**, 78-80.
2. Colinge, J.-P. (1991) *Silicon-On-Insulator Technology: Materials to VLSI*, Kluwer Academic Publishers, Dordrecht.
3. Maszara, W.P., Dockerty, R., Gondran, C.F.H. and Vasudev, P.K. (1997) SOI materials for mainstream CMOS technology, in S. Cristoloveanu (ed.), *Silicon-on-insulator technology and devices VIII*, The Electrochem. Soc. Proc. Series Vol. 97-23, The Electrochem. Soc., Pennington, pp. 15-26.
4. Arnold, E., Pein, H. and Herko, S.P. (1994) Comparision of self-heating effects in bulk-silicon and SOI high voltage devices, in *IEDM Technical Digest*, IEEE, New York, pp 813-816.
5. Edholm, B., Söderbärg, A., Olsson J. and Johansson E. (1995) Transient measurements of heat distribution in devices fabricated on silicon-on-diamond material, *Jpn. J. Appl. Phys.* **34**, 4706-4714.

6. Annamalai, N.K. and Chapski, J. (1990) Novel dielectrics for SOI structures, in *Proc. 1990 IEEE SOS/SOI Technology Conf.*, Key West, IEEE, New York, pp 59-60.

7. Annamalai, N.K., Sawyer, J., Karulkar, P., Maszara, W.P. and Landstrass, M. (1993) Radiation response of silicon on diamond (SOD) devices, *IEEE Trans. Nucl. Sci.* **40**, 1780-1786.

8. Edholm, B, Söderbärg, A. and Bengtsson, S. (1996) Reliability evaluation of manufacturing processes for bipolar and MOS devices on silicon-on-diamond materials, *J. Electrochem. Soc.* **143**, 1326-1334.

9. Bergh, M., Tiensuu, S. and Edholm B. (1998) Diamond silicon-on-insulator using wafer bonding, Manuscript., to be published.

10. Bengtsson, S., Choumas, M., Maszara, W.P., Bergh, M., Olesen, C., Södervall, U. and Litwin, A. (1994) Silicon on aluminium nitride structures formed by wafer bonding, in *Proc. 1990 IEEE SOS/SOI Technology Conf.*, Nantucket Island, IEEE, New York, pp 35-36.

11. Bengtsson, S., Bergh, M., Choumas, M., Olesen, C., and Jeppson, K. (1996) Applications of aluminium nitride films deposited by reactive sputtering to silicon-on-insulator materials, *Jpn. J. Appl. Phys.* **35**, 4175-4181.

12. Matsumoto, S., Sato, Y., Kamo, M. and Setaka, N. (1982) Vapour deposition of diamond particles from methane, *Jpn. J. Appl. Phys.* **21**, L183-L185.

13. Diamonex Inc., Allentown, PA, USA.

14. Bengtsson, S. (1992) Semiconductor wafer bonding: A review of interfacial properties and applications, *J. Electron. Mater.* **21**, 841-862.

15. Tokura, H., Yang, C.-F. and Yoshikawa, M. (1992) Study on the polishing of chemically vapour deposited diamond films, *Thin Solid Films* **212**, 49-55.

16. Bhushan, B., Subramanian, V.V., Malshe, A., Gupta, B.K. and Ruan, J. (1993) Tribological properties of polished diamond films, *J. Appl. Phys.* **74**, 4174-4180.

17. Landstrass, M.I. and Fleetwood D.M. (1990) Total dose radiation hardness of diamond-based silicon-on-insulator structures, *Appl. Phys. Lett.* **56**, 2316-2318.

Low-Temperature Processing of Crystalline Si Films on Glass for Electronic Applications

R. B. BERGMANN, T. J. RINKE, L. OBERBECK, and R. DASSOW
Institut für Physikalische Elektronik, Universität Stuttgart
Pfaffenwaldring 47, D-70569 Stuttgart, Germany

Abstract: The present paper gives an overview of the material properties and the technology of the low-temperature preparation and modification of crystalline Si films on glass. Electronic properties of Si films strongly depend on the film structure, which thus determines possible areas of devices applications. In detail, we discuss i) high-throughput pulsed laser crystallization using a solid state laser that enables the formation of Si films with elongated grains having a length of several ten μm, ii) low-temperature epitaxy at temperatures around 600°C with a rate up to 0.5 μm/min using ion assisted deposition, and iii) the formation of quasi-monocrystalline Si films via crystallization of porous Si. This innovative thin film transfer technology permits reuse of Si wafers and produces films with a thickness-dependent hole mobility of up to 78 cm^2/Vs and effective internal light trapping.

1. Introduction

Low temperature deposition and processing of crystalline Si films is of great practical interest for electronic applications that require cheap, large area and/or transparent substrates such as glass. Examples are thin film solar cells [1], active matrix displays [2], and displays with driver electronics integrated on glass [3].

The present paper discusses low temperature techniques for the preparation of crystalline Si films for application to electronic devices, which can be broadly classified into *majority carrier* devices such as thin film transistors (TFTs) for display or logic applications and *minority carrier* devices such as solar cells or sensors. While majority carrier devices require high majority carrier mobility, minority carrier devices need a high minority carrier diffusion length. For both applications, a whole range of materials from amorphous or nanocrystalline [4] to monocrystalline Si films [5] are being studied intensively. In this paper, the following topics are discussed in detail:

i) The structural features of thin crystalline Si films can be classified mainly according to grain size and processing temperature. We describe the correlation of structural and electrical properties of Si films on glass, as the electronic properties of Si films strongly depend on the crystallography and, thus, define areas of possible device applications.

ii) Crystallization of sub-μm thick amorphous Si films on glass substrates is accomplished using a solid state laser. Grain size can, within certain limitations, be adjusted to

P.L.F. Hemment et al. (eds.),
Perspectives, Science and Technologies for Novel Silicon on Insulator Devices, 109–120.
© 2000 *Kluwer Academic Publishers. Printed in the Netherlands.*

suit the desired device application. Elongated grains of several tens of microns in length are obtained using sequential lateral solidification (SLS). Solid state lasers achieve a pulse repetition rate of up to 100 kHz and thus allow one to crystallize several 10 cm^2 of Si film per second.

iii) For some device applications such as bipolar transistors one needs pn-junctions formed by epitaxy. Crystalline Si thin film solar cells require the deposition of films with a thickness of several microns in order to absorb a sufficient fraction of sunlight. For certain electronic devices on glass, we therefore need epitaxial growth of Si at a high deposition rate. A comparison of various deposition techniques shows that ion-assisted deposition (IAD) is a key technology for high-rate epitaxial deposition at temperatures below 650°C. We systematically investigate the quality of films formed by IAD as a function of deposition temperature and post-deposition treatment.

iv) Thermal crystallization of porous Si creates Si films with a crystallographic orientation determined by the substrate. The formation of such *quasi-monocrystalline* Si films and their transfer to glass is expected to overcome limitations of the electrical properties of polycrystalline Si films imposed by the presence of grain boundaries. We discuss the formation and possible applications of quasi-monocrystalline Si films.

2. Crystalline Si Films on Glass

The crystallographic features of thin crystalline Si films can be classified mainly according to grain size and processing temperature. Here, we review the current understanding of the correlation of structural and electrical properties of Si films on glass.

2.1. STRUCTURAL AND ELECTRONIC PROPERTIES OF CRYSTALLINE Si FILMS DEPOSITED ON GLASS

Table 1 presents a basic classification for the properties of crystalline Si on glass according to processing temperature and grain size. The following sections discuss deposition, crystallization (where applicable), technological constraints and electrical properties as well as availability of glass substrates in the respective temperature regimes.

Nanocrystalline Silicon: At low deposition temperatures T<450°C, the deposition of Si leads to the formation of either hydrogenated amorphous Si (a-Si:H) with its well known light-induced degradation [6], or (hydrogenated) nanocrystalline Si films (nc-Si or nc-Si:H) with a grain size on the order of 10 nm and a process-dependent fraction of amorphous material within the film [7]. The layers can be deposited on a large variety of substrates such as inexpensive soda-lime float-glass, see Ref. [8].

Electronic transport in nc-Si:H is considerably different from transport in poly-crystalline Si (poly-Si) [9] and is not yet understood in detail [7]. Recent experimental investigations, assisted by two-dimensional transport simulations, indicate that there is no significant band bending at grain boundaries of grains with 10 to 20 nm diameter. The Debye screening length in Si is within the range of 100 nm and thus much higher than the average grain size. The carrier transport in nc-Si:H is therefore dominated by trapping and recombination and not by potential barriers at grain boundaries. A comparison of simulated and calculated IV-curves of pin-junctions indicates that amorphous regions within the material are randomly distributed rather than located at the grain boundaries of the crystallites [10]. Majority carrier mobility is of the order of a few cm^2/Vs [7] and the minority carrier diffusion length is generally well below 1 μm [11]. Minority carrier devices are realized using pin structures in order to benefit from drift fields for carrier extraction. The deposition rate of minority carrier device grade material is, up to now, fairly low [12].

Microcrystalline or fine-grained polycrystalline Silicon: Micro- or polycrystalline Si films deposited or treated in the medium temperature range (450 ... 700°C) can be obtained, for example, by solid phase crystallization [13] or laser crystallization [1]. A number of glass substrates are available nowadays which extend the process temperature range to near 700°C [14 - 16].

Majority carrier transport is governed by potential barriers at grain boundaries, while minority carrier recombination may be dominated by either grain boundaries or intra-grain defects. Material quality can, to some extend, be improved by hydrogen passivation [17]. The majority carrier mobility of these films may approach that of monocrystalline Si, especially for devices fabricated in single crystalline areas of such films [18]. However, very few studies are concerned with measurements on minority carrier diffusion length in fine grained poly-Si [19], which are complicated by the broad grain size distribution observed in such films [20]. Diffusion lengths are usually in the range of only a few μm even for monocrystalline films deposited at temperatures around 600°C, see section 4. The fabrication of minority carrier devices such as solar cells using fine grained poly-Si is therefore still very challenging.

Large grained polycrystalline Silicon: Large grained Si films may be obtained in the high-temperature regime preferably by using zone melting processes [21]. Material properties may approach that of monocrystalline Si as far as majority and minority carrier properties are concerned and the material can be processed using standard high temperature device processing, deposition rates can be as high as 10 μm/min. High temperature processing of Si, however, requires the use of ceramics [22] or special glass substrates [23], which may add substantial cost to a fabrication process. However, during the last years the scope for comparatively cheap high-temperature glass substrates has significantly widened [24, 14], although the development of new glass substrates is considerably hindered by the extreme mismatch between the initial (experimental) need and the huge production volume of glass plants.

2.2. TRANSFER TECHNIQUES FOR THE FORMATION OF MONOCRYSTALLINE Si FILMS ON GLASS

Monocrystalline Si films appear to be ideally suited for use in thin film devices, if cost issues due to expensive processing can be overcome. Within the last few years, two innovative concepts for the fabrication of crystalline Si films have been developed: i) The "Smart cut" process, which relies on the separation of a Si film by implantation and subsequent annealing [5], and ii) epitaxial growth of Si on sacrificial, porous Si layers [25] and subsequent separation of the epi-layer from the starting wafer [26, 27]. Both processes, however, require potentially expensive processing steps such as implantation and/or epitaxy. A simpler and therefore presumably more cost effective process for the fabrication of thin monocrystalline films is described in section 5 of this paper.

TABLE 1: Classification of polycrystalline Si films on glass, explanation see text. GB = grain boundary, CVD = chemical vapor deposition, g = grain size, IAD = ion-assisted deposition.

film structure grain size temperature regime	large grained poly $g \rightarrow$ mm ... cm high temperature (> 800°C)	fine grained poly $g \approx 0.1 ... 10\,\mu m$ medium temperature (450°C ... 700°C)	nanocrystalline $g \approx 10$ nm low temperature (< 450°C)
charge carriers majority carrier mobility minority carriers: transport recombination	quasi-monocryst. behavior diffusion intra-grain	GB potential barriers medium diffusion grain boundaries	barrier height $\rightarrow 0$ low (percolation) drift grain boundaries
technology junction formation surface passivation	solid state diffusion thermal oxidation	heterojunction SiN_x, SiO_x coating	pin structure
deposition rates	(rapid) thermal CVD 1 ... 10 µm/min	plasma CVD IAD and others 0.01 ... 0.5 µm/min	plasma CVD hot-wire CVD 0.01 .. 0.2 µm/min
(re-)crystallization **processes** thermal budget	zone melting $T_{melt} \approx 1400°C$ high (s ... min)	laser crystallization $T_{laser} \gg 1400°C$ low (µs)	none
suitable glass **substrates**	high temperature glass, fused silica	boro-, alumino silicate glass	soda-lime-glass

3. Laser Crystallization for High Throughput and Large Grain Size

Different types of lasers and processing schemes have been investigated within the last few years [28 -31] resulting in grain sizes from some nm to some tens of microns. We apply the so-called sequential lateral solidification (SLS) process [29, 30] investigated by Im et al. [29]. In this process, small areas, a few microns wide and more than 100 microns in length, of the initial Si film are sequentially melted by successive laser pulses. Between laser pulses, the substrate is moved a distance of approximately one micron. By this procedure, grains grow from the interface of the previously processed area into the molten zone. The SLS process enables the crystallization of high quality poly-Si with grain sizes of several tens of microns. In order to obtain a cost effective production of large area electronic devices, we optimize the process to achieve large area throughput [32].

3.1. SEQUENTIAL LATERAL SOLIDIFICATION

In this section, we investigate the choice of the optimum laser parameters for high energy efficiency and high processing speed. The translation of the substrate between subsequent pulses is limited by the lateral length of the grains grown during the resolidification to less than 1 µm. The scanning speed is therefore determined by the repetition rate of the laser. The laser beam energy density for complete melting of a 100 nm thick a-Si film on glass is approximately 0.5 J/cm^2, the width of the processed area is around 5 µm. Conventional optical imaging systems limit the aspect ratio of the slit-like focussed laser beam to about 1,000 or probably 10,000 using purpose-designed optics. The length of the focus is thus limited between 5 mm and 50 mm. For complete melting of a Si film by a single laser beam, a laser pulse energy of 1.25 mJ is needed if one assumes a focus height of 50 mm and a focal width of 5 µm.

Therefore, a laser with a relatively low pulse energy of around 1 mJ and a high pulse repetition rate is favorable for a high-throughput SLS process. We apply a newly developed all solid state laser system (diode pumped Nd:YVO$_4$ laser), which is capable of emitting laser pulses at a rate up to 100 kHz with a pulse energy of 0.1 mJ. With such a laser we achieve a throughput of 2.8 cm^2/s. The pulse energy of this particular laser system is well beyond the limit of 1.25 mJ. However, systems with significantly higher pulse energy are available. A laser system with a pulse energy of 1.25 mJ and a repetition rate of 100 kHz would allow a throughput of 35 cm^2/s.

3.2. LASER CRYSTALLIZATION SETUP

We use a diode pumped, frequency doubled Nd:YVO$_4$ laser, operating at variable repetition rates up to 100 kHz. The laser wavelength is 532 nm, the pulse width 10 ns and the pulse energy is 100 µJ at 10 kHz and 5 µJ at 100 kHz. Depending on the thickness of the a-Si:H layer, the required pulse energy density varies between 600 mJ/cm^2 and 1000 mJ/cm^2. We crystallize 100 nm and 300 nm thick amorphous silicon films, which are deposited by Very High Frequency Plasma Enhanced Chemical Vapor Deposition (VHF-PECVD) on Corning 1737F glass [14]. Furnace annealing of the films at 550 °C for 45 minutes decreases the hydrogen content in the film to less than 3 %. Further details on the experimental setup and film deposition are described in [33].

3.3. CHARACTERIZATION OF LASER CRYSTALLIZED Si FILMS ON GLASS

Figure 1 shows a scanning electron micrograph (SEM) of a laser crystallized 300 nm thick a-Si film after Secco etching [34]. Curved lines are attributed to large angle grain boundaries and straight, parallel lines to twins. Both evolve in the scanning direction of the laser beam. Analyzing larger portions of the sample surface indicates a length of the grains of more than 100 µm. By evaluating the average number of grain boundaries using profile scans perpendicular to the scanning direction, we estimate the grain width

114

to exceed 3 μm. These findings are supported by previously presented [32] transmission electron microscopy (TEM) investigations.

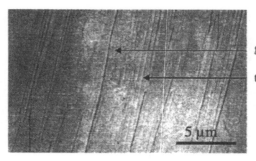

grain boundary

twin

Figure 1. Laser crystallized, 300 nm thick Si film on glass. Secco etching reveals the location of extended crystallographic defects. Scanning electron micrograph.

Electrical characterization of crystallized poly-Si films is performed using room temperature Hall-effect measurements. We use a Van der Pauw geometry with a sample size of 7×7 mm^2. Highly boron doped films with a hole concentration of 2×10^{21} cm^{-3}, a thickness of 100 nm and a sheet resistance of 10 $\Omega/\square$ show a Hall-effect mobility of 28 cm^2/Vs, which is near the value for bulk silicon at the same doping level. The large grain size of the films makes them interesting for TFT applications and solar cells. Highly doped films are in addition suitable as contact layers for solar cells.

4. High-Rate, Low-Temperature Epitaxy of Si Films

A great variety of techniques based on chemical vapor deposition (CVD) enable low-temperature epitaxy of Si. Most of these techniques, however, suffer from low deposition rates. Figure 2 compares epitaxial deposition rates as reported during the last years by various authors.

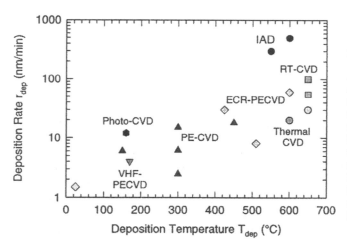

Figure 2: Comparison of various deposition techniques suitable for low-temperature Si epitaxy: Ion-assisted deposition (IAD) achieves deposition rates r_{dep} of up to 0.5 μm/min at a deposition temperature T_{dep} of 600°C. CVD-techniques: RT: rapid thermal, ECR: electron cyclotron resonance, PE: plasma enhanced, VHF: very high frequency. For references to data of deposition rates see [36].

To overcome the rate limitation, we have investigated the potential of ion-assisted deposition (IAD) [35] for rapid epitaxial growth at low temperatures. This technique is capable of epitaxial deposition rates as high as 0.3 µm/min at deposition temperatures as low as 550°C [36] and presently achieves a deposition rate of 0.5 µm/min at 600°C. In contrast to molecular beam epitaxy (MBE), IAD makes use of a small fraction of ionized Si atoms, which are accelerated towards the substrate by moderate acceleration voltages V_{acc} of around 10 to 50 V. These ions provide additional energy to the Si atoms on the substrate surface and enable the use of lower substrate temperatures [37].

4.1. ELECTRICAL PROPERTIES OF EPITAXIAL Si FILMS OBTAINED FROM ION-ASSISTED DEPOSITION

The *majority carrier mobility* of holes in monocrystalline epitaxial Si films deposited by IAD is determined by means of room-temperature Hall measurements. Figure 3a shows hole concentration dependent mobility data obtained from in-situ boron and gallium doped Si films [38]. The majority carrier mobilities μ_h in boron doped layers almost reach the values obtained for crystalline bulk Si [39] over a wide doping range. The hole mobility in the case of gallium doping is a little lower. Low Hall mobilities of gallium-doped films arise from the comparatively low activation of electrically active gallium acceptors due to the high ionization energy of 72 meV [40]. In contrast, at 600°C boron can be fully activated at concentrations greater than 1×10^{20} cm^{-3} [41].

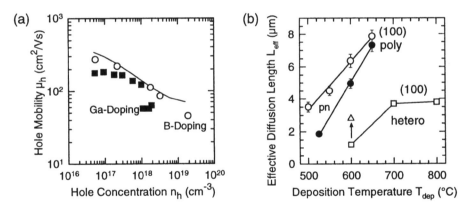

Figure 3. a) Hole mobility as determined by Hall-effect measurements of in-situ B- and Ga-doped epitaxial layers deposited by IAD at 600°C. Boron-doped layers show higher hole mobility μ_h as compared to Ga doped layers. Solid line indicates theoretical values for B-doped Si [39]. b) Effective diffusion length L_{eff} increases with deposition temperature T_{dep}. Highest diffusion lengths are observed in films with a diffused pn-junction, diffusion lengths of heterojunction devices stay at significantly lower levels, for explanation see text. Open symbols: films deposited on monocrystalline Si, filled symbols: films deposited on poly-Si, circles: pn-junction formation by phosphorous diffusion, squares and triangle: heterojunction. B-doping of heterojunction devices, Ga-doping otherwise, triangle: film post-deposition annealed at 700°C.

In order to characterize the *minority carrier diffusion length* in mono- and polycrystalline epitaxial Si films deposited by IAD, we fabricated two different types of pn-junctions: (i) A high-temperature pn-junction process [42] involving a phosphorous diffusion at 830°C and thermal oxidation at 1000°C an (ii) a low-temperature hetero-junction process [43] with the deposition of amorphous Si layers at 220°C which has been developed for compatibility with glass substrates. The minority carrier diffusion length is determined by the evaluation of internal quantum efficiency measurements [44].

Figure 3b shows that in the case of pn-junction devices, the effective diffusion length L_{eff} increases with increasing deposition temperature T_{dep} from about 3.5 to 8 µm on monocrystalline, (100)-oriented substrates and from 2 to 7.5 µm on polycrystalline substrates. In this study so called SILSO material [45] with a grain size of a few mm is used as a polycrystalline substrate, as this material is well suited to reveal grain boundary effects on the devices with an area of 1cm^2. The diffusion length of electrons in hetero-junction devices on polycrystalline substrates increases from 1 to 2.5 µm and on mono-crystalline substrates from about 1 to 4 µm in the temperature range investigated.

Gettering of impurities during the phosphorous diffusion and/or annihilation of crystallographic defects during the high-temperature oxidation may be responsible for the improved diffusion length of the diffused pn-junction devices. A film deposited at 700°C yields a diffusion length of 3.7 µm and a film deposited at 600°C and subsequently annealing at 700°C has a diffusion length of 2.8 µm. This observation indicates that intrinsic defects, which are partly annihilated at higher temperatures, rather than impurities limit the diffusion length in the epitaxial films. Further investigations are on the way to determine the mechanism that limits the performance of the films and to improve the diffusion lengths.

5. Quasi-monocrystalline Si films

In this section we present an alternative approach for the fabrication of mono-crystalline Si films. Instead of using porous Si films only as a substrate for epitaxy, as demonstrated by other authors in Refs. [26, 27], we avoid epitaxy and use porous Si directly for device fabrication: A thermal annealing process serves two purposes namely to crystallize the porous Si and to separate it from the substrate. As a consequence, the porous Si film is not wasted as a sacrificial layer and the process does not require epi-taxy.

5.1. FORMATION AND CRYSTALLIZATION OF POROUS Si

Porous silicon is formed by anodic etching in a two-electrolyte etching cell [46]. Details of the process are given elsewhere [47]. We use (100)-oriented, boron-doped, polished, 4 inch silicon wafers with a resistivity between 0.01 and 0.05 Ωcm. In order to demonstrate the reusability of the wafers, we have already obtained 30 layers from one 500 µm thick Czochralski (Cz)-grown wafer. In a first step, a porous film with a thick-ness ranging from sub-µm to up to 10 µm is formed with a porosity of around 20%. In a second step, a layer with high porosity is formed.

A high-temperature annealing step in high vacuum at a temperature of 1050°C for 2 h serves to crystallize the porous Si films. As a result of this process, we obtain a sandwich structure of i) the silicon wafer, ii) a separation layer, and iii) a quasi-monocrystalline, transferable Si film, which is still fixed to the substrate by the separation layer. Figure 4a shows a cross-sectional scanning electron (SEM) micrograph of the sandwich structure. All standard processes used in bulk Si technology such as diffusion, oxidation and photolithography are applicable without modification. The transfer of the quasi-monocrystalline film to a foreign substrate, if so desired, can be done at the end of device processing. Figures 4b and c show SEM micrographs of films of different thickness which are separated from the wafer.

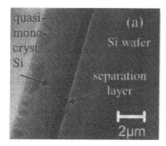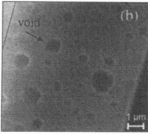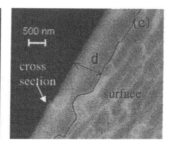

Figure 4. Quasi-monocrystalline Si films obtained from the crystallization of porous Si, shown in cross-sectional scanning electron micrographs. a) High-temperature annealing creates a separation layer between the quasi-monocrystalline film and the Si-wafer. b) Cross-sectional view of an 8-μm thick, free standing quasi-monocrystalline p-type Si film with a resistivity of 0.05 Ωcm. c) Sub-μm thick, p-type, monocrystalline film with a resistivity of 0.02 Ωcm. Out-diffusion of internal voids to the surface causes surface roughness.

The high-temperature treatment of porous silicon induces an essential change in the structure of the Si film. The inner surface decreases by several orders of magnitude by thermally activated diffusion of Si atoms. However, the recrystallized layer preserves the orientation of the original crystal lattice, as verified by X-ray diffraction [47]. The structure of the crystallized films strongly depends on the sample thickness. Several μm thick films contain well-defined void volumes, see Figure 4b, while sub-μm thin films are completely dense, see Figure 4c. In the case of sub-μm thin films, voids grow until they reach the surface where they are annihilated but cause surface roughness.

5.2. ELECTRICAL AND OPTICAL PROPERTIES OF QUASI-MONO-CRYSTALLINE Si FILMS

The carrier mobility of quasi-monocrystalline Si films is determined from room-temperature Hall-effect measurement. Starting wafers are boron-doped in the range of 6×10^{17} to 7×10^{18} cm^{-3}, corresponding to a resistivity of 0.01 - 0.05 Ωcm. Quasi-mono-crystalline Si films made of bulk Si with a resistivity of 0.05 Ωcm achieve a hole mobility of 78 cm^2/Vs. Figure 5a shows the dependence of the carrier mobility on the resistivity. Thinner films with a lower concentration of voids have a higher mobility, probably due to a lower probability of charge carrier scattering at inner surfaces.

Figure 5b shows the absorption coefficient of a 4 μm thick quasi-monocrystalline Si film compared to the absorption coefficient of crystalline Si. Over a wide spectral range the absorption coefficient exceeds that of bulk Si by a factor of 10. We assume this high absorption to be caused by internal light trapping, see inset of Fig. 5b. This phenomenon

118

is particularly interesting for thin film solar cells, as a 4 µm thick solar cell made from this material would have less than 10% transmission losses. First test devices fabricated from a phosphorous diffused pn-junction on a 4 µm thick, quasi-monocrystalline Si film on a Si wafer with a resistivity of 0.05 Ωcm result in a short circuit current density of 11.1 mA/cm^2 at one sun illumination without any anti-reflection coating or film texture.

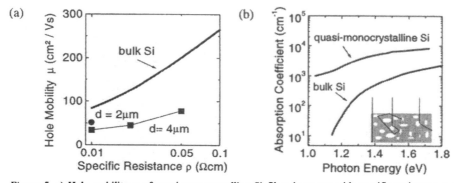

Figure 5. a) Hole mobility µ of quasi-monocrystalline Si films increases with specific resistance ρ and decreasing film thickness d. b) Absorption coefficient of a quasi-monocrystalline, 4-µm thick sample, derived from transmission and reflection measurements with an integrating sphere. The absorption coefficient is an order of magnitude larger than the one of bulk crystalline silicon (c-Si line). Inset shows a simple model of internal light scattering that leads to enhanced absorption.

6. Conclusions

Crystalline Si films on glass can be deposited with a great variety of properties mainly depending on deposition temperature and grain size. Silicon films with majority *and* minority carrier properties close to bulk Si can, up to now, only be fabricated with monocrystalline or large grained polycrystalline Si films deposited at high temperature. Three routes for low cost and/or rapid fabrication of device quality, low-temperature crystalline Si films are described in this overview:

Sequential lateral solidification using a pulsed solid state laser enables the fabrication of large grained Si films with a throughput on the order of several 10 cm^2/s.

Several µm thick Si films can be epitaxially deposited by ion-assisted deposition at a maximum rate of 0.5 µm/min at growth temperatures around 600°C with minority carrier diffusion lengths between 1 and 8 µm depending on deposition temperature and post-deposition treatment.

Crystallization of porous Si films at temperatures exceeding 1000°C creates an ultrathin, quasi-monocrystalline Si film on a separation layer and presents a novel route for device fabrication. These films achieve a hole mobility of 78 cm^2/Vs at a resistivity of 0.05 Ωcm. Due to internal light trapping, the absorption coefficient of this material is within a broad energy regime significantly higher as compared to bulk Si.

Acknowledgment: The authors thank M. Grauvogl, R. Hausner, N. Jensen, J. Köhler, U. Rau, F. Renner, J. Schirmer, M. B. Schubert, M. Vetter, and B. Winter for technical assistance and valuable discussions and are indebted to J. H. Werner for his continuous support. Part of this work is supported by the German ministry for education, research, science and technology (BMBF) under contract no. 0329818.

7. References

1. Bergmann, R. B., Köhler, J., Dassow, R., Zaczek, C., and Werner, J. H. (1998) Nucleation and Growth of Crystalline Silicon Films on Glass for Solar Cells, *Physica Status Solidi (a)* **166**, 587.
2. Boyce, J. B., Mei, P., Fulks, R. T., and Ho, J. (1998) Laser Processing of Polysilicon Thin-Film-Transistors: Grain Growth and Device Fabrication, *Physica Status Solidi (a)* **166**, 729.
3. Street, R. A. (1998) Large Area Electronics, Applications and Requirements, *Physica Status Solidi (a)* **166**, 695.
4. Schubert, M. B. (1999) Low Temperature Silicon Deposition for Large Area Sensors and Solar Cells, *Thin Solid Films* **337**, xxx, in press.
5. Bruel, M. (1996) Application of Hydrogen Ion Beams to Silicon On Insulator Material Technology, *Nuclear Instruments and Methods in Physics Research B* **108**, 313.
6. Fritzsche, H. (1997) Search for explaining the Staebler-Wronski effect, *Mat. Res. Soc. Symp. Proc.* **467**, 19.
7. Carius, R., Finger, F., Backhausen, U., Luysberg, M., Hapke, P., Houben, L., Otte, M., and Overhof H. (1997) Electronic properties of microcrystalline Silicon, *Mat. Res. Soc. Symp. Proc.* **467**, 283.
8. Pfaender, H. G. (1997) Schott Guide to Glass, Chapman & Hall, London.
9. Kamins, T. (1988) *Polycrystalline Silicon for Integrated Circuit Applications*, Kluwer Academic Publishers, Dordrecht.
10. Zimmer, J., Stiebig, H., Hapke, P., and Wagner, H. (1998) Study of the electronic transport in pin solar cells based on microcrsytalline silicon, in: *Proc. 2nd World Conf. on Photovoltaic Solar Energy Conversion*, in press.
11. Brüggemann, R. (1998) Improved steady-state photocarrier grating in nanocrystalline thin films after surface-roughness reduction by mechanical polishing, *Appl. Phys. Lett.* **73**, 499.
12. Torres, P., Meier, J., Goetz, M., Beck, N., Kroll, U., Keppner, H., and Shah, A. (1997) Microcrystalline silicon solar cells at higher depsoition rates by the VHF-GD, *Mat. Res. Soc. Symp. Proc.* **452**, 883.
13. Bergmann, R. B., and Krinke, J. (1997) Large grained polycrystalline silicon films by solid phase crystallization of phosphorus doped amorphous silicon, *J. Crystal Growth* **177**, 191.
14. Fehlner, F. P. (1997) Thin films on glass for liquid crystal displays, *J. Non-Crystalline Solids* **218**, 360.
15. Moffatt-Fairbanks D. M., Tennent, D. L. (1997) Substrate issues for advanced display technologies, *Mat. Res. Soc. Symp. Proc.* **471**, 9.
16. Schott Corp. (1997), Product information on glass, BOROFLOAT nos. 33, AF37 and 40
17. Plieninger, R., Wanka, H. N., Kühnle, J., and Werner, J. H. (1997) Efficient defect passivation by hot-wire hydrogenation, *Appl. Phys. Lett.* **71**, 2169.
18. Im, J. S., Crowder, M. A., Sposili, R. S., Leonard, J. P., Kim, H. J., Yoon, J. H., Gupta, V.V., Jin Song, H., and Cho H. S. (1998) Controlled Super-Lateral Growth of Si Films for Microstructural Manipulation and Optimization, *Physica Status Solidi (a)* **166**, 603.
19. Brendel, R., Bergmann, R. B., Fischer, B., Krinke, J., Plieninger, R., Rau, U., Reiß, J., Strunk, H. P., Wanka, H., and Werner J. H. (1997) Transport analysis for polycrystalline silicon solar cells on glass substrates, in: *Proc. 26th IEEE Photovoltaic Specialists Conf.*, IEEE, Picataway, p. 635.
20. Bergmann, R. B., Shi F. G., and Krinke, J. (1998) Non-coarsening origin of log-normal size distributions during crystallization of amorphous films, *Physical Review Letters* **80**, 1011.
21. Robinson, R. D., Miaoulis, I. N. (1994) Thermal parameters affecting low temperature zone-melting recrystallization of films, *J. Appl. Phys.* **75**, 1771.
22. Hebling, C., Glunz, S. W., Schuhmacher, J. O., and Knobloch J. (1997) High-efficiency (19.2%) silicon thin-film solar cells with interdigitated emitter and base front-contacts, in: *Proc. 14th Europ. Photovoltaic Solar Energy Conf.*, H. A. Ossenbrink, P. Helm, and H. Ehmann (eds.), Stephens & Assoc., Bedford, p. 2318.
23. Bergmann, R. B., Hebling, C., Ullmann, I., Bischoff, E., and J. H. Werner, (1997) Zone melt recrystallization of silicon films on glass, in: *Proc. 14th Europ. Photovoltaic Solar Energy Conf.*, H. A. Ossenbrink, P. Helm, and H. Ehmann (eds.), Stephens & Assoc., Bedford, p. 1464.

120

24. Bergmann, R. B., Darrant, J. G., Hyde, A. R., Werner, J. H. (1997) Crystalline Silicon films on a novel high temperature glass for applications in microelectronics and photovoltaics, *J. Non-Crystalline Solids* **218**, 388.

25. Sato, N., Sakaguchi, K., Yamagata, K., Fujiyama, Y., and Yonehara, T. (1995) Epitaxial growth on porous Si for a new bond and etchback silicon-on-insulator, *J. Electrochem. Soc.* **142**, 3116.

26. Tayanaka, H., Yamauchi K., and Matsushita, T. (1998) Thin-film crystalline silicon solar cells obtained by separation of a porous silicon sacrificial layer, in: *Proc. 2nd World Conference on Photovoltaic Solar Energy Conversion*, in press.

27. Brendel, R., Artmann, H., Oelting, S., Frey, W., Werner, J. H., and Queisser, J. H. (1998) Monocrystalline Si waffles for thin solar cells fabricated by the novel perforated-silicon process, *Appl. Phys. A.* **67**, 151.

28. Kuriyama, H., Nohda, T., Ishida, S., Kuwahara, T., Noguchi, S., Kiyama, S., Tsuda, S., Nakano, S. (1993) Lateral grain growth of poly-Si films with a specific orientation by an excimer laser annealing method, *Jpn. J. Appl. Phys.* **32**, 6190.

29. Im, J. S., and Sposili, R. S. (1996) Crystalline Si films for integrated active-matrix liquid crystal displays, *Mater. Res. Bull.* **21**, 39.

30. Im, J. S., Spossli, R. S. and Crowder, R. S. (1997) Single-crystal Si films for thin film transistor devices, *Appl. Phys. Lett.* **70**, 3434.

31. Plais, F., Legagneux, P., Reita, C., Huet, O., Petinot, F., Pribat, D., Godard, B., Stehle M. and Fogarassy, E. (1995) Low temperature polysilicon TFT's: A comparison of solid and laser crystallization, *Microelect. Eng.* **28**, 443.

32. Köhler, J. R., Dassow, R., Bergmann, R. B., Krinke, J., Strunk H. P., and Werner, J. H. (1998) Large grained polycrystalline silicon on glass by copper vapor laser annealing, *Thin Solid Films* xxx, in press.

33. Dassow, R., Köhler, J. R., Grauvogl, M., Bergmann, R. B., and Werner, J. H. (1998) Laser-crystallized polycrystalline silicon on glass for photovoltaic applications, in: *Polycrystalline Semiconductors V*, Werner, J. H., Strunk, H. P., Schock, H. W., eds., in Series *Solid State Phenomana*, Scitech Publ., Uettikon am See, Switzerland, 1999, to be published.

34. Secco d´Aragona, F. (1972) Dislocation etch for (100) planes in silicon, *J. Electrochem. Soc.* **119**, 948.

35. Oelting, S., Martini, D., Köppen, H., Bonnet, D. (1995) Ion assisted deposition of crystalline thin film silicon solar cells, in: *Proc. 13th European Photovoltaic Solar Energy Conf.* (H. S. Stephens, Bedford, 1995), p. 1681.

36. Bergmann, R. B., Zaczek, C., Jensen, N., Oelting, S., Werner, J. H. (1998) Low-temperature Si epitaxy with high deposition rate using ion-assisted deposition, *Appl. Phys. Lett.* **72**, 2996.

37. Rabalais, J. W., Al-Bayati, A. H., Boyd, K. J., Marton, S., Kulik, J., Zhang, Z., and Chu, W. K. (1996) Ion-energy effects in silicon ion-beam epitaxy, *Phys. Rev. B* **53**, 10781.

38. Kühnle, J., Bergmann, R. B., Oelting, S., Krinke, J., Strunk, H. P., and Werner, J. H. (1997) Poly-crystalline silicon films on glass for solar cells by ion-assisted deposition, in *Proc. 14th Europ. Photovoltaic Solar Energy Conf.*, Stephens & Assoc., Bedford, p. 1022.

39. Klaassen, D. B. M. (1992) A unified mobility model for device simulation I. Model equations and concentration dependence, *Solid-State Electronics* **35**, 953.

40. Arch, J. K., Werner, J. H., and Bauser, E. (1993) Hall effect analysis of liquid phase epitaxy silicon for thin film solar cells, *Solar Energy Materials and Solar Cells* **29**, 387.

41. Parry, C. P., Kubiak, R. A., Newstead, S. M., Whall, T. E., and Parker, E. H. C. (1991) Temperature dependence of incorporation processes during heavy boron doping in silicon molecular beam epitaxy, *J. Appl. Phys.* **71**, 118.

42. Blakers, A. (1990) High efficiency crystalline silicon solar cells, in: *Festkörperprobleme / Advances in Solid State Physics* Vol. 30, Vieweg, Braunschweig, p. 403.

43. Hausner, R. M., Jensen, N., Bergmann, R. B., Rau, U., and Werner, J. H. (1998) Heterojunctions for polycrystalline silicon solar cells, to be published, see [33].

44. Brendel, R., Hirsch, M., Stemmer, M., Rau, U., and Werner, J. H. (1995) Internal quantum efficiency of thin epitaxial silicon solar cells, *Appl. Phys. Lett.* **66**, 1261.

45. Watanabe, H. (1993) Overview of cast multicrystalline silicon solar cells, *MRS Bulletin* **18**, 29.

46. Lang, W. (1996) Silicon microstructuring technology, *Mater. Sci. Engin.* **R17**, 55.

47. Rinke, T. J., Bergmann, R. B., Brüggemann, R., and Werner, J. H. (1998) Ultrathin quasi-monocrystalline silicon films for electronic devices, to be published, see [33].

β-SiC ON SiO₂ FORMED BY ION IMPLANTATION AND BONDING FOR MICROMECHANICS APPLICATIONS

C. Serre[1], A. Pérez-Rodríguez[1], A. Romano-Rodríguez[1], J.R. Morante[1], L. Fonseca[2], M.C. Acero[2], J. Esteve[2], R. Kögler[3] and W. Skorupa[3].

[1] E.M.E., Dept. d'Electrònica, Unitat Associada CNM-CSIC, Universitat de Barcelona, Avda. Diagonal 645-647, 08028 Barcelona, Spain.
[2] Centre Nacional de Microelectrònica CNM-CSIC, Campus UAB, 08193 Bellaterra, Spain.
[3] Forschungzentrum Rossendorf, P.F. 510119, D-01314 Dresden, Germany.

Abstract:

β-SiC on SiO₂ multilayer structures have been fabricated by ion implantation into Si substrates and thermal bonding. This process involves three steps: i) multiple energy C⁺ implants into Si, to obtain a broad buried β-SiC layer, ii) selective oxidation of the top Si layer, and iii) bonding and etch-back of Si. These are processes compatible with Si processing technology, and permit high crystalline quality β-SiC films on SiO₂ to be formed without using expensive bulk SiC or Silicon-On-Insulator wafers. The structures have been characterised after the different process steps mainly by Fourier Transform Infrared Spectroscopy, X-Ray Photoelectron Spectroscopy, Secondary Ion Mass Spectroscopy and Atomic Force Microscopy. The analysis of samples processed after the different steps has allowed the key parameters for fabricating high quality structures for electronic devices and sensors applications to be defined.

1- Introduction

SiC, as a wide band gap semiconductor, constitutes a promising material for power and high temperature electronics as well as for visible optoelectronics, while its high stiffness, high mechanical strength and extreme chemical inertness make it suitable for the development of electronic devices or micromechanical applications which have to operate in chemically or physically aggressive environments and at high temperatures. Among the wide variety of SiC polytypes, the cubic (3C-SiC or β-SiC) presents very interesting electronic properties. However, this phase cannot be grown as bulk material, and is usually obtained as thin films deposited on Si substrates. High temperature ion implantation into Si followed by annealing [1.2], has also been shown to produce high quality crystalline β-SiC layers. Nevertheless, the fabrication of devices in these layers requires high electric isolation between the active SiC layer and the Si substrate.

121

P.L.F. Hemment et al. (eds.),
Perspectives, Science and Technologies for Novel Silicon on Insulator Devices, 121–126.
© 2000 *Kluwer Academic Publishers. Printed in the Netherlands.*

Therefore, the development of SiC on Insulator (known as SiCOI) technologies represents an attractive route to achieve low cost SiC devices and systems in Si substrates. One of the most promising techniques is the wafer bonding and etch-back approach. The Bond and Etch-back technique [3] was originally developed for the synthesis of Silicon On Insulator (SOI) structures, leading to the so called BESOI technique which is one of the main technologies for the fabrication of SOI wafers [4]. It has been successfully used to transfer CVD SiC layers grown on Si to insulating substrates such as oxidised silicon or sapphire [5] and more recently in combination with the smart cut process [6].

In this context, we report the fabrication of SiCOI structures by a bond and etch-back technique of β-SiC layers onto oxidised Si wafers. The SiC layers were obtained by high temperature multiple energy C$^+$ ion implantation into Si followed by high temperature annealing. The structural analysis indicates the formation of a β-SiC buried layer. Oxidation experiments have been performed in order to insure complete oxidation of the top Si layer while minimising the penetration into the SiC layer. Afterwards, bonding onto a Si wafer covered with a thick thermal oxide has been performed by a mechanically assisted room temperature bonding procedure followed by a high temperature anneal to strengthen the bonding. After removal of the Si substrate by etching in TMAH (CMOS compatible etchant), a SiC/SiO$_2$/Si structure was obtained with an average RMS surface roughness as low as 5 nm. In this etching process, the buried ion beam synthesised SiC layer itself acts as etch-stop layer.

2- Ion beam synthesis of the SiC layers

(001) B-doped Si wafers (16-24 Ωcm) were implanted at 500°C, at energies of 100, 120, 150 and 195 keV, and doses of 2.6×10^{17}, 3.3×10^{17}, 4.7×10^{17} and 10^{18} cm^{-2}, respectively. These parameters were determined using TRIM simulations [7] to achieve a flat stoechiometric carbon profile. These implantations were performed at 500°C, in order to avoid amorphisation of the implanted layer. The samples were subsequently annealed at 1150°C for 6 hours in a nitrogen atmosphere. Similar to the results reported in [2] for a single implantation at 500°C, the structural analysis revealed the direct formation of a β-SiC layer with abrupt SiC/Si interfaces. This layer consists of SiC precipitates with the same crystalline orientation as the Si substrate. The high crystalline quality of the layer after annealing can be seen in figure 1, where the FTIR spectra from an annealed sample exhibits a lorentzian absorption peak centred at 796 cm^{-1}, with a FWHM of about 50 cm^{-1}, characteristic of crystalline β-SiC.

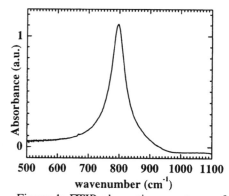

Figure 1: FTIR absorption spectrum of a β-SiC buried layer synthesised by multiple C ion implantation at 500 °C and annealed at 1150 °C, 6 h (centre: 797 cm^{-1}, FWHM: 49 cm^{-1}).

XPS measurements (curve (a) in figure 2) show a broad flat carbon profile, stable under annealing, with an SiC stoechiometric concentration. The thickness of the SiC layer is about 3000 Å, while that of the top Si layer is about 2500 Å. Finally, the layers have proved excellent etch-stop properties, as can be seen from the XPS analysis (curve (b) in figure 2) of a sample etched one hour in TMAH 25% wt at 80 °C. These samples will be referred to as "SiC samples".

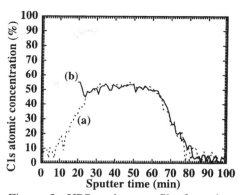

Figure 2: XPS carbon profile from ion beam synthesised SiC layers: curve (a) as implanted sample, and curve (b) sample annealed and etched.

3- Pre-bonding surface preparation

First of all, in order to obtain a SiCOI structure, the top Si layer of the SiC samples must be eliminated. The easiest approach would be to remove it by selective etching. However, we found in a previous etch-stop study [8] that the resulting RMS roughness of the SiC surface left by etching in TMAH is of the order of 50 to 60 Å as measured by AFM. This is much higher than that of a bare Si wafer, which is usually about 1.5 to 2 Å, and would seriously decrease the efficiency of subsequent bonding. In view of this, we finally chose to oxidise this top Si layer. The surface roughness of the SiC samples after wet oxidation was found to be around 15 Å. This is lower than the 29 Å for the RTCVD SiC layers that were successfully used for bonding by Tong et al. [5].

On the other hand, emphasis should be put on the importance of a good adjustment of the oxidation depth. Any residual Si layer between the SiO_2 and the SiC in the final SiCOI structure has to be eliminated. At the same time the penetration of the oxidation front into SiC has to be minimised in order to limit the loss of SiC. The finally retained set of oxidation conditions corresponds to an equivalent SiO_2 thickness of 6000 Å in Si, which consumed less than 100 Å of SiC.

The next step in the preparation of the samples consists of a careful clean of the oxidised Si wafers and SiC samples. This is a standard RCA cleaning, which in addition to removing any contamination, inclusions or particles of dust, leaves an hydrophilic surface. Hydrophilicity of the surfaces has been found to have a crucial influence on the efficiency of the bonding [9].

4- Thermal bonding

The bonding of two mirror-polished, flat and clean wafers usually involves a room temperature bonding step (weak bonding) followed by a high temperature "strengthening" step. The roughness and cleanliness of the two surfaces are highly critical factors for the success of the weak bonding step. In general, two wafers that fail to bond at room temperature will also fail to bond at high temperatures. In the case of two Si wafers, with or without oxide, they spontaneously adhere to each other when brought into contact at room temperature. However, this is not true when one the

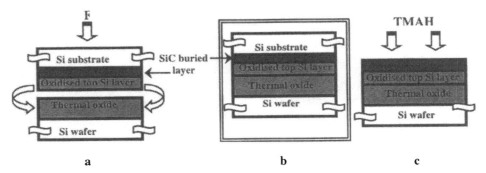

Figure 3: Overall scheme of the bonding process. a) mechanically assisted room temperature bonding. b) High temperature strengthening. c) Etch-back of the implanted sample's subtrate.

contact surfaces is made of SiC: even with the most careful surface preparation our SiC samples still failed to adhere to the Si oxidised wafers. Finally, room temperature bonding could only be managed with the application of an external force as described in [5].

The subsequent high temperature strengthening process was then performed, in conditions that are currently used for standard Si wafer bonding: 2 hours dry oxidation followed by 8 hours annealing in N_2 at 1100°C.

The wafer bonding is followed by the thinning or removal of one of the substrates by chemical etching, which is often controlled by the presence of a buried etch-stop layer. However, in our case, the process is simpler, since the active ion beam synthesised SiC buried layer itself acts as the etch-stop layer, owing to the excellent etch-resistant properties of our implanted layers as stated in section 2.

The overall bonding process is schematically represented in figure 3. According to this, the original SiC buried layers have been successfully transferred onto the surface of the oxidised Si wafer, leading to an SiC/SiO$_2$/Si structure. The resulting SiCOI structure is shown in figure 4a: the interface between the two oxide layers cannot be distinguished after bonding, and neither voids nor bubbles are observable. As for the SiC/SiO2 interface, we can see in figure 4b a 20-25 nm thick layer with SiC precipitates in an oxide matrix, which are assumed to correspond to the tail of the implantation profile. Further assessment will be required to determine the electrical characteristic of this layer. The sample presents an average RMS surface roughness as low as 45 to 50 Å, with a remarkably flat stoechiometric C profile as shown by SIMS analysis (fig. 5).

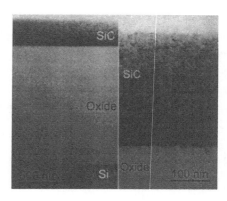

Figure 4: cross section TEM image of the SiCOI structure.

Bonding of about 75 % of the initial area has been accomplished. This coverage ratio could be increased mainly by improving the

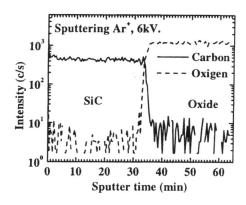

Figure 5: SIMS analysis of the final SiCOI structure. Ar$^+$ sputtering was used in order to be able to measure the oxygen concentration and detect the interface SiC/SiO$_2$.

efficiency of the room temperature bonding step. Two critical issues related to this are treatments to enhance hydrophilicity and a better homogeneity of the applied external force. On the other hand, further experiments are also needed to characterise the electrical behaviour of the SiC/SiO$_2$ interface, since it has been observed that fixed charge and interface traps densities in SiC/oxide interfaces strongly depend on the oxidation method as well as on the quality of the starting material [10]. These features will control the performance of the electronic devices to be made in the structures. Furthermore, previous mechanical assessment of the SiCOI structures has already demonstrated the viability of these SiCOI substrates for micromechanical applications.

5- Conclusions

High quality crystalline β-SiC buried layer have been successfully transferred by wafer bonding onto oxidised Si. The SiC layers were produced by high temperature multiple energy ion implantation and annealing, and the bonding was performed after wet thermal oxidation of the top Si layer. The resulting SiCOI structures have a surface roughness as low as 45 Å and a sharp SiC/SiO$_2$ interface. Further experiments are planned to improve the efficiency of the bonding process and to carry out electrical characterisation.

References

[1] Martin, P., Daudin, B., Dupuy, M., Ermolieff, A., Olivier, M., Papon, A.M. and Rolland, G. (1990) High temperature in beam synthesis of cubic SiC, J. Appl. Phys. **67** (6), 2908-2912.

[2] Serre, C., Pérez-Rodríguez, A., Romano-Rodríguez, A., Morante, J.R., Kögler, R. and Skorupa, W. (1995) Spectroscopic characterisation of phases formed by high-dose carbon ion implantation in silicon, J. Appl. Phys. **77** (6), 2978-2984.

[3] Maszara, W.P. (1991) Silicon-on-insulator by wafer bonding: a review, J. Electrochem. Soc. **138** (1), 341-347.

[4] Harendt, C., Hunt, C.E., Appel, W., Graf, H.G., Höfflinger, B. and Penteker, E. (1991) Silicon on insulator material by wafer bonding, J. Electron. Mater. **20** (3), 267-277.

[5] Tong, Q.-Y., Gösele, U., Yuan, C., Steckl, A.J. and Reiche, M. (1995) Silicon Carbide Wafer Bonding, J. Electrochem. Soc. **142** (1), 232-236.

[6] Di Cioccio, L., Le Tiec, Y., Letertre, F., Jaussaud, C. and Bruel, M. (1996) Silicon carbide on insulator formation using the Smart Cut process, Electronics Letters **32** (12), 1144-1145.

[7] Ziegler, J.F., Biersack, J.P. and Littmark, U. (1985) The stopping and range of ions in solids, Vol. 1, Pergamon Press, New York.

[8] Serre, C., Pérez-Rodríguez, A., Romano-Rodríguez, A., Calvo-Barrio, L., Morante, J.R., Esteve, J., Acero, M.C., Skorupa, W. and Kögler, R. (1997) Synthesis of SiC microstructures in Si technology by high dose carbon implantation: etch-stop properties, J. Electrochem. Soc. **144** (6), 2211-2215.

[9] Tong, Q.-Y. and Gösele, U. (1989) A model of low-temperature wafer bonding and its applications, J. Electrochem. Soc. **143** (5), 1773-1779 (1996).

[10] Chaudry, M.I. (1989) A study of native oxides of β-SiC using Auger electron spectroscopy" , J. Mater. Res. **4** (2), 404-407.

LASER RECRYSTALLIZED POLYSILICON LAYERS FOR SENSOR APPLICATION: ELECTRICAL AND PIEZORESISTIVE CHARACTERIZATION

A.A.DRUZHININ, I.I.MARYAMOVA, E.N.LAVITSKA,
Y.M.PANKOV, I.T.KOGUT

"Lviv Polytechnic" State University, Lviv, Ukraine
Kotlarevsky street 1, Lviv 290013, Ukraine

1. Introduction

The increasing needs in polycrystalline silicon films which are suitable for fabrication of the low-cost microelectronic sensors stimulate studies aimed to improve the properties of polysilicon. A possible way to obtain the high-quality poly-Si layers is the microzone laser recrystallization (MLR) technique. The aim of the studies was to obtain the material with improved characteristics for fabrication of piezoresistive mechanical sensors. The microzone laser recrystallization of a polysilicon layer using the heating of a substrate represents an easily realized technology of SOI structures for their further application in fabrication of IC and microelectronic sensors [1,2]. The microzone laser recrystallization changes the microstructure of poly-Si layers modifying their electrical and piezoresistive properties. The main   these properties, being the most important in the mechanical sensors technology, are studied.

A set of theoretical and experimental investigations was carried out to reveal the possibilities of MLR in the technology of piezoresistive mechanical sensors.

2. Theory

The theoretical analysis of the electrical and piezoresistive characterization of poly-Si layers is based on consideration of both the bulk silicon and the barrier regions. The carrier transport through the potential barrier is supposed to be due to the thermionic emission combined with diffusion [3,4].

It was established [4] that the potential barrier height does not change under an external strain and the reason of piezoresistance at the grain boundaries (GB) is the change in carrier mobilities which influence diffusion and recombination velocities in the potential barrier region.

P.L.F. Hemment et al. (eds.),
Perspectives, Science and Technologies for Novel Silicon on Insulator Devices, 127–135.
© 2000 *Kluwer Academic Publishers. Printed in the Netherlands.*

For numerical computations of the electrical and piezoresistive properties of the poly-Si layers the characteristics of the potential barrier were evaluated in the wide range of impurity concentrations and temperatures.

For p-type poly-Si layers, the change of the electrical conductivity σ_b of the barrier due to the strain applied was estimated using the method proposed in [4]

$$\frac{\Delta\sigma_b}{\sigma_b} = \frac{(B_1 - B_2)(\Delta E_2 - \Delta E_1)}{2kT(B_1 - B_2)} + \frac{(\Delta B_1 + \Delta B_2)}{(B_1 + B_2)}, \tag{1}$$

where $\Delta E_2 - \Delta E_1$ is the strain-induced energy gap between the light- and heavy-hole subbands,

$$B_i = m_i \Big/ \Big[1 + \big(2kTm_i / \pi q^2 E_b^2 \tau^2\big)\Big]^{1/2}, \tag{2}$$

$$\Delta B_i = \frac{\Delta m_i}{1 + \big(2kTm_i / \pi q^2 E_b^2 \tau^2\big)^{1/2}} \left\{ 1 - \frac{\big(2kTm_i / \pi q^2 E_b^2 \tau^2\big)^{1/2}}{2\Big[1 + \big(2kTm_i / \pi q^2 E_b^2 \tau^2\big)^{1/2}\Big]} \right\}, \tag{3}$$

m_i is the hole effective mass in the i-th subband, Δm_i is the strain-induced change of m_i, E_i is the maximum electrical field at the grain boundary and τ is the effective time between collisions. For rigorous consideration the details of the valence-band structure of silicon were taken into consideration for calculations of the electrical and piezoresistive properties both of the grain volume and of the grain boundary [5].

Assuming the potential barriers at the grain boundaries and the grains as consequently connected linear resistances, the resistivity of poly-Si layer, ρ, and longitudinal gauge factor, G_L, may be estimated according to the expressions [4]:

$$\rho = \frac{(L - 2w)}{L}\rho_g + \frac{2w}{L}\rho_b, \tag{4}$$

$$G_L = 1 - \sum_j \frac{S_{ij}'}{S_{ii}'}\big(1 - \delta_{ij}\big) + \frac{\rho_g \pi_{lg}}{\big[\rho_g + 2w\rho_b / (L - 2w)\big]S_{ii}'} + \frac{\rho_b \pi_{lb}}{\big[\rho_b + (L - 2w)\rho_g / 2w\big]S_{ii}'} \tag{5}$$

where π_{lg} and π_{lb} are the longitudinal piezoresistance coefficients of the grain and the grain boundary, S_{ij}' are the elastic constants of silicon; the thickness of the grain boundary, L is an average grain size, w is the width of the depleted region; the width of the grain δ is supposed to be substantially less than L and w and, thus, neglected. Indices g and b here define the values related to the grain and the grain boundary correspondingly.

The laser recrystallization enlarges the average grain size and, at the same time, increases the carrier concentration due to reducing the total surface of the grain boundaries and traps' passivation. Therefore, the laser recrystallization increases contributions of diffusion and tunnelling to the electrical conductivity of the GB's. On the other hand, the contribution of the GB's themselves to the total resistivity of polycrystalline silicon strongly depends on the average grain size. For the large-grained material the volume of the grains dominates in the electrical conductivity and piezoresistance.

For boron-doped p-type polysilicon the software has been developed and for the wide concentration and temperature ranges electrical and piezoresistive properties of the polysilicon layers were calculated with the average grain size as a parameter. Electrical resistivities of the grain and the grain boundary (ρ_g, ρ_b) as well as their change due to the strain were calculated separately.

Fig.1 and 2 show the calculated dependencies of resistivity and carrier mobility for boron-doped poly-Si *vs* doping impurity concentration for several grain sizes.

Calculations of the longitudinal gauge factor, G_L, show the distinct maxima at concentrations about 1×10^{19} cm^{-3} for the large grain size (Fig.3). For the smallest grain structures there were no extrema. One can note that the maxima of the gauge factor correspond approximately to the maxima of effective mobilities. This behaviour of G_L *vs* impurity concentration dependence becomes evident since the carrier mobility is known to be a factor determining piezoresistance mechanisms both in the grain volume and at the GB.

3. Experimental

The experimental studies of the electrical conductivity, carrier mobility and piezoresistivity of boron-doped polysilicon-on-insulator patterns both before and after the laser recrystallization were carried out.

For experimental measurements the boron-doped LPCVD poly-Si on SiO$_2$ layers were used before and after the treatment by CW YAG-laser (λ=1.06 μm, power 10-20 W). Experimental SOI test patterns were specially designed for these studies. The cross section of the experimental test structure is presented in Fig.4.

For the optimum temperature profile in the laser-irradiated region the combined capping layer was formed at the film surface. This layer contained the 0.5 μm thick SiO$_2$ layer and 0.15 μm thick Si$_3$N$_4$ strips. More detailed information about the method is presented in [6]. Depending on the regime of the laser recrystallization, on whether or not SiO$_2$/Si$_3$N$_4$ have been used we obtained after the laser treatment different kinds of recrystallized layers: with mosaic-shaped grains (an average grain size about 1 μm), chevron-like grains (L from few micrometers to 10 μm) and elongated monocrystalline grains (L from few tens of μm up to 100 μm).

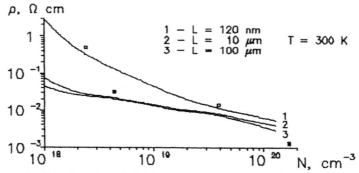

Figure 1. Calculated resistivities of poly–Si films (curves 1–3) as compared with experimental data: □ – before and ▪ – after the laser recrystallization

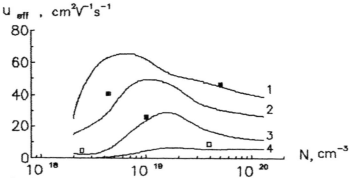

Figure 2. Calculated carrier mobility of poly–Si films (curves 1–4) as compared with experimental data: □ – before and ▪ – after the laser recrystallization. Average grain size: 1 – 100 μm, 2 – 10 μm, 3 – 1 μm; 4 – 20 nm.

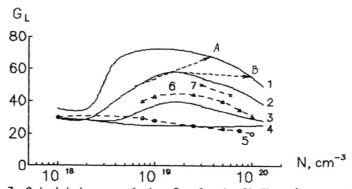

Figure 3. Calculated gauge factor G_L of poly–Si films (curves 1–4) for the grain size: 1 – 100 μm; 2 – 10 μm; 3 – 1 μm; 4 – 20 nm. Experimental data: 5 – fine-grained poly–Si [4]; 6 – after thermal anneal [8]; 7 – after the laser recrystallization [8].

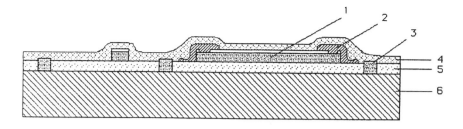

Figure 4. Cross section of a test pattern: 1 - polysilicon piezoresistor; 2 - Al contacts; 3 - seed regions; 4 - protective glass layer; 5 - insulating SiO_2 layer; 6 - monocryslalline Si.

The test patterns, due to their special layout, provide the study of the influence of laser recrystallization on the longitudinal and transversal resistances and gauge factor as related to the scanning direction of the laser beam. Poly-Si layers in the test patterns were boron-doped by the ion implantation method in the wide concentration range.

Some results of the measurements are presented in Figs.5,6 and Table 1.

Table 1. Electrical and piezoresistive characterization of SOI test patterns

p, cm^{-3}	G-factor longitudinal	G-factor transversal	TCR, deg^{-1}	Longitudinal TCGF, deg^{-1}	Transversal TCGF, deg^{-1}	Recrystallization yes/no
2.4×10^{18}	23.5	-5.2	-3.6×10^{-3}	-2.6×10^{-3}	-2.5×10^{-3}	no
4.4×10^{18}	38.2	-11.1	-2.0×10^{-4}	-2.1×10^{-4}	-2.3×10^{-3}	yes
3.9×10^{19}	19.6	-3.4	-2.5×10^{-4}	-1.7×10^{-3}	-4.9×10^{-3}	no
1.7×10^{20}	19.7	-4.8	$+1.2\times10^{-3}$	-1.3×10^{-4}	-1.9×10^{-4}	yes

Experimental measurements of the longitudinal gauge factor prove our theoretical predictions: when the carrier concentration in the initial samples was less than 1×10^{19} cm^{-3}, a significant rise of G_L was achieved due to the laser recrystallization. In the heavy doped samples there was no improvement in G_L or even drop of the gauge factor has been found [2]. For $N<1\times10^{19}$ cm^{-3} after the laser recrystallization the longitudinal gauge factor G_L of poly-Si piezoresistors increases by a factor of 1.5-2.3, whilst G_T increases by a factor of 2.1-2.4. At the same time resistances of poly-Si samples after the recrystallization strongly decrease, sometimes this decrease achieves a decade.

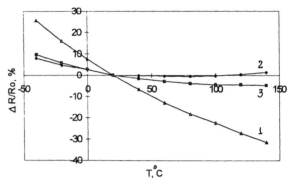

Figure 5. Related change of the resistance of the test patterns: 1 - initial non-recrystallized resistors with carrier concentration 2.4×10^{18} cm^{-3}; 2,3 - after recrystallization (carrier concentration becomes 4.4×10^{18} cm^{-3} (2 - in longitudinal direction; 3 - in transversal direction as related to the longitudinal axis of the resistors).

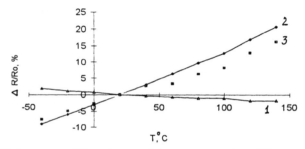

Figure 6. Relative change of the resistance of the test patterns: 1 - initial non-recrystallized resistors with carrier concentration 3.9×10^{19} cm^{-3}; 2,3 - after recrystallization (carrier concentration becomes 1.7×10^{20} cm^{-3}, 2 - in longitudinal direction; 3 - in transversal direction as related to the longitudinal axis of the resistors.

Temperature dependencies of poly-Si resistivity also change after the laser recrystallization; these changes significantly depend on the scanning direction of the laser beam. Figs.5 and 6 show the results of laser recrystallization for the patterns with the initial carrier concentrations equal to 2.4×10^{18} cm^{-3} and 3.9×10^{19} cm^{-3}, correspondingly. The best results as to the temperature coefficient of resistance was obtained for the sample with the initial carrier concentration $N = 2.4 \times 10^{18}$ cm^{-3} where resistivity after the laser recrystallization very slightly depends on the temperature in the range 20-140°C (Fig.5). This proves the possibility to achieve the minimum zero drift with strongly improved strain sensitivity of the mechanical sensors with recrystallized strain gauges for this doping level. At the same time there was no improvement both in the gauge factor and the temperature coefficient of resistance for the carrier concentration in the initial samples with $N=3.9\times10^{19}$ cm^{-3} (Fig.6). The difference in the resistances (and resistivities) in longitudinal and transversal directions as related to the scanning direction confirms the strongly textured microstructure of the laser recrystallized patterns [7]. It was shown that the temperature coefficient of the longitudinal gauge factor (TCGF) reduces after the laser recrystallization.

Non-recrystallized poly-Si resistors with carrier concentration of about 2×10^{18} cm^{-3} due to their relatively high TCR value ($\approx -3.6\times10^{-3}$ deg^{-1}) could be used as temperature sensors in the range -40 - +180 °C. The possibility is considered to create a multifunctional SOI sensor for pressure and temperature measurement. It contains laser-recrystallized piezoresistors on a diaphragm and properly doped non-recrystallized poly-Si thermoresistors out of the diaphragm area. Use of the poly-Si thermoresistors in multifunctional sensors also makes it possible both to correct the temperature error and improve the measurement accuracy in the wide temperature range.

4. Discussion

The possible results of laser recrystallization in respect to the gauge factor are shown in Fig.3 by the dotted lines A and B. The carrier concentrations and the grain size before and after recrystallization were chosen arbitrary to explain clearly the above speculations. The transition A shows the most desirable way of MLR when both enlarging of the grain size and increasing of the carrier concentration result in increasing the value of the longitudinal gauge factor. The transition B corresponds to the case when increasing of the carrier concentration diminishes the results of the grain size enlarging. In case B there is no advantage in the gauge factor achieved.

Other important parameters of SOI piezoresistors are the temperature coefficient of resistance (TCR) and temperature coefficient of the gauge factor (TCGF). TCR of boron-doped poly-Si increases with the carrier concentration in the range 1×10^{18} - 2×10^{20} cm^{-3} changing its sign from negative values to positive ones at approximately 2×10^{19} cm^{-3}. Therefore, one can expect a substantial reduction of TCR after recrystallization by its magnitude for relatively small carrier concentrations. For $N>1\times10^{19}$ cm^{-3} small positive TCR values are expected in the recrystallized SOI layers that is typical for a bulk material.

TCGF for the samples studied has a slight dependence from the carrier concentration and its reduction after recrystallization is expected mainly due to an increase of the carrier concentration. These expectations were confirmed by our measurements (see the next section).

It is necessary to note the dual action of the laser recrystallization on the SOI layers: enlarging of the grain size and increasing of the carrier concentration. The second effect causes an additional increasing of G_L at the low dopant levels and an opposite effect for $N>1\times10^{19}$ cm^{-3}. That is why we have obtained for the high doping levels no improvement of G_L after the recrystallization: increase due to the grain enlarging have been compensated by its reduction because of the concentration rise. At the same time for smaller doping levels ($<1\times10^{19}$ cm^{-3}) increasing of the carrier concentration does not cause the G_L decrease. Therefore, the doping levels 1×10^{18} - 1×10^{19} cm^{-3} are the most recommended for laser recrystallization in order to obtain the maximum gauge factor.

It was shown experimentally that after the laser recrystallization G_L increased 1.5-1.7 times, whilst G_T increased 2.1-2.4 times for boron concentrations 1×10^{18} cm^{-3} - 5×10^{18} cm^{-3}. At the same time resistances of poly-Si samples after MLR strongly decrease, sometimes this decrease achieves a decade. Taking into account the temperature-dependent characteristics of the polysilicon samples (temperature coefficient of resistivity TCR and the emperature coefficient of the gauge factor TCGF) it was shown that the laser recrystallization is the most useful, as concerning parameters of poly-Si piezoresistors, at a dopant level near $(1-5)\times10^{18}$ cm^{-3}. One might consider these results as concentration-dependent restrictions of the laser recrystallization. Similar results were obtained experimentally by other authors [8,9].

As concerning the application of laser recrystallization for piezoresistive sensors fabrication one should note stabilizing of the SOI layer parameters after MLR. This factor combined with the sensitivity increasing and possibility to control temperature-dependent parameters allows to recommend MLR for the sensor technology. Nevertheless, it is worth noting that MLR can induce an additional scattering of SOI resistor's parameters within a SOI wafer (and even a chip) due to the non-uniform action of the laser beam on different parts of it. This limitation becomes the most important in the batch technology of sensors. That is why the question whether apply laser recrystallization or not remains an actual one. Generally, this method could be recommended for the sensor applications when a high sensitivity and stability are needed.

5. Conclusion

Results of the theoretical simulation are in a good agreement with experimental measurements. They allowed to develop technological recommendations, namely, related to the optimum doping levels of the polysilicon layers and were applied to optimize the performances of the piezoresistive sensors on the basis of laser-recrystallized SOI-structures [1].

6. References

1. Druzhinin, A., Lavitska, E., Maryamova, I. and Voronin, V. (1997) Mechanical sensors based on laser-recrystallized SOI structures, *Sensors and Actuators* **61A**, 400-404.
2. Druzhinin, A., Lavitska, E., Maryamova, I. and Kogut I. (1997) Laser recrystallization of polysilicon in sensor technology: possibilities and restrictions, in S.Cristoloveanu

(ed.), *Silicon-on-Insulator Technology and Devices*, Electrochemical Soc. Proc., **97-23**, Pennington, NJ, pp. 92-97.

3. Seto, J. (1975) The electrical properties of polycrystalline silicon films, *J.Appl.Phys.* **46**, 5247-5254.

4. French, p.J. and Evans, A.G. (1989) Piezoresistance in polysilicon and its application to strain gauges, *Solid State Electronics* **32**, 1-10.

5. Lavitskaya, E.N., Maryamova, I.I., Druzhinin, A.O. and Pankov, Yu.M. (1996) Effects of microstructure on the piezoresistance of thin polysilicon layers, *Inorganic Materials* **32**, 1016-1018.

6. Druzhinin, A., Kostur, V., Kogut, I., et.al. (1995) Microzone laser recrystallized polysilicon layers on insulator, in: J.P.Colinge et al. (eds.), *Phys. and Tech. Problems of SOI Struct. and Devices, NATO ASI Series*, Kluver Acad. Pub., pp. 101-105.

7. Voronin, V.A., Maryamova, I.I., Druzhinin, A.A., Lavitska, E.N. and Pankov, Y.M. (1995) SOI pressure sensors based on laser recrystallized polysilicon, in: J.P.Colinge et al. (eds.), *Phys. and Tech. Problems of SOI Struct. and Devices, NATO ASI Series*, Kluver Acad. Pub., pp. 281-286.

8. Suski, J., Mosser, V., Goss, J. and Obermeier, E. (1989) Polysilicon SOI pressure sensor, *Sensors and Actuators* **A17**, 405-414.

OPTICAL SPECTROSCOPY OF SOI MATERIALS

Alejandro Pérez-Rodríguez, Christophe Serre, Joan Ramón Morante
E.M.E., Departament d'Electrònica, Unitat Associada CNM-CSIC.
Universitat de Barcelona, Avda. Diagonal 645-647, 08028 Barcelona,
Spain

Abstract:

The application of Raman scattering and FTIR spectroscopies for the structural characterisation of multilayer semiconductor on insulator structures is presented. These are vibrational techniques well suited for the characterisation of the different layers (Si, SiO_2) in Silicon on Insulator structures. Examples of the microstructural analysis of SOI wafers formed by high dose oxygen implant (SIMOX) are given, and the ability of Raman scattering to characterise the top Si film from as-implanted structures in terms of average stress and correlation length is shown. For high quality SOI, the analysis of thermal effects on the Raman spectra can provide additional information on the presence of defects such as Si interstitials. These data are complemented with the structural characterisation of the SiO_2 buried layer, which is performed by FTIR. Finally, the ability of these techniques for the analysis of other systems such as SiC on Insulator is discussed.

1- Introduction

Optical characterisation techniques have interest for the analysis of multilayer semiconductor on insulator structures, such as Silicon on Insulator (SOI), due to their non destructive character. Among them, Raman and Fourier Transform Infrared (FTIR) spectroscopies are especially suited for the analysis of SOI materials. Raman spectroscopy [1-3] provides information on the structure and crystalline quality of the top semiconductor layer. The Raman spectra are sensitive to features such as strain, damage and the presence of structural defects, crystalline structure, composition of the layer, density of carriers (for high doping levels) and the presence of secondary phases. In addition, a high lateral resolution can be achieved when working in the MicroRaman configuration, which makes the technique well suited to the assessment of the active layer in devices under operation. FTIR [4] complements well this technique for the analysis of the dielectric layer in the SOI structure (usually SiO_2), which has a low Raman efficiency. Moreover, FTIR has been demonstrated as an efficient tool for the characterisation of other semiconductor on insulator systems, such as SiC on Insulator (SiCOI).

This paper reviews the application of Raman scattering and FTIR techniques to the characterisation of SOI structures. As an example, the detailed analysis of structures

P.L.F. Hemment et al. (eds.),
Perspectives, Science and Technologies for Novel Silicon on Insulator Devices, 137–148.
© 2000 *Kluwer Academic Publishers. Printed in the Netherlands.*

obtained by oxygen ion implantation (SIMOX) is given. SIMOX is a commercially mature technology for the fabrication of SOI wafers, combining high dose oxygen implantation (with doses of the order of 10^{18} cm^{-2}) and a very high temperature anneal. This is required to relieve the structure of damage and residual stress in the top Si film, which is specially severe for this technology where a high dose of oxygen ions is used. The non destructive analysis of the implantation induced damage in terms of structural defects and stress is interesting, since both features can be related to the final density of defects in the annealed structures. For these, information can also be provided from the analysis of the dependence of the spectra on the power of the incident laser radiation. Finally, the ability of these techniques for the analysis of other systems such as SiCOI is also discussed.

2- Silicon on Insulator. Raman scattering analysis

Raman spectroscopy is based on the inelastic scattering of photons with elemental excitations of the material [1-3]. For first order processes, the main interactions correspond to the creation (Stokes process) or annihilation (Antistokes process) of a phonon. The comparison between both spectra allows one to determine the effective temperature of the scattering volume. Due to the conservation of energy, the wavenumber of the scattered photon is shifted in relation to that of the incident photon by an amount which corresponds to the energy of the interacting phonon. Conservation of momentum implies that only zone centre phonons can be Raman active .

The first order Raman spectrum of crystalline Si has a single line at about 520 cm^{-1} which corresponds to a triply degenerated phonon. This line has a lorentzian shape, with a Full Width at Half Maximum (FWHM) of about 3 cm^{-1}. The position and shape of this line is sensitive to features such as strain, structural defects and damage in the lattice, temperature, chemical composition and concentration of carriers.

In crystalline Si, Raman spectra are usually measured in a backscattering geometry. In a MicroRaman configuration, excitation of the sample and collection of scattered ions is performed through an optical microscope, which facilitates a high lateral resolution. According to the Rayleigh diffraction criteria, the diameter of the light spot on the sample is given by 1.22 λ/NA, where λ is the wavelength of the incident light and NA the numerical aperture of the microscope objective. Using a X100 objective with NA=0.95, the size spot in the sample can be as low as 0.6 μm (λ= 457.9 nm). Thus, submicron lateral resolution can be achieved.

In this configuration, the penetration depth of the scattered light is determined by optical absorption. For a multilayer structure such as SOI, the interpretation of the spectra requires previous knowledge of the contribution from the different layers in the structure to the spectra. In principle, if the thickness of the top Si layer is much bigger than the absorption length, the depth of the scattering volume can be roughly estimated by 1/2α, where α is the absorption coefficient. However, when the thickness of the layers, the light penetration depth and the excitation wavelength are all of the same order of magnitude this assumption can lead to gross miss-interpretation [5]. Usually, the excitation in the Raman spectrometer is provided by the different visible lines from an Ar^{+} laser (the main ones being at 457.9 nm, 488 nm and 514 nm). This leads to problems in the interpretation of the spectra when measuring SOI structures with layers with typical thicknesses in the range 100-400 nm. In this case, the coherent nature of the excitation light and the multilayer structure of the system determines the appearance of strong interference effects in the distribution of the light intensity in the layers, which depend on their geometrical features (thickness of layers, nature and quality of the interfaces) and optical parameters.

Such effects can be clearly seen in Figure 1, which shows a simulation of the light distribution in a Silicon on

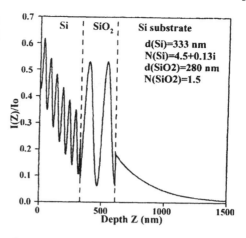

Fig. 1. Light intensity profile calculated in a SOI structure (Si thickness 333 nm, SiO_2 thickness 280 nm) [5].

Insulator structure assuming multiple reflections and neglecting optical absorption in the SiO_2 buried layer [5]. This is especially critical for as-implanted structures, where the high level of damage in the layers can give rise to significant changes in the refraction index and optical absorption of the different regions in the structure. The high oxygen content and oxide inclusions in the top Si layer can lead to an enhanced substrate contribution to the Raman spectra. In this case, the Raman contribution from the Si substrate can be determined by comparing the spectra measured on the SOI structure with those directly measured on the buried oxide, from samples where the top Si layer is removed by a chemical treatment (for example, a KOH etch). Then, the convenient subtraction of this spectrum from the one measured on

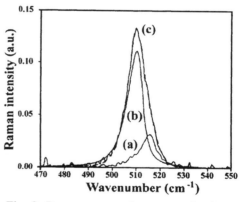

Fig. 2. Raman spectra from an as-implanted SIMOX structure: (a) substrate contribution, (b) top Si contribution, (c) experimental spectrum from the whole structure [6].

the full structure enables the spectral contribution from the top Si layer to be determined. This can be seen in Figure 2, which shows the top Si and substrate contribution in the spectrum from a SIMOX as-implanted structure.

2.1. STRESS AND DISORDER EFFECTS. CORRELATION LENGTH MODEL

The first order Si Raman line measured from as-implanted SIMOX shows an asymmetric broadening and a shift towards lower frequencies [7-10]. This is due to the high level of damage and residual strain in the top Si layer after the high dose implant. The high density of structural defects in the scattering volume determines the appearance of disorder effects in the spectra, which are related to the breaking of the conservation of momentum due to spatial confinement of phonons by the defects. This can be modelled by introducing a correlation length L, which corresponds to the average size of the crystalline domains in which phonons are confined [6,11]. Assuming a constant correlation length L in the scattering volume, the intensity of the first order Si Raman band is given by:

$$I(\omega) = \int\limits_{0}^{2\pi/a_0} \frac{|C(q)|^2 4\pi q^2 dq}{(\omega - \omega(q))^2 + (\Gamma_0/2)^2} \qquad (1)$$

where a_0 is the lattice constant of Si, q is the wave vector , Γ_0 is the intrinsic linewidth of crystalline silicon and $C(q)$ is an appropriate multiplying function that localises the mode inside a finite space. For spherical shape confinement, $C(q)$ is gaussian:

$$|C(q)|^2 \propto \exp\left(-\frac{q^2 L^2}{8}\right) \qquad (2)$$

and the phonon dispersion relationship can be taken as $\omega(q) = \omega_p - 120\,(q/q_0)^2$, where ω_p is the wavenumber of the first order Raman band in the absence of disorder effects and $q_0 = 2\pi/a_0$ [12]. Figure 3 shows the spectra simulated for different values of L according to these expressions: for L < 200 Å, a red shift and asymmetric broadening of the peak occur, which increases as L decreases.

On the other hand, the shape and position of the peak are also sensitive to stress in the scattering volume. For uniform stress, the shape of the peak is not affected, and the only effect of the stress is an additional shift of the mode. The stress induced shift is defined as $\Delta\omega_s = \omega_p - \omega_0$, being ω_0 the wavenumber of the first order Raman line in the absence of both disorder and stress effects. For biaxial stress, the stress is given by $\sigma = 250$ MPa/cm^{-1} $\Delta\omega_s$ [13].

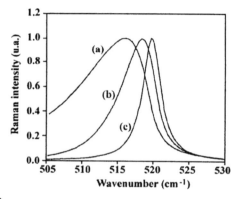

Fig. 3. First order Si Raman spectrum simulated assuming spherical confinement: (a) L =60Å, (b) L=100Å, (c) L=200Å.

The fitting of the spectra with the above expressions enables an estimation to be made of both the correlation length and the stress, assuming that both parameters are uniformly distributed in the scattering volume. However, for an as-implanted SIMOX structure some depth variation of these parameters is to be expected. In this case, the parameters estimated by the model correspond to effective average values in the scattering volume.

This model has been applied for the structural analysis of samples as-implanted, where the implantation process was performed under different surface conditions leading to different concentration of dislocations in the top Si layer after annealing [6]. In spite of the simplicity of the model, a clear correlation between these parameters and the final density of defects (in the corresponding annealed samples) has been observed. This can

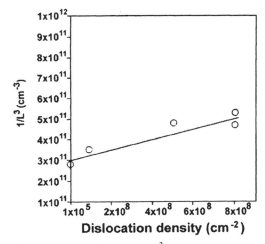

Fig. 4. 1/(correlation length)^3 from as-implanted SIMOX samples versus dislocation density from the corresponding annealed ones [6].

correlation of these features with the dislocation density after annealing clearly demonstrates the value of Raman scattering for the diagnosis of the final density of defects in the top Si film. In this sense, the Raman spectra from the as-implanted structures are sensitive to final dislocation densities higher than 5×10^4 cm^{-2}.

2.2. LASER BEAM INDUCED THERMAL EFFECTS

For high quality state-of-the-art SOI wafers, the first order Si Raman line from the top Si film becomes indistinguishable from virgin single crystal Si. In this case, further information can be obtained by performing the measurements at higher laser power densities [10,15]. Overheating of the samples by the exciting beam produces an anharmonic shift and broadening of the Raman line, which depend on the presence and nature of defects in the scattering volume.

Figure 5 presents the frequency shift of the first order Raman peak from an annealed SIMOX structure in relation to that from bulk silicon and the FWHM versus the laser excitation power. It is evident that the response of the Raman parameters to the excitation power is

be seen in figure 4, where $1/L^3$ (before annealing) versus the final dislocation density (after annealing) is plotted. The increase in the density of defects in the scattering volume -which is dependent upon $1/L^3$- is also accompanied by a significant increase of tensile stress. This suggests that dislocations in the top Si after annealing are determined by both tensile stress and defects induced by the implant process. These are probably related to the ability of the surface of the sample to accommodate the flux of Si interstitials coming from the oxidised buried layer during the implantation process [14]. The

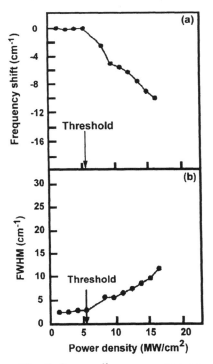

Fig. 5. Raman line parameters vs. excitation power from a SIMOX annealed wafer (dislocation density 4×10^5 cm^{-2})

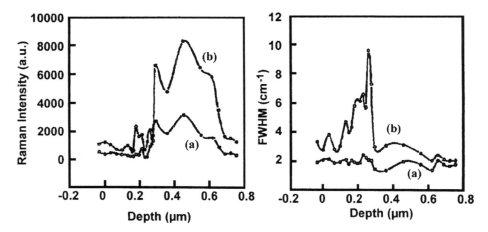

Fig. 6. Intensity and low frequency half width of Raman spectra versus position in bevelled surface, with excitation power below and above threshold, curves (a) and (b), respectively.

characterised by the presence of a threshold: for excitation power density values below this threshold, the Raman band only shows a weak deviation from bulk Si. Above this threshold, a negative frequency shift and broadening of the Raman peak is evident. This is related to the increase of the temperature in the scattering volume during laser heating, which depends on the ability of the sample to dissipate the heat due to the non radiative recombination of excess photogenerated electron-hole pairs. Accordingly, higher frequency shifts and FWHM values are observed for samples with higher density of defects.

On the other hand, the spatial resolution of the Raman microprobe allows depth studies of the SOI structure to be carried out, by rastering the laser microbeam across a low angle bevelled surface. By doing this, the structures can be profiled in terms of the Raman spectral parameters for different laser power densities, revealing the local response to the laser excitation. This also provides information on the structural quality of the Si/SiO$_2$ interface regions.

Figure 6 shows the low frequency half-width and the intensity of the Raman spectra measured at different possitions along the low angle bevelled surface of the same SIMOX structure as refered to in figure 5, at excitation power densities below and above the threshold. These features are characterised by the presence of an interference pattern superimposed upon the Raman spectra parameters. This effect is due to the changes in the reflectivity as the thickness of the top Si layer decreases, which determines an oscillation of the effective excitation power in the scattering volume. As shown in the figure, the changes in the Raman parameters increase when the laser beam approaches the top Si/buried SiO$_2$ interface. In principle, this can be related to the low thermal conductivity of the SiO$_2$ buried layer. However, no clear correlation of these effects has been found with the oxide layer thickness. This indicates that other factors, such as the concentration of defects at the Si/SiO$_2$ interface and its roughness, have also to be considered.

For samples prepared with the surface covered with an oxide capping film during the whole implant process, there is a strong increase in the intensity of the Raman line measured at points close to the surface of the sample. This can be seen in figure 7, which plots this parameter versus depth for a SIMOX structure obtained by implanting the oxygen ions trough a 190 nm thick capping film with standard implant (dose 1.8×10^{18} cm^{-2}, energy 190 keV, temp. 650°C) and annealing (1320°C, 6 hours) conditions. This enhancement of the intensity of the Raman line is not observed if the implantation is fully or partially performed without the capping film. This is likely related to the accumulation of Si interstitials in the top Si layer during the implant [16], which cannot be released in the free surface. The analysis of the Raman spectra measured at low excitation conditions from the corresponding as-implanted samples revealed the presence of a significant compressive stress in the top Si film [6]. After annealing, a significant relaxation of stress is observed, when the Raman line measured

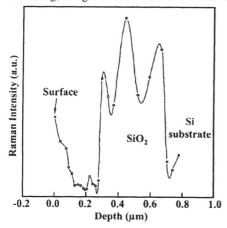

at low excitation power is similar to that from virgin Si. However, the enhancement in the intensity of the Raman line at higher excitation power densities shown in Figure 7 suggests the presence of a significant residual content of interstitial Si in the surface region. This behaviour agrees with that observed in the interface region between the buried oxide layer and the Si substrate, where a significant increase of the intensity, which is higher than that expected from the change of optical reflectivity, is also found. This feature is related to a compressive stress component.

Fig. 7 Intensity of first order Si Raman line at high excitation power versus position in bevelled surface, from a sample implanted with a capping film.

Thus, the analysis of the response of the Raman spectra with increasing excitation power densities allows information on the thermal behaviour of the structure to be determined. This is related to the presence of defects in the layers and their interfaces which are not observed when making the measurements under standard low power conditions. However, and in contrast with the analysis of the spectra obtained at low excitation power densities, quantification of the data is a complex issue, as it requires the detailed modelling and calculation of the temperature profile in the scattering volume and heat dissipation through the buried oxide and Si substrate regions. In any case, the observation of higher thermal effects for samples with higher densities of dislocations (up to 2.5×10^{9} cm^{-2}) is related to their lower effective thermal conductivity, which is an important feature for the performance of the devices to be made on these structures.

3- Silicon on Insulator. FTIR

FTIR is an optical technique well suited for the structural analysis of dielectric films in multilayer structures, as SiO$_2$ in SOI [4]. The FTIR spectrum from SiO$_2$ is charaterised by the presence of 4 transverse optical (TO) and longitudinal optical (LO) vibrational modes of the Si-O-Si unit, which are sensitive to the stoichiometry, stress and structural

144

features in the SiO$_2$ matrix. Typical FTIR transmission spectra from SOI structures are plotted in Figure 8, which includes the TO$_1$, TO$_2$ and TO$_3$ vibrational modes [17]. As is shown in the figure, the higher absorption band corresponds to the asymmetric stretching motion of the Si-O-Si unit (TO$_3$), located around 1070 cm^{-1}.

Stoichiometry of the SiO$_x$ film from SIMOX structures has been studied from the broadening and shift of the TO$_3$ vibrational mode towards lower wavenumbers [7]. Moreover, the position of the mode is also stress dependent [18]. This is due to the dependence of the vibrational frequency in the average Si-O-Si bond angle: compressive stress in the oxide determines a decrease in the average bond angle and, hence, a reduced

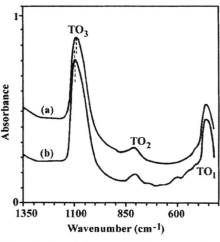

Fig. 8. Transmission FTIR spectra from SIMOX samples (a: sequentially implanted and annealed, (b) standard process) [17].

vibrational frequency. In the same way, tensile stresses lead to an increase in the frequency of the TO$_3$ mode. Distortions in the bond angle distribution, related to stress gradients in the oxide, also give rise to a broadening of the peak. In a more general way, changes in the IR spectrum are related to changes in the dielectric function, which reveal structural differences in the oxide network. In this sense, disorder effects in oxide layers have been related to an increase in the contribution of the TO$_4$-LO$_4$ vibrational modes [19].

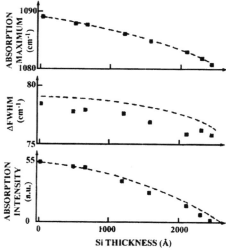

Fig. 9. Theoretical and experimental IR features of TO$_3$ mode versus thickness of the top Si film [20].

However, the analysis of the oxide layers by FTIR has also to take into account the dependence of the spectra on the geometrical arrangement of the structure. This is specially important for multilayer structures such as SOI. Figure 9 shows the features of the TO$_3$ mode from SIMOX wafers as a function of the thickness of the top Si layer (buried oxide thickness 390 nm), together with their theoretical fitting [20]. In this case, the interpretation of the spectra calls for the optical modelling of the structures, which must take into account multiple reflection and transmission events at the different interfaces, as well as the parameters of the dielectric function of the different layers. For the data shown in Figure 9, the spectra have been fitted to a model [21] that incorporates the measured dielectric function of a thermally grown SiO_2 (950°C). In this example, the position of the TO_3 vibrational mode has been shifted by 3 cm^{-1}. The need for this shift is attributed to the high temperature annealing in SIMOX technology (typically 1350°C, several hours), which produces a relaxation of the average Si-O-Si bond angle in the SiO_2 network, together with a decrease in the bond angle distribution. This supposition is supported by the lower values of the experimental FWHM in relation to the theoretical ones.

The dependence of the spectral IR features on the geometrical arrangement can be minimised if measurements are performed in transmission (normal incidence) mode directly on the buried oxide layer. This has been done for the samples from Figure 9, after chemical (KOH) etching of the top Si layer. The spectra obtained showed differences in the TO_4-LO_4 region which correlated with the density of Si inclusions in the buried oxide layer close to the back SiO_2/Si interface. This was interpreted in terms of disorder in the SiO_2 matrix in this region.

Attempts have also been made to develop a depth profiling technique using an FTIR spectrometer coupled with an Infrared microscope [22]. In this case measurements are made along a low angle bevelled surface, when spectra from thinner regions of the oxide layer can be obtained, and the detailed analysis of the spectra provides information on the depth homogeneity of the film. However, this is limited by the relatively poor lateral resolution of the IR microscope (about 20 μm).

4- SiC on Insulator (SiCOI)

The combination of Raman scattering and FTIR spectroscopies has also been applied to the analysis of the semiconductor top layer of silicon carbide on insulator (SiCOI) structures. In this case, the FTIR spectrum is characterised by the absorption band at about 800 cm^{-1} related to the Si-C stretching mode. The shape and position of this band is strongly dependent on the crystalline nature of SiC: for crystalline SiC, the band has a lorentzian shape with a FWHM of about 50 cm^{-1}. For amorphous SiC, the band has a gaussian shape, being centred at lower wavenumbers (in the range 700-800 cm^{-1}) and with higher values of FWHM. This has allowed studies of the crystallisation of thin SiC layers produced by different techniques, such as ion implantation and sputtering deposition, to be carried out [23]. For these structures, Raman scattering is helpful in order to detect the different SiC phases within the layers. Raman scattering is also very sensitive to the presence

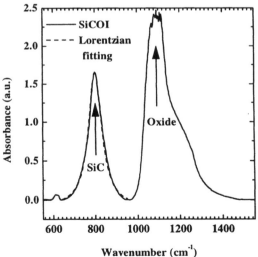

Fig. 10: FTIR spectrum from a SiCOI structure.

of secondary phases, such as graphitic carbon domains in grain boundary regions.

Figure 10 shows the FTIR transmission spectrum measured from a SiCOI wafer fabricated by ion implantation and thermal bonding [24]. This spectrum is characterised by the Si-C absorption band at about 800 cm^{-1}, in addition to the SiO$_2$ vibrational peaks. The SiC film has been synthesised by a multiple C$^+$ implant into Si at 500°C, forming a 300 nm thick β-SiC film buried in the Si wafer. Annealing at 1200°C for 6 hours has produced an improvement of the crystalline quality of the film, which is observed from the decrease in the FWHM of the Si-C stretching mode in the IR spectra. Complementary X-Ray Diffraction measurements also show a significant reduction of the residual strain in the film. Selective oxidation of the top Si film allows the SiO$_2$/SiC/Si structure to be achieved. Then, this structure is thermally bonded to a previously oxidised Si wafer, and the original Si substrate is etched back. Monitoring of the Si-C stretching band during the whole process has confirmed the absence of any structural degradation of the initial SiC film during subsequent processing. The high structural quality of the SiC film is confirmed by the similar shape and position of the Si-C band in relation to that theoretically simulated for a β-SiC monocrystalline film.

5- Conclusions

In conclusion, this work reviews the characterisation of Silicon on Insulator (SOI) and Silicon Carbide on Insulator (SiCOI) structures by Raman scattering and Fourier Transform Infrared (FTIR) spectroscopies. For SOI, the detailed characterisation of samples obtained by high dose oxygen implant (SIMOX) has been performed for both as-implanted and annealed structures. In the first case, modelling of the first order Si Raman band allows to estimate both the correlation length (which is given by the average distance between structural defects breaking translational symmetry of the crystal) and average stress in the scattering volume. Both features are found to correlate with the density of defects in the top Si layer after annealing, for densities higher than 5×10^4 cm^{-2}. For high quality annealed structures, the first order Si Raman band measured under standard low excitation power conditions becomes indistinguishable to that from reference single crystal Si. In this case, the analysis of the response of the Raman spectra with increasing excitation laser power allows information on the thermal behaviour of the structure to be determined, which is related to the presence of defects in the layers and their interfaces. In this sense, an enhancement on the intensity of the Raman line has been observed in the surface region from samples implanted with the surface covered with a capping layer, as well as in the region close to the back Si/SiO$_2$ interface. This has been interpreted as related to the presence of Si interstitials which are not fully removed during the high temperature anneal step.

The analysis of the Si regions in the SOI structure is complemented with the study of the dielectric layer, which can be performed by FTIR. The FTIR spectrum from SiO$_2$ is characterised by the presence of the 4 tranverse optical (TO) and longitudinal optical (LO) vibrational modes of the Si-O-Si unit, which are sensitive to features such as stoichiometry of the SiO$_2$ layer, stress and structural disorder effects in the oxide matrix. However, the FTIR characterisation of the buried layer from SOI wafers calls for the optical modelling of the structures, which must take into account geometrical features such as the thickness of the different layers and the roughness of the interfaces, as well as structural features related to the parameters of the dielectric function of the different regions in the wafer. In principle, this makes more difficult the interpretation of the FTIR spectra. However, the development of suitable fitting programmes allows to deepen in the structural analysis of the wafers, obtaining information directly related to the dielectric function of the oxide layer.

Finally, the structural analysis of SiCOI wafers synthesised by ion implantation, selective oxidation and thermal bonding is presented. In this case, FTIR can be simultaneously used for the characterisation of both the top SiC layer and the buried SiO_2 one. This is due to the strong dependence of the features of the Si-C stretching absorption band on the structure of the SiC matrix. In this case, Raman scattering can provide additional information related to the possible presence of secondary phases such as graphitic carbon in polycrystalline SiC, as well as to identify the SiC polytypes in the layer.

References

[1] Cardona M, and Güntherodt G. (eds) (1975 to 1989), *Light Scattering in Solids, Vols. I to V*, Springer-Verlag, Berlin, Heidelberg.

[2] Prévot B. and Wagner J. (1991), Raman characterization of semiconducting materials and related structures, *Prog. Crystal Growth and Charact.*, **22**, 245-319.

[3] Pollak F.H. and Tsu R. (1984), Raman characterization of semiconductors revisited, in *Spectroscopic Characterization Techniques for Semiconductor Technology*, Proc. SPIE vol. 452, pp. 26-43

[4] Ferraro J.R. and Krishnan K. (eds) (1990), *Practical Fourier Transform Infrared Spectroscopy*, Academic Press, San Diego

[5] Macía J. (1996), *Caracterización estructural de tecnologías SOI-SIMOX mediante espectroscopía Raman*, PhD Thesis, University of Barcelona

[6] Macía J, Martín E., Pérez-Rodríguez A., Jiménez J., Morante J.R., Aspar B. and Margail J. (1997), Raman microstructural analysis of silicon-on-insulator formed by high dose oxygen ion implantation: as implanted structures, *J. Appl. Phys.* **82**, 3730-3735.

[7] Harbeke G., Steigmeier E.F., Hemment P.L.F., Reeson K.J. and Jastrzebski L. (1987), Monitoring of SIMOX layer properties and implantation temperature by optical measurements, *Semicond. Sci. Technol.* **2**, 687-690.

[8] Takahashi J. and Makino T. (1988), Raman scattering measurement of silicon-on-insulator substrates formed by high-dose oxygen-ion implantation, *J. Appl. Phys.* **63**, 87-91.

[9] Olego D.J., Baumgart H. and Celler G.K. (1988), Strains in Si-on-SiO_2 structures ormed by oxygen implantation: Raman scattering analysis, *Appl. Phys. Lett.* **52**, 483-485.

[10] Pérez-Rodríguez A., Cornet A., Morante J.R., Jiménez J., Hemment P.L.F. and Homewood K.P. (1991), Raman scattering and photoluminescence analysis of silicon on insulator structures obtained by single and multiple oxygen implants, *J. Appl. Phys.* **70**, 1678-1683.

[11] Fauchet P.M. and Campbell I.H. (1988), Raman spectroscopy of low-dimensional semiconductors, *CRC Crit. Rev. Solid State Mater. Sci.* **14**, S79-S101.

148

[12] Sui Z., Leong P.P., Herman I.P., Higashi G.S. and Temkin H. (1992), Raman analysis of light-emitting porous silicon, *Appl. Phys. Lett.* **60**, 2086-2088.

[13] Anastassakis E. (1985), Stress measurements using Raman scattering, in Kassabov J. (ed), *Physical Problems in Microelectronics*, Proc. 4th Int. School ISPPME, Varna, Bulgaria, pp. 128-153.

[14] Stoemenos J., Aspar B. and Margail J. (1994), Mechanisms of SIMOX synthesis and related microstructural properties, in Cristoloveanu S. (ed), *Silicon on Insulator Technology and Devices*, The Electrochemical Society Proc. Vol. 94-11, The Electrochem. Soc. Inc., Pennington, pp. 16-27.

[15] Pérez-Rodríguez A., Morante J.R., Martín E., Jiménez J., Margail J. and Papon A.M. (1992), Screen oxide effects on the SIMOX material quality observed by Raman microprobe measurements, in Bailey W.E. (ed), *Silicon on Insulator Technology and Devices*, The Electrochemical Society Proc. Vol. 92-13, The Electrochem. Soc. Inc., Pennington, pp. 228-236.

[16] Van Ommen, A.H., Kook, B.H and Viegers, M.P.A. (1986), Amorphous and crystalline oxide precipitates in oxygen implanted silicon, *Appl. Phys. Lett.* **49**, 628-630.

[17] Pérez, A., Samitier, J., Cornet, A., Morante, J.R., Hemment, P.L.F. and Homewood, K.P. (1990), Optical characterization of silicon-on-insulator material obtained by sequential implantation and annealing, *Appl. Phys. Lett.* **57**, 2443-2445.

[18] Lucovski G., Manitini M.J., Srivastava J.K. and Irene E.A. (1987), Low-temperature growth of silicon dioxide films: a study of chemical bonding by ellipsometry and infrared spectroscopy, *J. Vac. Sci. Technol.* B **5**, 530-537.

[19] Lange P. (1989), Evidence for disorder-induced vibrational mode coupling in thin amorphous SiO_2 films, *J. Appl. Phys.* **66**, 201-204.

[20] Samitier J., Martinez S., Pérez-Rodríguez A., Garrido B., Morante J.R., Papon A.M. and Margail J. (1993), Buried oxide layers formed by oxygen implantation on screened oxide silicon wafers: structural analysis, *Nucl. Intrum. and Meth.* B **80/81**, 838-841.

[21] Naiman M.L., Kirk C.T., Aucoin R.J., Terry F.L., Wyatt P.W. and Senturia S.D. (1984), Effect of nitridation of silicon dioxide on its infrared spectrum, *J. Electrochem. Soc.* **131**, 637-640.

[22] Pérez-Rodríguez A., Martín E., Samitier J., Jiménez J., Morante J.R., Hemment P.L.F. and Homewood K.P. (1991), In depth resolved analysis of SIMOX materials by optical characterization techniques, in *1991 IEEE Inter. SOI Conf. Proc.*, The Institute of Electrical and Electronics Engineers Inc., Piscataway, pp. 110-111.

[23] Serre, C., Calvo-Barrio, L., Pérez-Rodríguez A., Romano-Rodríguez A., Morante J.R., Pacaud, Y., Kögler, R., Heera, V. and Skorupa, W. (1996) Ion-beam synthesis of amorphous SiC films: structural analysis and recrystallization, *J. Appl. Phys.* **79** (9), 6907-6013.

[24] Serre, C., Pérez-Rodríguez A., Romano-Rodríguez A., Morante J.R., Fonseca, L., Acero, M.C., Kögler, R. and Skorupa, W. (1998) β-SiC on SiO2 formed by ion implantation and bonding for micromechanics applications, *Sensors & Actuators A*, to be published.

COMPUTER SIMULATION OF OXYGEN REDISTRIBUTION IN SOI STRUCTURES

V.G. LITOVCHENKO, A.A. EFREMOV,
Institute of Semiconductor Physics, NASU, Kiev, Ukraine
45, Prospect Nauki, 252028 Kiev-28, Ukraine
Phone: (380-44) 265 62 90; Fax: (380-44) 265 83 42;
E-mail: Lvg@div9.semicond.kiev.ua

Abstract Physical mechanisms of oxygen transport and precipitation in silicon during the synthesis of buried oxide are reviewed. Important role of influence of subcritical nuclei of SiO_2 phase on oxygen diffusion is emphasized. Different effects caused by interaction of weakly bonded oxygen with mobile point defects and static defect complexes are analysed. Different methods of computer simulation of oxygen redistribution during SOI synthesis are discussed. The computer modeling based on kinetic quasichemical description of evolution of the system is considered in detail. The results of calculations are compared with the experimental data obtainded in the framework of low-dose approach combined with defect engineering. Some new effects in kinetics of oxygen redistribution are revealed and discussed.

1. Introduction

The synthesis of a buried oxide (BOX) layer in Si by high dose O^+ ion implantation for Silicon-On-Insulator (SOI) applications is controlled in large degree by complex point defects dynamics. Oxygen ions are generally implanted at elevated temperatures ($\geq 500°C$) to promote dynamic annealing of defects, that are generated during implantation [1]. Very high doses ($2 \cdot 10^{17}$ to $1.8 \cdot 10^{18}$ $O^+.cm^{-2}$) are used for stoichiometric SiO_2 buried oxide formation. For the highest doses, however, one can form a thick (~400 nm) BOX but the crystalline quality of the Si overlayer is not enough for commersial device fabrication [2]. Post implantation annealing at temperatures above 1300°C is still needed, to produce device quality material [1,2].

The high doses used in the standard process is the main reason for the high cost of SIMOX [3]. Synthesis of a relatively thin (~80 nm) and continuous buried SiO_2 layer can be achieved by low dose oxygen implantation (and the same high temperature post implantation annealing) only within the optimal dose range of about $(3.5-4.0) \cdot 10^{17}$ cm^{-2} [4,5]. Beyond this "dose window" the continuity of the buried SiO_2 layer becomes dramatically bad; numerous Si pipes or Si islands are left in the BOX synthesized using lower or high doses, respectively [5]. On the other hand, crystalline quality of the top silicon is higher for lower doses [3]. There are several other low dose approaches, namely: low energy SIMOX formation, two (or multi)-energy implantation, consecutive implantation of oxygen and nitrogen by varying the temperature ramp rates of the anneals and so on [3,6]. In most of these cases high temperature post implantation annealing, mentioned above, should also be used. As a result complex defects

P.L.F. Hemment et al. (eds.),
Perspectives, Science and Technologies for Novel Silicon on Insulator Devices, 149–161.
© 2000 *Kluwer Academic Publishers. Printed in the Netherlands.*

(dislocation loops, etc.) appear in the silicon overlayer. The need for very high temperatures, is a further reason for the high cost of SIMOX technology.

In the standard SIMOX technology the knowledge about details of oxygen redistribution is not quite so important: a more or less perfect buried oxide with sharp interfaces may be formed «without any tricks» just due to the high temperature anneal. Estimations show, however, that if we use the formula for oxygen diffusitivity commonly accepted for crystalline silicon [7]:

$$D(T) = D_o \exp(-2.5 \text{ eV}/kT); \qquad (1)$$

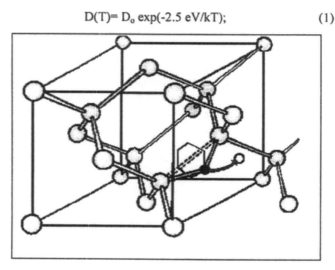

Figure 1. The mechanism of oxygen diffusion via weakly bonded interstitials. The arrow shows a jump from one position to another. Hexagon corresponds to six equivalent positions of oxygen atom around Si-Si bond.

where $D_o = 0.23$ cm^2/s, we shall obtain (at T=1200° C) the value of $D(T) = 6.42 \cdot 10^{-10}$ cm^2/s, which corresponds to diffusion length L_D=10µm, for t=13 min. Therefore, it seems that only several tens of minutes should be enough to gather all oxygen in a silicon wafer into the buried layer. Experiments show this not to be the case and one must conclude that in reality the redistribution of oxygen during SOI formation is a very complicate process even for standard technology. In the low dose approaches (both outlined above and considered below) control of the accumulation of oxygen in a very thin oxide layer buried in the silicon becomes an important issue. So, special studies and a deep understanding of this key point are quite necessary for the development of technological procedures. On the other hand, the kinetics and dynamics of oxygen redistribution can be investigated by *in situ* experimental techniques (for example, IR spectroscopy). They have, however, their own restrictions, first of all, inadequite depth resolution. A computer simulation, which gives us information about the *qualitative behaviour* of the O/Si system is an alternative approach that can be used as a variant of the *in situ* technique for the study of the physical mechanisms and evolution of the SOI structure. In this case the choise of an appropriate theoretical background is very important. The present paper is devoted to a consideration of some aspects of this problem. After theoretical analysis of the main mechanisms of oxygen transport and precipitation in silicon, we shall discuss different methods of computer simulation of these processes. We shall show that computer simulation is a powerful tool to solve this circle of problems. In conclusion comparison of simulations with experimental data will be presented [6,8].

2. Physical Mechanisms of Oxygen Redistribution

2.1. OXYGEN TRANSPORT: DIFFUSION AND QUASICHEMICAL REACTIONS

The mechanisms of oxygen transport in silicon are well studied only for perfect crystalline matrices at low concentrations of native defects and impurities, such as oxygen. Under these «idealized» conditions the diffusion of oxygen is realized via weakly bonded interstitials as shown in figure 1 [9]. Jumping from one position to another an oxygen atom forms temporary Si-O-Si chains with an Si-O-Si angle of about 109° (in comparison with 120-144° for Si-O-Si bridge in SiO_2 [10]). Such an unusual value for the bond angle results in weak bonding of oxygen atoms. This mechanism is responsible for high values of oxygen diffusivity in silicon and the sensitivity of its diffusion to external stress. Under these conditions oxygen may form also a large variety of complexes with point defects (yet without SiO_2 phase formation), in particular, with vacancies. The simplest is the well-known A-centre, which is unstable and may be easily destroyed at elevated temperatures [11]. There are several models which take into account, also the interaction between oxygen and self-interstitials, in particular, the Si-O quasimolecule is assumed to have a very high diffusivity along open channels in crystalline structure [12].

The conditions during oxygen ion implantation and subsequent annealing are quite different from the picture above, because of the high concentrations of oxygen, vacancies and polyvacancies. All these components have nonuniform and different spatial concentration distributions, changing with time both due to diffusion, and multiple quasichemical reactions. High level of crystalline lattice damage inside an implanted layer, allows us to exclude all mechanisms, connected with ultra fast diffusion of Si-O quasimolecules. Hence, the main processes, determining oxygen transport are (i) precipitation due to strong oversaturation and (ii) interaction of weakly bonded oxygen interstitials with mobile point defects (vacancies), static vacancy clusters and with nuclei for the SiO_2 phase. Therefore, we can conclude that with time the *weakly bonded oxygen* concentration depends on (i) the quasi chemical reactions between implanted species and defects, (ii) their diffusion and pairing and (iii) SiO_2 precipitates growth/decomposition. In turn, the final spatial distribution of *bounded oxygen* is determined by the evolution of stable sinks and hence, prove to be sensitive to the initial stage of SiO_2 nucleation. It should be emphasized that the conditions for SiO_2 phase growth/decomposition depend themselves strongly on concentration both of oxygen and point defects in the local vicinity of inclusions. Therefore, the transport and bonding of oxygen influence each other in a rather complicate manner, which is considered in section 2.3.

2.2. INTERACTION WITH DEFECTS: TRAPPING/DETRAPPING

High concentrations both of vacancies and oxygen during BOX formation follow the law of mass action and shift the equilibrium between weakly bonded oxygen and oxygen trapped by vacancies to a higher formation rate of A-centers and other similar complexes. Under these conditions all V_nO_m clusters (otherwise being unstable) are dinamically stabilized by the favourable environment. Their multiple decay and recombination leads to intensive trapping/detrapping of weakly bonded oxygen. As a result for oxygen transport we have a diffusion of atoms in the presence of shallow traps. It is well known that shallow traps (and only shallow traps) result in a dramatical suppression of the effective diffusion coefficient [13]:

$$D_{eff} = \frac{D}{1+\dfrac{C_t}{(1-C_t)P}} \qquad (2)$$

where $P = \exp(-\Delta U/kT)$, ΔU is an energetic depth of a shallow trap, and C_t is a normalized concentration of the traps. At

$$C_a/(1-C_t) << P << C_t/(1-C_t),$$

where C_a is the concentration of mobile atoms, C_t is the concentration of shallow traps, we have a strong inequality, namely $D_{eff} << D$. This occures at high values of C_t, *relatively* low concentration of moving atoms and low energetic depth of a trap. Points defects and vacancy complexes, however, are not the only kind of shallow traps in the diffusion zone during buried layer growth. Another important factor is the presence of small (critical and subcritical) SiO_2 nuclei, discussed in the next section.

2.3 OXYGEN PRECIPITATION: DYNAMICAL RETARDATION OF DIFFUSION

In crystalline silicon even at very high temperatures oxygen has a low solubility limit [7] ($<10^{18}$ cm^{-3}). So, it is expected to be easy to achieve oversaturation during oxygen implantation. However, in non-uniformly damaged silicon the process of SiO_2 inclusion formation depends also on oversaturation of defects. In order to accommodate the inclusion of the SiO_2 phase in the silicon matrix an excess free volume must be available. It may be created by either emission of silicon interstitials into the matrix or by accumulation of vacancies in the vicinity of the growing inclusion. Indeed, according to [14] the critical radius (r_c) of the SiO_2 nucleus in defective silicon may be described as follows:

$$r_c = 2\sigma/(\Delta G_{ch} - \Delta G_{el}) \qquad (3)$$

where σ is the interface energy per unit area, $\Delta G_{ch} = [(1 - \varepsilon)^{-3}xk_BT/\Omega_p] \ln(C_{ox}/C^*_{ox})$ $(C_V/C^*_V)^\beta (C^*_I/C_I)^\gamma$ are the general relations for the chemical energy of supersaturation per unit volume, C_{ox}, C_V, and C_I are concentrations of oxygen, vacancies and self-interstitials in silicon matrix, C^*_{ox}, C^*_V, and C^*_I are the solubility limits of oxygen and defects, respectively. The second term in the denominator, $\Delta G_{el} = 6\mu\delta\varepsilon$, is the change in the total elastic energy of a spherical nucleus related to the change of the unit volume.

Here μ is shear modulus of silicon, $\delta = [(\Omega_P/\Omega_M) - x(\beta+\gamma)]^{1/3} - 1$ is the linear misfit, ε is the constrained strain. Ω_P and Ω_M are the volumes per silicon atom in the precipitate and in the matrix respectively, x is the oxygen/silicon ratio in the SiO_x precipitate, γ and β are the numbers of self-interstitials and vacancies emitted and absorbed (per oxygen atom) by a precipitate for partial strain relief, respectively. After creation of a supercritial nucleus it begins to grow according to the law of diffusion [15]. A radius, ρ, of isolated spherical inclusion as a function of time (in the steady state approximation) obeys the simple equation [9,15]:

$$d\rho^2/dt = 2D\,(C-C_e(\rho))/\,(C_{np}-C_e(\rho)) \qquad (4)$$

where C_e is an equilibrium concentration of oxygen at the inclusion - matrix interface, C_{np} is the concentration of oxygen in the SiO_x phase. We can see that growth or decomposition of the new phase depends on the relation between C and $C_e(\rho)$. It is usually assumed that C_e decreases with an increase of ρ [15], so small inclusions should decay even at a high concentration of mobile atoms. In the case of an ensemble of interacting inclusions of the new phase their growth proceeds slower than predicted by equation (4) due to their mutual competition for weakly bonded oxygen [15]. The two formulas (3) and (4) are valid strictly only under quasi steady state-conditions. In the case of BOX formation we have observed from computer simulations, the effect of dynamically hampering the growth and decomposition due to the existence of negative feedbacks in the system involving the chain of events «subcritical nucleus $\leftrightarrow$ mobile atoms $\leftrightarrow$ defects». Creation of an inclusion due to oversaturation of oxygen and vacancies in some regions of the system leads to an emission of interstitials and rapid consumption of rather large amount of weakly bonded oxygen. As a result the critical radius, r_c, according to the equation (3), increases. At the same time the condition $C<C_e(\rho)$ becomes valid and in conformity with equation (4) the inclusion begins to dissolve. Because of this mechanism, small SiO_2 inclusions become very effective shallow traps for mobile oxygen in the diffusion zone.

3. Computer Simulation of the Processes

There are three possible ways of describing the physical processes outlined above: (a) Monte-Carlo, (b) Molecular Dynamics and (c) Finite-difference (grid) methods, using the kinetics approach [16]. The last method in turn, is subdivided into explicit and implicit techniques. Each approach has its own advantages and disadvantages, which, however, are not absolute [16,17]. Published results, as well as our own experience, allow us to conclude that the problems, outlined above, may best be solved successfully in the framework of explicit finite-difference methods. The last are the most simple, economic and flexible [17] and may be easy implemented on a modern PC.

3.1.   MODELS AND EQUATIONS

In order to handle the processes outlined above we have written several versions of code "SYNTHESYS", which considers the problem of ion beam induced new phase formation in the framework of the phenomenological quasichemical description. The code, based on the mathematical technique of finite differences, calculates the solution of a set of differential equations describing the diffusion and reactions of all those components of the system under investigation, which are important for the process studied.

$$\partial C_k/\partial t = D_k \partial^2 C_k/\partial z^2 + F_k(C_1, C_2, ...,C_n) + V_s \partial C_k/\partial z + g_k(z) \quad (5)$$

Here C_k is the volume concentration of the k-th component of a system (foreign or matrix atoms or defects) which can diffuse with diffusitivity D_k and interact with other components according to the non-linear function $F_k(C_1, C_2, ...,C_n)$. Here V_s is a sputtering rate (or drift velocity). The origin of the k-th component, due to either implantation or other reasons (e.g. Frenkel pair generation), is described by a distributed power function $g_k(z)$. The respective functions are available, for example, from Burenkov's et al Tables [18]. In the general case, F_k is an arbitrary function of the following type:

$$F_k(C_1,... C_i,... C_l,.... C_n) = \Sigma_{il} \Sigma_{\mu v} K_{il\mu v} C_i{}^{\mu} C_l{}^{v} \quad (6)$$

Here the subscripts i, l and k identify the components of the system; $K_{il\mu v}$ is the rate constant for the reaction:

$$\mu C_i + v C_l \rightarrow C_k \quad (7)$$

3.2 ALGORITHMS

Commonly used explicit algorithm may be presented as [17]:

$$C_j{}^{n+1} = (1 + \mathbf{D}^{\wedge} + \mathbf{R}^{\wedge} + \mathbf{V}^{\wedge}) C_j{}^{n} \quad (8)$$

where $C_j{}^{n}$ is the concentration of any component of the physical system (subscript k is omitted) in a {j,n} node of the grid and $C_j{}^{n+1}$ is a corresponding concentration at $t=t+\tau$. Operators $\mathbf{D}^{\wedge}$, $\mathbf{R}^{\wedge}$, and $\mathbf{V}^{\wedge}$, which act on an initial field of concentrations, generate the values of the concentration change, caused by diffusion, quasi chemical reactions and drift (or/and sputtering), respectively during the time τ. For example:

$$\mathbf{D}^{\wedge} C_j{}^{n} = (D\tau/h^2) (C_{j-1}{}^{n} - 2C_j{}^{n} + C_{j+1}{}^{n}) \quad (9)$$

$$\mathbf{R}^{\wedge} C_j{}^{n} = F_i{}^{n}\tau + G_i{}^{n}\tau \quad (10)$$

$$\mathbf{V}^{\wedge} C_j{}^{n} = (V\tau/h) (C_{j+1}{}^{n} - C_j{}^{n}) \quad (11)$$

In most cases of practical importance (e.g., under a strong non linearity of function F), the computational scheme described by equations (9)-(11) proved to be inadequate. This is because in this scheme the influences of various processes on the concentrations (at low τ) are assumed to be mutually independent and additive (see equation (8)). Thus we use another modification of the explicit scheme, which (ignoring drift) may be presented as,

$$C_j^* = (1 + \mathbf{R}^\wedge) C_j^n \tag{12}$$

$$C_j^{n+1} = (1 + \mathbf{D}^\wedge)C_j^* = (1 + \mathbf{R}^\wedge + \mathbf{D}^\wedge + \mathbf{D}^\wedge\mathbf{R}^\wedge)C_j^n) \tag{13}$$

Here C_j^* is an auxiliary intermediate concentration, which is used to calculate the changes induced by diffusion. Instead of the Euler-like addend (10), we use for the calculation of $\mathbf{R}^\wedge C_j^{\ n}$ the much more precise 4-th order Runge-Kutt method. The equation (13) differs from (8) by an additional cross-member $\mathbf{D}^\wedge\mathbf{R}^\wedge$. The last allows us to take into account the mutual influence of diffusion and reactions during time τ. The stability and the conservativity of the scheme (12), (13) are determined by the respective characteristics of each stage (12) and separately (13). During diffusion in a closed system the total number of each species of atom is conserved with an accuracy about 0.1%. Moreover, in order to increase the stability and accuracy of the calculations we use in turn two different space grids: "standard" and "fine", the latter with a half-step $h/2$. Switching between these grids allows us to remove short period distortions of the spatial distributions (which are typical for explicit schemes [17]), as well as avoiding the difficulties which arise when diffusion-induced evolution of the concentration profile with a sharp front is included.

3.3. CODE

The initial programme code was written in C programming language. We have worked out also a version of the programme for Windows-95 (in Visual C/C++®). The programme has been tested by simulating the well-studied analytically and numerically non-linear models for self-organization, which have been developed in biophysics («Brusselator», etc.) [19].

4. Comparison with the Experimental Results

4.1. LOW DOSE APPROACH

Of particular interest is the low-dose SOI fabrication technology known as Low-Dose Approach Combined with Optimized Defect Engineering - LDACODE [6,8]. This approach is based on the use of multiple low dose oxygen ion implantation and intermediate annealing cycles. After the first O^+ implantation a second H^+ implantation

156

(or another impurity: isovalent or inert gas species) is used to create at a specific depth in the silicon centres for oxygen precipitation with subcequent enhanced growth of the SiO_2 phase. In this manner we form a «seed» layer. Quite moderate temperatures for both the intermediate and final anneals are used in LDACODE. This method of SOI preparation allows us to obtain a BOX with thickness about 50 nm using a total dose of oxygen of about $2 \cdot 10^{17}$ cm^{-2}.

In figures 2, 3 and 4 we show the results of computer simulations of a low dose and moderate anneal temperature to stage process involving C^+ or H_2^+ implantations to form an ultra-thing buried oxide layer.

The computer simulations are compared with experimentalresults in Sections 4.2, 4.3.

4.2 CARBON-INDUCED SOI SYNTHESIS

The efficiency of oxygen accumulation during the first stage of the process nonmonotonously depends on the C^+ ion energy (Figure 2). The most complete accumulation is observed in the optimum case. As shown in a previous publication [6] this effect depends also on the dose of C^+ and on the anneal temperature. The process of

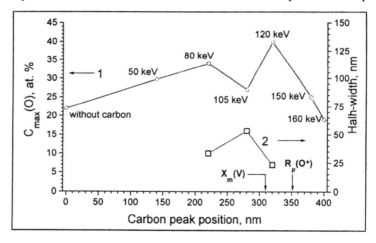

Figure 2. Dependence of (1) the maximum of the oxygen concentration and (2) the half-width of the concentration depth profile on the depth of the implanted carbon after the first implantation and anneal.

seed layer formation is found to be very sensitive both to the dose D_o and the temperature T_{imp} of oxygen implantation. If D_o and T_{imp} are higher than some critical values, the enhanced growth of isolated precipitates occurs over a broad depth range. The second-stage oxygen implantation and anneal at a temperature of 1250°C results in the formation of an ultra-thin stoichiometric (~67% of oxygen) SiO_2 buried layer. The

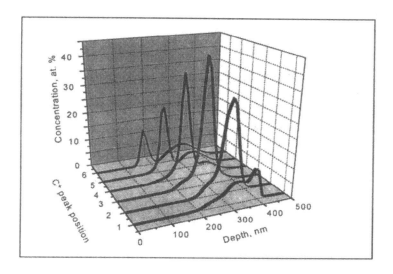

Figure. 3. Spatial distribution of bound oxygen ($R_p(O^+) = 350$ nm) vs. depth of implanted carbon (1) $R_p(C^+) = 400$ nm; (2) $R_p(C^+)=$ 350 nm, (3) $R_p(C^+)=300$ nm, (4) $R_p(C^+) =250$ nm, (5) $R_p(C^+)=200$ nm, (6) $R_p (C^+) = 150$ nm.

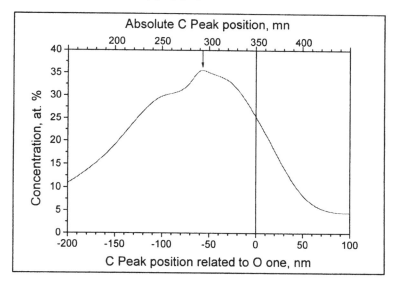

Figure. 4. Calculated maximum oxygen concentration vs. the depth of the implanted C^+

layer thickness, d, is about 15 nm. Both above and below this layer, in the silicon overlayer and substrate, respectively small SiO_2 precipitates have been observed in band of thickness about 50 nm. In the case of O^+ implantation only (no carbon implantation), the silicon overlayer contains high tensile stress ($\sigma \sim 2 \cdot 10^8$ Pa). In contrast, samples implanted with both O^+ and C^+ exhibit a practically full stress compensation ($\sigma < 3 \cdot 10^7$

Pa). This important observation supports the main concept of carbon serving to enhance SiO$_2$ growth. We propose that carbon atoms create additional free volume which accomodates the SiO$_2$ nuclei and precipitates in the silicon matrix. Indeed, Si-C particles in silicon produce a volume contraction of about 50%. [12]. Moreover, this free volume gives rise to a decrease of the critical radius, r$_c$, of SiO$_2$ nuclei [14]. On the other hand C-O complexes as the most probable nuclei [12] are also assumed to be involved in precipitation. Another stimulating factor which acts in our samples are vacancy clusters formed during the evolution of the primary ion beam induced vacancies. [12]. These clusters, as well as Si-C, play an important role in the accomodation of the volume misfit associated with precipitate growth. The third important agent involved in the processes are silicon interstitials. They are generated both during ion stopping and during growth of SiO$_2$ inclusions. We speculate that a further source and sink of self-interstitials are the silicon wafer surfaces [20]. Self-interstitials result in dissolution of vacancy complexes and dissolution of SiO$_2$ precipitates [14,20]. This complicated interaction between different factors acting in opposite directions creates the necessary conditions for spatially non-uniform SiO$_2$ growth, which is promoted in some local regions and suppressed in other regions.

We have included in the model the reactions described above and have obtained the dependences of bound oxygen distributions upon the seed layer and the position of the peak of the carbon implantation R$_p$(C$^+$), shown in Figure 3. One can see that if the carbon implantation is situated far in front of oxygen distribution R$_p$(O$^+$), or far behind it, the oxygen distribution splits into two broad peaks. If carbon is implanted into the region near the maximum of the O distribution these two peaks coincide and narrowing of the oxygen profile is observed. Some ill-defined maximum attributed to pure "vacancy induced" oxidation is observed also (curves 1,2,3) in front of the main peak. The most substantial narrowing and oxygen accumulation is observed at R$_p$(C$^+$) = (0.8 - 0.9) R$_p$(O$^+$) in good agreement with experiment [8]. The maximum value of the oxygen concentration as a function of the relative position of R$_p$(C$^+$) and R$_p$(O$^+$) is shown in Figure 4. The optimum peak position for the carbon, inferred from the computer simulations is practically the same as experimentally observed. The model also predicts an approximate value for the reduction of the maximum value of the concentration at R$_p$(C$^+$) ≈270 nm (Figures 2 and 4).

4.3 HYDROGEN -INDUCED SOI SYNTHESIS

Hydrogen ion implantation is often incorporated into silicon processing, for example to create nano or microvoids in silicon for impurity gettering [21] and as a fundamental step in Smart Cut® SOI technology [22]. In our experiments H$_2^+$ implantation into O$^+$ implanted structures (at the depth of about 320 nm) leads to the formation of a very narrow seed layer(d = 15-25 nm) upon which the buried oxide grows (Figure. 5). In order to simulate this process we have taken into account the formation of nanocavities and their possible influence upon oxygen precipitation. We have included the nucleation of nanoblisters [23] in the model, but simulate this process in a somewhat simplifying form. Only formation of elementary nanobubble (ENB) with simultaneous self interstitial injection into the silicon matrix is considered. The introduction of ENB allows us to apply the quasichemical approach to the problem of bubble formation under oxidation without a detail description of nanobubble growth, which is out of the scope of the present paper. The interaction of oxygen with silicon proceeds in this model by two processes: "ordinary" oxidation, accompanied by stressed SiO_2 precipitation growth and Si interstitial injection with an enhanced oxidation rate in the local vicinity of the nanobubbles. In this case unstressed growth of precipitates takes place without (stress or interstitial induced) impedance. We have considered also the generation and redistribution of Si interstitials during annealing (including temperature rump-up [3,20]).

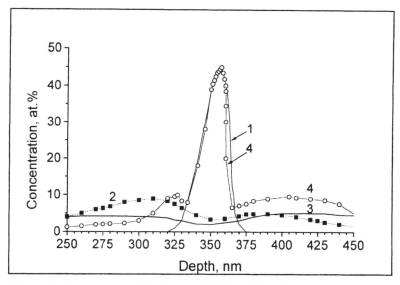

Figure 5. Comparison of the theory (1,2,3) and the AES depth profile (4) for O and H implantation and annealing 1- ultra thin unstressed oxide SiO_2, (2) SiO_2 in ordinary (stressed) precipitates, (3) bonded O interstitials.

In our case at least two processes - SiO_2 precipitation and ENB formation are accompanied by the injection of additional interstitials. We include interstitial emission from the surface during temperature rump up [20]. The results of computer simulations of this annealing of H_2^+ and O^+ implanted samples, taking into account diffusion of Si interstitials, vacancies, O, H and H_2 are shown in Figure 5. One can see rather good agreement with the experimental oxygen profiles (curve 4) both in width and shape. Simulations show, however, that the final result of synthesis is very sensitive to the intensities of different reactions treated in the model. So, a very narrow buried seed layer may be formed only under optimum conditions of both implantation and annealing. The most important feature of the initial stage of hydrogen enhanced synthesis is a rapid growth of ordinary (stressed) SiO_2 inclusions accompanied by self-interstitial generation with a subsequent appearance of relaxed SiO_2 phase in a narrow layer near the peak of the hydrogen regime profile. The optimum condition for this case is superposition of the oxygen and hydrogen concentration depth profiles. Indeed, according to equation (2) for strain-free SiO_2 nucleation (which takes place in the region of silicon containing nanocavities) r_c has its minimum value at the depth where C_{ox} has its maximum i.e. at $R_p(O^+)$. The oxygen content within this narrow layer continues to increase without increase of its thickness if the rate of ENB formation is low enough in comparison with the rate of enhanced oxidation in the vicinity of nanobubbles.

5. Conclusion

For the further development of a relatively cheap Low Dose Approach to SIMOX technology we have modelled the controlling processes in detail and determined the resulting oxygen redistributions. Theoretical analysis of the main mechanisms of oxygen transport and precipitation in silicon show the importance of a thorough consideration of all of the reactions between weakly bonded oxygen, point defects and SiO_2 nuclei. For this reason computer simulation, using a quasichemical kinetic approach, based upon a sound theoretical background has proved to be a powerful and adequate tool for the solution of this circle of problems. The codes developed for the computer modeling of ion -beam synthesis of buried layers have shown their reliability both in the investigation of the physical processes and in the search for the optimum processing conditions. The results of calculations have been compared with experimental data from samples implanted with a low O^+ ion dose and incorporating defect engineering. Not only good agreement with experiment but also a sound understanding of fine details of the physical mechanisms have been achieved.

6. References

1. Pantelides S. T. and Ramamurthy M. (1997) Theory of the nucleation and growth of SiO_2 in Si: Application to SOI, in S. Cristaloveanu et al (eds.), Proc.of VIII Int. Symp. *Silicon on Insulator Technology and Devices* (Pennington, NJ: The Electrochemical Society) p. 39-42.
2. Auberton -Herve J.A., Aspar B., Pelloie J.L. (1994) Low dose SIMOX for UILSI applications, in J.P. Colinge, A.N. Nazarov et al. (eds), *Physical and Technical Problems of SOI structures and Devices*, Kluwer Academic Publishers, Dortrecht, p. 3
3. Cerofolini G.F., Bertoni S., Meda L. and Spaggiari C. (1996) The fluence spectrum allowing the formation of a connected buried SiO_2 in silicon by oxygen implantation *Semicond. Sci. Technol.* 11 p.398-415
4. Jablonski J., Saito M., Miyamura Y., and Katayama T. (1997) Internal Thermal Oxidation of Discontinous Buried SiO_2 Layers in Silicon, in S. Cristaloveanu et al (eds.) Proc. of VIII Int. Symp. *Silicon on Insulator Technology and Devices* (Pennington, NJ: The Electrochemical Society) p. 51-54.
5. Ogura A. (1997) Extention of Dose Window for Low-Dose SIMOX in S Cristaloveanu et al (eds) Proc.of VIII Int. Symp. *Silicon on Insulator Technology and Devices* (Pennington, NJ: The Electrochemical Society) p. 57-63.
6. Litovchenko V., Romanyuk B., Efremov, M.Klyui, A., Melnik V, (1996) in P.L.F. Hemment et al (eds.) Proc.of VII Int. Symp. *Silicon on Insulator Technology and Devices*. Low Dose SIMOX Approach and Stimulating Factors, (Pennington, NJ: The Electrochemical Society) p. 117-121.
7. Hu S.M., (1973) Diffusion in silicon and Germanium in D. Show (ed.) *Atomic Diffusion in Semiconductors*, Plenum Press London-New York, p 248-396

8. Litovchenko V.G., Efremov A.A., Romanyuk B.N., Melnik V.P., Claeys C. (1998) Processes in Ultrathin Buried Oxide Synthesis Stimulated by Low Dose Ion Implantation. *J.Electrochem. Soc..* **145**, No 8, p.2964-2969

9. Flynn C.P. (1972) *Point Defects and Diffusion* Clarendon Press-Oxford

10. Liebau F.(1985) *Structural Chemistry of Silicates*, Springer-Verlag, Berlin-Heidelberg.

11. Lannoo M., Friedel J. (1981) *Point Defects in Semiconductors I. Theoretical Aspect* Springer-Verlag, Berlin-Heidelberg-New York.

11a. Watkins G. D., (1964) in P. Baruch (ed.) *Radiation Effects in Semiconductors* - Dunod-Paris, p. 97.

12. Borghesi A., Pivac B., Sasela A. and Stella A. (1995) Oxygen Precipitation in Silicon *J. Appl.Phys.* **77**(9), p. 4169

13. Geguzin Ya. E.(1979) *Diffusion Zone* (in Russian), Nauka-Press, Moscow.

14. Vanhellemont J. and Claeys C. (1987) A Theoretical study of precipitates and its application to silicon oxide in silicon *J Appl. Phys.* **62**, No 9, pp 3960-1371

15. Lyubov B.Ya.(1981) *Diffusion Processes in Inhomogeneous Solid Media* (in Russian), Nauka-Press, Moscow.

16. Kirsanov V.V., Suvorov A.L., Trushin Yu. V (1985) *Processes of irradiation induced defect generation in metals* (in Russian) Energoatomizdat, Moscow.

17. Nikitenko N. I.(1983) *Theory of Heat and Mass Transfer* (in Russian), Naukova Dumka-Press, Kiev.

18. Burenkov A.F., Komarov F.F., Kumachov M.A. Temkin M.M. (1985) *Spatial Distribution of Energy Deposited within a Cascade of Atomic Collisions in Solids* (in Russian), Energoatomizdat, Moscow.

19. Volkenshtein M. V. (1981) *Biophysics* (in Russian), Nauka-Press, Moscow.

20. Abbe H., Suzuki I., Koya H. (1997) The effect of hydrogen annealing on oxygen precipitation Behavior and Gate Oxide Integrity in Cz-Si wafers. *J. Electrochem. Soc.* v **144**, No 1, pp.306-311

21. Zhang M., Lin Ch., Hemment P.L.F, Gutjar K., Gosele U. (1997)Gettering of Cu to voids induced by H^+ implantation in SIMOX substrate. *In Meeting Abstracts of the1997 Joint International Meeting*, vol. 97-2, Pennington, NJ: The Electrochemical Society) p 2344.

22. Malewille C., Aspar B., Poumeirol T. Moriceau H. et al (1996) Physical Phenomena Involved in the Smurt Cut Process P.L.F. Hemment et al (eds.) Proc. of VII Int. Symp. *Silicon on Insulator Technology and Devices VII* (Pennington, NJ: The Electrochemical Society) p. 34-46.

23. Cerofolini G., Balboni R., Bisero D. et al (1995) Hydrogen Precipitation in Highly Oversaturated Single Crystalline Silicon *Phys. Stat. Sol.(a)* **150**, pp. 539-578.

ELECTRICAL INSTABILITIES IN SILICON-ON-INSULATOR STRUCTURES AND DEVICES DURING VOLTAGE AND TEMPERATURE STRESSING

A.N. NAZAROV, I.P. BARCHUK, V.I. KILCHYTSKA
Institute of Semiconductor Physics, NAS of Ukraine
Prospect Nauki 45, 252028, Kyiv, Ukraine

1. Introduction

Electrical instabilities in silicon-on-insulator (SOI) materials and devices during voltage and thermal stressing are fundamentally due to the movement and trapping of charge in the buried oxide (BOX), this being electrically the least robust part of the SOI structures. These processes will be controllerd both by the intrinsic structure and the impurity composition of the BOX and by the properties of the BOX/semiconductor interfaces and charge trapping into the BOX after the charge injection from outer layers.

In this paper we will review the voltage and temperature dependent processes and show how the structure and defects in the amorphous SiO_2 (a-SiO_2) give rise to this SIMOX BOX peculiarities.

2. Defects in amorphous SiO_2

2.1. STRUCTURE OF a-SiO_2 NETWORK

Structure investigations of crystalline and amorphous silicon dioxides by different research techniques [1-3] have shown that the main component of the amorphous network of the crystalline lattice of SiO_2 is the SiO_4 tetrahedral block. Thus in a-SiO_2 short-range order exists, which controls the main electronic and optical properties of the oxide. However, the mechanical and electrical properties of a-SiO_2 are determined by long-range order which, to a great extent, is related to the Si-O-Si bridging bond, linking the neighboring SiO_4 tetrahedra, and to the bond angle, ϕ, as shown in Figure 1a.

In a-SiO_2 the bond angle, ϕ, can vary over a wide range [4]. However, Revesz and Gibbs [5] demonstrated that the energy of the bridging bond is only weakly dependant upon the bond angle over the range from $120°$ to $180°$ (Fig. 1b). The angle providing the minimal bond energy for the bridging bond was found to be $144°$, in a good agreement with experimental determination of the bond angle in different modifications

P.L.F. Hemment et al. (eds.),
Perspectives, Science and Technologies for Novel Silicon on Insulator Devices, 163–178.
© 2000 *Kluwer Academic Publishers. Printed in the Netherlands.*

of crystalline SiO$_2$. It is worth noting that the bond energy for a bond angle below 120°
rises rapidly, and, as a result, the bonds are chemically active in respect to interactions
with different atoms and are considered to be «strained».

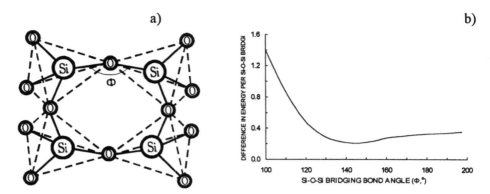

Figure 1. a) A schematic presentation of the 4 neighboring rings in SiO$_2$ (φ is the angle for the
Si-O-Si bridging bond); b) The calculated energy for Si-O-Si bridging bond as a function of the
bond angle [4].

The SiO$_4$ tetrahedra can be arranged into rings [3] with 6 members in the relaxed
dioxide and having from 3 to 8 members in the transition layers of Si/SiO$_2$ interfaces
[6].
Defects in the amorphous oxide can be classified in the following manner [3]:
channels, microcrystals, micropores, broken bonds and impurities. The first and second
classifications involve ordering of the amorphous network, the last three are local
disturbances. In addition, the channels, microcrystals, micropores and some broken
bonds are associated with native defects in a-SiO$_2$.

2.2. NATIVE DEFECTS

Ordered epitaxial dioxide growth on crystalline silicon can lead to a situation when the
SiO$_4$ tetrahedral rings are aligned one with another. In these materials channels or
«pipes» are formed which provide paths for electrical breakdown and increased leakage
currents through the oxide films. These defects may also provide paths for easy
movement of ions in the a-SiO$_2$. The «pipes» have been observed in thin thermal oxides
and pipe like defects have been observed in the BOX produced by low-dose (from
2×10^{17} to 5×10^{17} O$^+$/cm^2) oxygen ion implantation (low-dose SIMOX) [7].
In the BOX fabricated by a (standard) single energy implantation (D=(1.7-2)$\times 10^{18}$
O$^+$/cm^2) the oxide inclusions are localized in the vicinity of the BOX/substrate
interface [8]. These inclusions can be delineated by etching as they exhibit a lower etch
rate in HF than the «bulk» oxide and subsequently can be studied by atomic force
microscopy (AFM) [9]. It was reported in [8], that these oxide inclusions have a
coesite-like crystalline structure and may contain silicon inclusions in the form of a-Si
clusters or crystalline platelets. The heights of the protrusions, measured by AFM after

special HF etching of the BOX in this standard SIMOX material [9], are close to 100 nm, which is typically 1/4 part of the BOX thickness (Fig.2a).

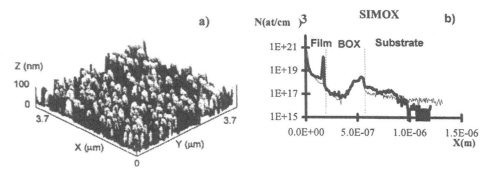

Figure 2. a) AFM image of the BOX / substrate interface in a SIMOX structure implanted under standard conditions (1.8×10^{18} O^+/cm^2; E=180 keV) [9]; b) Deuterium concentration profiles in SIMOX samples deuterated by RF plasma treatment at 250 °C, for 30 minutes at a power density of 2 W/cm^2 (thick line) and 1 W/cm^2 (thin line) [10].

Use of the photoinjection technique reveals the presence of hole and electron traps with large capture cross-sections (10^{-14} - 10^{-13} cm^2) and with densities of up to 5×10^{12} cm^{-2} located mainly at the BOX/substrate interface and correlating with Si enrichment. These traps exhibit an amphoteric behavior and show sensitivity to optical ionization [11]. The low-dose SIMOX BOX does not show this kind of traps.

The appearance of Si enrichment and a strong micro- non-uniformity of the BOX/silicon interfaces leads to a decrease of the effective barrier heights during Fowler-Nordheim electron injection into the BOX. For single energy implanted SIMOX structures the effective barriers are 0.9 - 1.3 eV and 1.4 - 1.6 eV for the BOX/substrate and the BOX/silicon film interfaces, respectively [12]. For the thermal oxide the barrier is 3.1 eV [13].

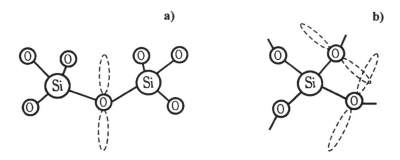

Figure 3. Models for EPR self-trapped hole centers: STH1 (a) and STH2 (b) [14].

The presence of Si inclusions in the a-SiO_2 network also introduces strain in the structure. It has been shown in [15], that the «strained» (Si-O-Si)* bonds can trap holes.

Such self-trapping hole centres (STH1 and STH2) are EPR active and plausible structures for these defects are shown in Figure 3. The STH1 and STH2 centres are thermally unstable at temperatures higher than 300K.

In the a-SiO$_2$ network the simplest broken bond defects are three bonded Si (O$_3$≡Si) and nonbridged oxygen (O$_3$≡Si-O^0), that can result from «strained» (Si-O-Si)* bonds as a result of ionizing irradiation [15] or from Si-Si bonds as a result of hole trapping [16]

$$(O_3≡Si-O-Si≡O_3)* + E \rightarrow O_3≡Si^0 \quad {}^0O-Si≡O_3 \tag{1}$$

$$(O_3≡Si-Si≡O_3)* + h \rightarrow O_3≡Si^0 \quad {}^+ Si≡O_3 \tag{2}$$

In equations (1) and (2) the reaction products can be considered as precursors for E'-centres [17]. The classical E$_\gamma$' centre has been observed in crystalline SiO$_2$ [18], in thermal a-SiO$_2$ [17, 19] and in buried oxides [20, 21] subjected to ionizing radiation and has been attributed to a positively charged asymmetrically relaxed oxygen vacancy (Si^0 $^+$Si≡) (Fig.4a). The E$_\gamma$'-centre EPR signal had a good correlation with the well-defined optical bond at 5.85 eV [22]. Careful study of E$_\gamma$'-centre formation in irradiated thermal oxides by EPR and C-V techniques lead to the conclusion that the centre is associated with positive charge in the dielectric [17].

The presence of the E$_\delta$'-EPR centres is a characteristic of the buried oxide [23]. It has been shown [7, 24] that the appearance of the centre is related to a «confinement» effect, that occurs during a high-temperature long-time anneal of oxide covered by a silicon layer. The nature of the centre seems to be related to silicon enrichment or to an oxygen deficiency in the BOX. Recent first-principles quantum-mechanical calculations of the structure of this defect [25] have shown that the E$_\delta$'-centre is a symmetrical relaxed oxygen monovacancy (Fig.4b) rather than a five-centre tetrahedral Si cluster (Fig.4c), as was proposed by Vanheusden and Stesmans [23].

a)   b)   c)

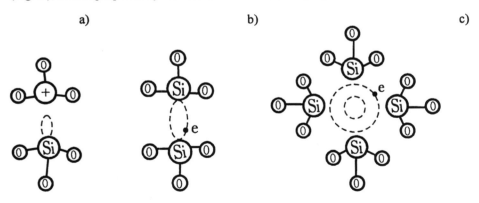

Figure 4. Models: a) of an asymmetrically relaxed oxygen vacancy (E$_\gamma$'-centre) [18]; b) symmetrically relaxed oxygen vacancy (E$_\delta$'-centre) [25]; c) five centre tetrahedral Si cluster (E$_\delta$'-centre) [23].

The E'-centres are thermally stable up to 300°C [16]. The cross-section for hole trapping is ≥ 10^{-13} cm^2 [26]. Electron traps with an activation energy of 1.17 eV and

1.30 eV have been reported by Strzalkowski et al. [27] who used a thermal depopulation technique to investigate ion bombarded thermal oxide. These traps were supposed to be attributed to the E'-centre. In a number of theoretical and experimental papers [16, 27, 28] it has been suggested that the oxygen vacancy in the dioxide can trap both holes and electrons.

It is worth noting that the E'-centre concentration is considerably higher in the BOX produced by the SIMOX technique than in a thermal oxide [26]. In addition, the positive charge, associated with the E'-centers, is compensated by negatively charged electron traps. These electron traps have been assosiated with amphoteric defects ($\sigma = 10^{-14} - 10^{-13}$ cm^2), and also with relatively deep neutral traps with a capture cross section of 3×10^{-16} and 2×10^{-17} cm^2, respectively[29].

2.3. IMPURITY-RELATED DEFECTS

2.3.1. *Hydrogen*
Hydrogen is the most frequently observed impurity in silicon dioxide. Fast silicon oxidation in the presence of the water vapour, widely used in silicon technology [30], naturally leads to hydrogen and hydroxile incorporation into the dioxide network. During silicon dioxide growth two kinds of reactions may occur [31]. The first one is oxide growth:

$$Si + OH (H_2O) \rightarrow SiO_2 + H_2\uparrow \qquad (3)$$

and the second reaction is the interaction of water with the dioxide network

$$O_3\equiv Si\text{-}O\text{-}Si\equiv O_3 + H_2O \rightarrow O_3\equiv Si\text{-}OH + HO\text{-}Si\equiv O_3 . \qquad (4)$$

In addition, hydrogen will be incorporated into the SiO$_2$ network during annealing in a hydrogen ambient at elevated temperature (above 1000°C) of the oxide-silicon structure [31]. Thus a third reaction between hydrogen and SiO$_2$ can take place:

$$O_3\equiv Si\text{-}O\text{-}Si\equiv O_3 + H_2 \rightarrow O_3\equiv Si\text{-}OH + H\text{-}Si\equiv O_3 \qquad (5)$$

The Si-OH and Si-H bonds are candidates for electron trapping centres with capture cross sections ranging from 10^{-18} to 10^{-17} cm^2 [32]. Since the hydrogen can be captured by «strained» and dangling bonds, its presence can serve as an indication of the quality of the amorphous structure. Myers et al. [33] shown that the deuterium concentration in a SIMOX BOX layer after thermal annealing at 700 - 900°C in the deuterium ambient can exceed 10^{20} cm^{-3}. This is considered as evidence for a large concentration of «strained» bonds in SIMOX buried oxides. The maximum concentration of these strained bonds is located near the BOX/Si interfaces, which has been deduced from SIMS analysis of plasma deuterated SOI structures [10] (see Figure 2b). It should be noted that the increase of the deuterium concentration near the BOX/substrate interface is in a good agreement with the measured E'-centre distribution in similar SIMOX BOX layers [24] and with the heights of protrusions, obtained by HF etching [9].

Hydrogen interaction with the structural defects in the vicinity of the BOX/silicon interfaces at temperatures higher than 500°C leads to the formation of positively charged diamagnetic defects that can be observed by the C-V method [34]. It was suggested, that the positively charged overcoordinated oxygen centre may be responsible for this positive charge formation in the BOX [35]. In addition it has been speculated that the appearence of fast moving atomic hydrogen in the BOX [36] leads to large instability of the positive charge in the oxide at room temperature. It is worth noting that protons in the BOX layer are not discharged at the BOX/silicon interfaces and can be trapped in the BOX for a long time [36]. The activation energy for proton drift is 0.76 eV. Based on the latter effect, a new kind of radiation hard memory device has been proposed by Warren et al [37].

The EPR signals (74G doublet and 10G doublet) related to hydrogen-linked defects, such as $O_2=Si-H$ and $O_2=Si-OH$, respectively, have been found in the BOX in SIMOX structures formed by a single energy implantation [38].

2.3.2. Sodium, potassium

It is well known that sodium and potassium are unintentional impurities which can be incorporated in SiO_2 during oxidation processes and various chemical treatments of the wafers [39]. Sodium and potassium in the dioxide can exist in bonded and mobile states [40]. If various environmental conditions affect the dioxide, their concentration in these states is changed. There is reason to believe that sodium is captured in SiO_2 to form an ionic pair with unbonded oxygen. In addition, during «wet» oxidation the trapping occurs more intensively due to the formation of «weak» bonds, such as, $\equiv Si-O-H$ [40]. The reaction can be written as follows:

$$\equiv Si-O-H + Na^+ + Cl^- \rightarrow \equiv Si-O^-Na^+ + HCl \qquad (6)$$

These same arguments can be applied to potassium ions. The activation energies for fast and slow moving sodium were determined to be 0.66 - 0.75 eV and 1.1 eV, respectively [41]. Charge drift takes place in the temperature region from 50 to 150 °C for «fast» ion drift and from 250 to 450 °C for «slow» ion drift. It was suggested that the fast ion drift in the oxide is consistent with ions moving along channels. The larger size of the potassium ions (r_{K^+} =1.33 Å, r_{Na^+} =1.00 Å) leads to a higher migration temperature in the oxide for potassium as compared with «fast» moving sodium. The activation energy of the potassium drift process is 1.09 eV [42].

In SIMOX SOI structures there is usually a low concentration of sodium and potassium in the BOX [43]. However, these impurities can be found in the BOX of SOI structures fabricated by zone melting recrystallization (ZMR) or by wafer bonding (WB) techniques.

3. Bias-temperature (BT) processes in the BOX

3.1. METHODS OF STUDIES

Bias-temperature instability processes can be studied by making measurements of device operation characteristics at room temperature, after applying a voltage on the gate of a MOSFET or capacitor whilst it is held at a high temperature and during high temperature heating of the samples.

The most widely used method for studying the BT instability *at room temperature*, after applying the voltage on the gate at high temperature, is the <u>C-V method</u>. For SOI structures the C-V method is very useful, since it permits the total net charge, Q_{tn}, in the BOX and its centroid, $\bar{X}_0$, to be determined. As it was proposed in [44]:

$$Q_{tn} = Q_f + Q_{sub} = \frac{C_d \cdot (V_{FB}^f - V_{FB}^s)}{q \cdot S} \tag{7}$$

$$\bar{X}_0 = \frac{\int_0^d \rho(x)x\,dx}{\int_0^d \rho(x)\,dx} = \frac{|V_{FB}^f| \cdot d}{V_{FB}^f - V_{FB}^s} \tag{8}$$

where V_{FB}^f, V_{FB}^s are flat-band voltages related to the film/BOX and the substrate/BOX interfaces, respectively, $\rho(x)$ is the charge distribution in the BOX, d is the thickness of the BOX, C_d is buried insulator capacitance. The centroid is defined with respect to the substrate/BOX interface (Fig.5).

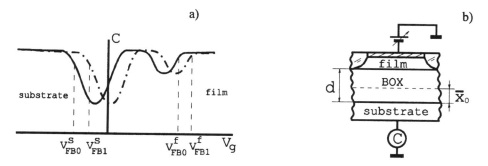

Figure 5. a) Low-frequency C-V characteristics before (1) and after (2) BT stress and b) a schematic diagram of a SOI capacitor

The amount of moving charge related to the BOX/silicon film interface, ΔQ_m^f, and to the BOX/substrate interface, ΔQ_m^s, after BT stress can be calculated as:

$$\Delta Q_m^f = \frac{C_d \cdot (V_{FB1}^f - V_{FB0}^f)}{q \cdot S} = \frac{C_d \cdot \Delta V_{FB}^f}{q \cdot S} \tag{9}$$

and

$$\Delta Q_m^s = \frac{C_d \cdot \Delta V_{FB}^s}{q \cdot S} \tag{10}$$

170

where V_{FB1}, V_{FB0} are flat-band voltages after and before BT stress, respectively (see Figure 5a). Thus, the change of the charge in the BOX is:

$$\Delta Q_{tr} = \Delta Q_m^f - \Delta Q_m^s \tag{11}$$

If $\Delta Q_{tr}=0$, we can consider the BOX/semiconductor interfaces as blocking electrodes, if $\Delta Q_{tr}\neq0$, they are nonblocking electrodes.

Thermally stimulated polarization/depolarization current method is used to investigate the charge transfer processes in a dielectric *during* linear heating of the structure, holding a fixed voltage across the capacitors and measuring the resulting current (Fig.6).

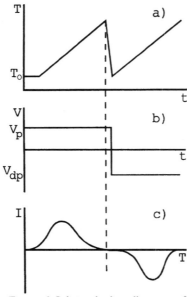

Figure 6. Schematic time diagrams of temperature (a) and applied voltage (b), and TSP/TSD current (c).

It has been shown, that charge moving in the outer circuit during polarization or depolarization processes, $Q_{TSP/TSD}$, is equal to the mirror charge at the blocking electrode [45]. So,

$$Q_{TSP/TSD} = \frac{1}{\beta \cdot S} \int_{T_0}^{T} I(T)dT = \Delta Q_m^f, \tag{12}$$

if the film/BOX interface is blocking, or

$$Q_{TSP/TSD} = \Delta Q_m^s \tag{13}$$

if the substrate/BOX interface is blocking (β is the heating rate). Therefore, the comparison of C-V with TSP/TSD data allows us to identify the blocking interface and also to determine the degree of blocking. In addition, the TSP/TSD method gives us the possibility of determining the activation energies of the polarization/depolarization processes, their energy distribution and the frequency factor or capture cross-section for traps taking part in the processes (see, for instance, [46]).

Source-gate characteristics of MOSFETs and the channel current variation at *high temperature* allow us to determine the polarization process parameters and to analyze the operational stability of the devices. By determining the source-drain current, I_{sd}, in a back-channel SOI MOSFET at high-temperature, after holding the back-gate voltage negative or positive (Fig. 7), it is possible to determine the change in the charge in the BOX layer relative to the BOX/film interface:

$$\Delta Q_m^f = C_d \cdot \frac{V_{th1} - V_{th2}}{q \cdot S} = \frac{C_d \cdot \Delta V_{th}}{q \cdot S} \tag{14}$$

where V_{th1}, V_{th2} are back-channel threshold voltage after negative and positive applied voltage to the back-gate, respectively.

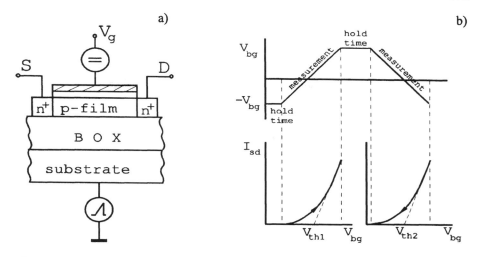

Figure 7. A schematic diagram of a SOI MOSFET (a) and the applied voltage at the back-gate (b) for BT instability measurements using the drain-gate characteristic technique.

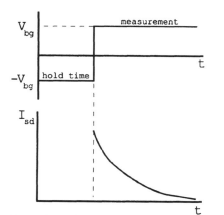

Figure 8. A schematic diagram for (a) back-gate voltage switching and (b) measurement of the relaxation current

If the polarity of the back-gate voltage is reversed and, as result, relaxation of the current occurs at different temperatures it is possible to determine the activation energy of the instability processes. This phenomena is shown schematically in Figure 8. In addition, in dependence on the nature of the active processes in the BOX layer (shift or changing of charge), we can determine either the frequency factor or the capture cross section of the traps, respectively (see, for example [47]).

Thus a combination of the C-V and TSP/TSD current methods with the source-gate characteristics of an MOSFET fabricated on the same SOI structure opens a wide range of possibilities for the study of BT processes in the BOX.

3.2. ZMR SOI STRUCTURES

Measurements of the C-V characteristics of ZMR SOI capacitors after TSP and TSD processes up to 400°C has led to the conclusion that the BOX/semiconductor interfaces in these materials are almost blocking (Table 1) [48].

172

TABLE 1. Total, Q_{tn}, mobile charges, Q_{TSP} and Q_{TSD}, and the charge centroids measured in the buried SiO_2 in ZMR SOI structures.

Q_{tn}, cm^{-2} (C-V)			$\overline{X}_0$, Å (C-V)			Q_{TSP}, cm^{-2}	Q_{TSD}, cm^{-2}
initial	after TSP	after TSD	initial	after TSP	after TSD		
$1 \cdot 10^{12}$	$8.3 \cdot 10^{11}$	$9.4 \cdot 10^{11}$	1070	1900	380	$5 \cdot 10^{11}$	$7 \cdot 10^{11}$

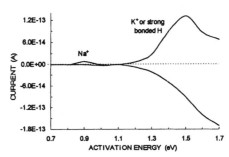

Figure 9. TSP/TSD current spectra measured in ZMR mesa structures (β=0.3 °C/sec) [48].

During charge motion from one interface to the other only 10% of the charge is lost. Comparison of the total net charge in the BOX obtained by the C-V method with the moving charge in the BOX, obtained by TSP/TSD current method, helps us to conclude that almost all of the positive charge, which is trapped in the BOX, participates in the observed movement of charge. Investigations of the thermal polarization/ depolarization processes (Fig. 9) [48] make it possible to suggest that the small low-temperature current peak with an activation energy ranging from 0.75 to 0.9 eV can be related to Na$^+$ ions whilst the high-temperature peak, with an activation energy from 1.2 to 1.7 eV, is due to the motion of K$^+$ ions or to strongly bonded hydrogen.

3.3. SINGLE IMPLANTED SOI SIMOX STRUCTURES

Investigation of the BT stability of the charge in the BOX layer in SIMOX and ZMR SOI structures using C-V and TSP/TSD current methods have shown [49] distinct differences between these two materials.

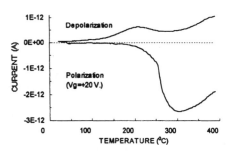

Figure 10. The TSP/TSD current spectra measured in SIMOX mesa structures [49].

Firstly, it should be noted that there is a significant difference, which is greater than a factor of 5, between the values of the polarization and depolarization current in SIMOX samples (Fig. 10). Secondly, the current peaks are completely asymmetric, from which it is concluded that different processes are involved in the BOX charging and discharging. Thirdly, after thermal polarization when a negative voltage is applied to the substrate positive charge accumulation in the BOX is observed; whilst a positive voltage on the substrate leads to negative charge accumulation which compensates the positive charge in the BOX. In each case a considerable variation in the charge close to the lower BOX/substrate interface is observed whilst a very small charge change takes place near the upper BOX/silicon film interface (Table 2). During biassing at temperatures above 250°C a small change of the charge in the BOX is observed but at the same time, a large TSP current peak with the maximum at 300°C occurs (Fig. 10).

TABLE 2. The charges, calculated from the C-V characteristics (ΔQ_{sub} , ΔQ_f , ΔQ_{tn} , ΔQ_m^s , ΔQ_m^f , ΔQ_{tr}) and the TSP current (Q_{TSP}) and the charge centroid ($\overline{X}_0$) determined for SIMOX structures subjected to thermal polarization with a negative bias applied to the substrate.

Kind of treatment	PARAMETERS							
	ΔQ_{sub} (cm^{-2})	ΔQ_f (cm^{-2})	ΔQ_{tn} (cm^{-2})	$\overline{X}_0$ (Å)	Q_{TSP} (cm^{-2})	ΔQ_m^s (cm^{-2})	ΔQ_m^f (cm^{-2})	ΔQ_{tr} (cm^{-2})
Initial	$-3\cdot10^{10}$	$6.8\cdot10^{11}$	$6.5\cdot10^{11}$	3600				
Polarization (250 °C)	$2\cdot10^{11}$	$6.4\cdot10^{11}$	$8.4\cdot10^{11}$	2680	$1.8\cdot10^{11}$	$1.7\cdot10^{11}$	$5.8\cdot10^{10}$	$1.1\cdot10^{11}$
Polarization (400 °C)	$9\cdot10^{10}$	$7.4\cdot10^{11}$	$8.4\cdot10^{11}$	3270	$1.1\cdot10^{12}$	$1.2\cdot10^{11}$	$6\cdot10^{10}$	$1.8\cdot10^{11}$

The activation energy of the polarization process with a negative bias applied to the substrate, determined by the fractional thermal cleaning method [50], is 1.2 eV. Charge moving during the polarization process can reach the value of 1.1×10^{12} cm^{-2} (see Table 2). In the case of depolarization we observed two approximately small current peaks from which the activation energy can only be estimated.

We suggested that the thermal polarization process is consistent with electron emission from the traps located near the BOX/substrate interface. It should be noted, that the flat-band voltage at the BOX/substrate interface is close to zero which is attributed to the complete compensation of the electrical charge.

3.4. HIGH-TEMPERATURE KINK-EFFECT IN SOI n-MOSFET

The processes of charging and discharging in the BOX at high temperatures can lead to some unusual effects in the MOSFETs. In the paper [51] a new high-temperature effect in the SIMOX SOI n-MOSFET is described which the aothors have called the high-temperature back-channel kink-effect. It appeared when a negative voltage is applied to the substrate at a temperature above 200°C when the substrate (back-gate) voltage, V_{bg}, is rapidly swept to a positive value. Under these conditions and with a bias voltage close to zero it was observed that a jumpoccurs in the source-gate current, I_{sd} (Fig.11). This current jump increases with increasing temperature and hold time but upon setting a constant positive voltage the excess current decays.

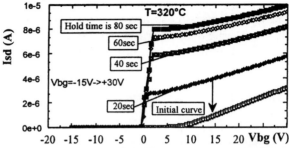

Figure 11. Dependence of the high-temperature kink-effect in n-MOSFET on the hold time of the negative voltage applied to the substrate of a SOI structure (measurement temperature is 320°C) [52].

The current jump is not related to the floating-body effect in the SOI MOSFET because the drain voltage is low during the measurement (0.1 V) and the substrate is held at a high temperature, when the floating body effects are suppressed. All the observed results attest to the current jump being linked with the positive charge accumulation in the BOX under conditions of a negative voltage applied to the substrate when the electrons localized in the negative charged traps and/or in the surface inversion layer in the silicon substrate compensate this positive charge. At zero voltage the hole generation in the substrate neutralizes the negative electron charge when the remaining positive charge in the deep traps will affect the channel current in the MOSFET. It is concluded that the increase in current is connected with the accumulated positive charge in the BOX. The authors estimate that the magnitude of the positive charge is about 2.5×10^{12} cm^{-2} with an activation energy of the accumulation process being 1.2 eV [52]. It should be noted that good agreement exists between these experimental results and values determined from thermal polarization measurements [49].

Analysis of the channel current relaxation at different temperatures and voltages, after switching the substrate bias from negative to positive (Fig. 12a) enables the activation energy of the discharging processes to be estimated as 0.66 eV and 1.00 eV. The dependence of the capture cross-sections on the electric field for the trapping process is shown in the Figure 12b. It is worth noting that these values are in good agreement with the photoinjection data for deep electron traps [29]. The authors conclude that these neutral electron traps control the discharging process at high temperatures.

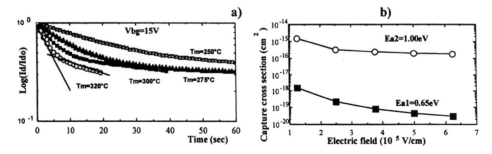

Figure 12. a)The normalized relaxation of the source-drain current in SIMOX n-MOSFET at different temperatures (back-gate voltage is switched from -15V to +15V); b) capture cross-sections, extracted from the current relaxation, vs. electric field applied to the BOX.

Since the activation energy and concentration of the observed positive charge accumulation are similar to the parameters for E'-centre [27] in SIMOX material, we suggest that E'-centre controls the high-temperature positive charge accumulation in the BOX.

4. Conlusions

The bias-temperature processes in the BOX layer of SOI structures can differ considerably from the bahaviour observed in thermal gate oxides. This differences is consistent with the confinement effect taking place in the BOX layer during high-temperature annealing of the structures.

The differences in defect composition in the BOX, as compared to a gate oxide, give rise to new phenomena in the SOI structures and SOI devices, such as, high-temperature kink-effect in n-MOSFETs.

The oxygen vacancy defects and their discharging in the BOX layer can give rise to high-temperature instability in the SIMOX SOI structures and devices.

Acknowledgments

The authors are indebted to Prof. V.S. Lysenko (ISP NASU, Ukraine) for helpful discussions. This research was supported by INTAS contract 93-2075ext, NATO Linkage grant HTECH.LG 951189 and Int. Soros Education Programme under Grant PSU 072035.

References

1. Litovchenko, V.G. and Gorban A.P. (1978) *The basis of MOS microelectronic system physics*, «Naukova Dumka», Kiev (in Russian).
2. Edelman, F.L. (1980) *Structure of LICs components*, Nauka, Novosibirsk (in Russian).
3. Revesz, A.G. (1980) The defect structure of vitreous SiO_2 films on silicon, part I, *Phys. Stat. Sol. A* **57**, 235-243.
4. Aujla, R., Dupree, R., Farnan, I. And Holland, D. (1988) A comparison of the structure of a-SiO_2 prepared by different routes, in R.A.B. Devine (ed.), *The Physics and Technology of Amorphous SiO₂*, Plenum, New York, pp.77-82.
5. Revesz, A.G. and Gibbs, G.V. (1980) Structure and bond flexibility of vitreous SiO_2 films, in G. Lukovsky, S.T. Pantelides and F.L. Galeena (eds.), *The Physics of MOS Insulators, Pergamon,* New York, pp.92-96.
6. Grunthaier, F.J., Levis, B.F., Zamini, N., Maserjian, I. And Madhukar, A. (1980), XPS studies of structure-induced radiation effects at the Si-SiO_2 interface, *IEEE Trans. Nucl. Sci.* **NS-27**, 1640-1646.

7. Revesz, A.G. and Hughes, H.L. (1995) The defect structure of buried oxide layers in SIMOX and BESOI structures, in J.P. Colinge at al. (eds.), *Physical and Technical Problems of SOI Structures and Devices*, Kluwer, Dordrecht, pp.133-156.

8. Afanas'ev, V.V., Stesmans, A. and Twigg, M.E. (1996) Epitaxial growth of SiO_2 produced in silicon by oxygen ion implantation, *Phys. Rev. Lett.* 77, 4206-4209.

9. Afanas'ev, V.V., Stesmans, A., Revesz, A.G. and Hughes, H.L. (1997) Structural inhomogenity and silicon enrichment of buried SiO_2 layers formed by oxygen ion implantation in silicon, *J.Appl. Phys.* 82, 2184-2199.

10. Boutry-Forveille, A., Nazarov, A. and Ballutaud, D. (1998) Deuterium diffusion, trapping, and stability in buried siicon oxide layers, in N.H. Nickel et al. (eds.), *Hydrogen in Semiconductors and Metals, V.513*, MRS, Pennsylvania, pp.319-324.

11. Fedoseenko, S.I., Adamchuk, V.K. and Afanas'ev, V.V. (1993) Silicon clusters as photoactive traps in buried oxide layers of SIMOX structures, *Microelectr. Eng.* 22, 367-370.

12. Ngwa, C.S. and Hall, S. (1994) Electron trapping studies in multiple- and single-implant SIMOX oxides, *Semicond. Sci. Technol.* 9, 1069-1079.

13. Revesz, A.G., Brown, G.A. and Hughes, H.L. (1993) Bulk electrical conduction in buried oxide of SIMOX structures, *J. Electrochem. Soc.* 140, 3222-3229.

14. Griscom, D.L. (1989) Self-trapped holes in amorphous silicon dioxide, *Phys. Rev. B* 40, 4224-4227.

15. Devine, R.A.B. and Arndt, J. (1989) Correlation defect creation and dose-dependent radiation sensitivity in amorphous SiO_2, *Phys. Rev.* B39, 5132-5138

16. Devine, R.A.B. (1994) The structure of SiO_2, its defects and radiation hardness, *IEEE Trans. Nucl. Sci.* 41, 452-460.

17. Conley, J.F. and Lenahan, P.M. (1996) A rewiew of electron spin resonance spectroscopy of defects in thin film SiO_2 on Si, in H.Z. Massoud et al. (eds.), *The Physics and Chemistry of SiO_2 and the Si-SiO_2 interface - 3, 96-1*, ECS, NJ, pp.214-249.

18. Weeks, R.A. (1963) Paramagnetic spectra of E_2'-centers in crystalline quartz, *Phys. Rev.* 130, 570-576.

19. Griscom, D.L. (1978) Defects and impurities in α-quartz and fused silica, in S.T. Pantelides (ed.), *The Physics of SiO_2 and its interfaces*, Pergamon Press, NY, 232-252.

20. Conley, J.F., Lenahan, P.M. and Roitman, P. (1991) Electron spin resonance study of E'-trapping centers in SIMOX buried oxides, *IEEE Trans. Nucl. Sci.* 38, 1247-1252.

21. Stesmans, A., Revesz, A.G. and Hughes, H.L. (1991) Electron spin resonance of defects in silicon-on-insulator structures formed by oxygen implantation: influence of γ-irradiation, *J. Appl. Phys.* 69, 175-181.

22. Weeks, R.A. and Sonder, E. (1963) The relation between the magnetic susceptibility, electron spin resonance and the optical absorption of the E'-center infused silica, in W. Low (ed.), *Paramagnetic Resonance II*, Academic Press, NY, pp.869-875.

23. Vanheusden, K. and Stesmans, A. (1993) Characterization and depth profiling of E' defects in buried SiO_2, *J. Appl. Phys.* 74, 275-283.

24. Vanheusden, K. and Stesmans, A. (1994) Similarities between separation by implanted oxygen and bonded and etchback silicon-on-insulator material as revealed by electron spin resonance, in S.Cristoloveanu et al. (eds.), *Silicon-on-Insulator Technology and Devices 94-11,* ECS, NJ, 197-202.

25. Chavez J.R., Karna S.P., Vanheusden, K., Brothers, C.P., Pugh R.D., Siugaraya, B.K., Warren W.L. and Devine, R.A.B. (1997) Microscopic structure of the E_δ'-center in amorphous SiO_2: A first principles quantum mechanical investigation, *IEEE Trans. Nucl. Sci.* **44**, 1799-1803.

26. Conley, J.F., Lenahan, P.M. and Roitman, P. (1992) Evidence for a deep electron trap and charge compensation in separation by implanted oxygen oxides, *IEEE Trans. Nucl. Sci.* **39**, 2114-2120.

27. Strzalkowski, I., Marczewski, M. and Kowalski, M. (1986) Thermal depopulation studies of electron traps in ion implanted silica layers, *Appl. Phys. A* **40**, 123-127.

28. Fowler, W.B. and Rudra, J.K. (1987) Oxygen vacancy and the E' center in crystalline SiO_2, *Phys. Rev. B* **35**, 8223-8230.

29. Afanas'ev. V.V., Revesz, A.G., Brown, G.A. and Hughes, H.L. (1994) Deep and shallow electron trapping in the buried oxide layer of SIMOX structure, *J. Electrochem. Soc.* **141**, 2801-2804.

30. VLSI Technology (1983) ed by S.M.Sze, McGraw-Hill Book Comp., NY.

31. Revesz, A.G. (1979) The role of hydrogen in SiO_2 films on solicon, *J.Electrochem.Soc.* **126**, 121-130.

32. Sugano, T. (1989) Carrier trapping in silicon MOS devices, *Acta Polytechn. Semicond. Electr. Eng. Sci.* **64**, 220-241.

33. Myers, S.M., Brown, G.A., Revesz, A.G. and Hughes, H.L. (1993) Deuterium interactions with ion-implanted SiO_2 layers in silicon, *J. Appl. Phys.* **73**, 2196-2206.

34. Vanheusden, K., Stesmans, A. and Afanas'ev, V.V. (1995) Combined electron spin resonance and capacitance-voltage analysis of hydrogen-annealing induced positive charge in buried SiO_2, *J. Appl. Phys.* **77**, 2419-2424.

35. Warren, W.L., Vanheusden, K., Schwank, J.R., Fleetwood, D.H., Winokur, P.S. and Devine, R.A.B. (1996) Mechanism for anneal-induced interfacial charging in SiO_2 thin films on Si, *Appl. Phys. Lett.* **68**, 2993-2995.

36. Vanheusden, K., Schwank, J.R., Warren, W.L., Fleetwood, D.M. and Devine, R.A.B. (1997) Radiation-induced H^+ trapping in buried SiO_2, *Microelect. Eng.* **36**, 241-244.

37. Warren, W.L., Fleetwood, D.M., Schwank, J.R., Vanheusden, K., Devine, R.A.B., Archer, L.B. and Wallace, R.M. (1997) Protonic nonvolatile field effect transistor memories in $Si/SiO_2/Si$ structures, *IEEE Trans. Nucl. Sci.* **44**, 1789-1798.

38. Conley, J.F. and Lenahan, P.M. (1992) Room temperature reactions involving silicon changing bond centers and molecular hydrogen in amorphous SiO_2 thin films on silicon, *IEEE Trans. Nucl. Sci.* **39**, 2186-2191.

39. Snow, E.H., Grove, A.S., Deal, B.E. and Suh, C.T. (1965) Ion transport phenomena in insulating films, *J. Appl. Phys.* **36**, 1664-1673.

40. Foukes, F.M. and Witherell, F.E. (1974) Sodium mobility in irradiated SiO_2, *IEEE Trans. Nucl. Sci.* **21**, 67-72.

41. Hofstein, S.R. (1967) Proton and sodium transport in SiO_2 films, *IEEE Trans. On Electr. Dev.* **ED-14**, 749-759.
42. Stagg, J.P. (1977) Drift mobility of Na^+ and K^+ ions in SiO_2 films, *Appl. Phys. Lett.* **31**, 532-533.
43. Dimitrakis, P., Papaioannou, G.J. and Cristoloveanu, S. (1996) Electrical properties of buried oxide-silicon interface, *J.Appl.Phys.* **80**, 1605-1610.
44. Nazarov, A.N., Mikhailov, S.N., Lysenko, V.S., Givargizov, E.I. and Limanov, A.B. (1992) The study of transportation and accumulation charge processes in the buried SiO_2 layers in SOI structures fabricated by zone melting recrystallization technique, *Microelectronica* **21**, 3-13 (in Russian).
45. Vertoprahov, V.N., Kuchumov, B.M. and Salman, E.G. (1981) Structure and properties of $Si-SiO_2-M$ systems, Nauka, Novosibirsk (in Russian).
46. Van Turnhaut, J. (1980) Thermally stimulated discharge of electrets, in G.M. Sessler (ed.), *Electrets*, Springer-Verlag, Berlin-New York, pp. 105-270.
47. Ioannou-Sougleridis, V., Papaioannou, G.J., Dimitrakis, P. and Cristoloveanu, S. (1993) Characterization of the buried oxide in SOI structures by a rate window method, *J. Appl. Phys.* **74**, 3298-3302.
48. Nazarov, A.N., Lysenko V.S., Gusev, V.A. and Kilchitkaya, V.I. (1994) C-V and thermally activated investigation of ZMR SOI meza structures, in S. Cristoloveanu (ed.), *Silicon-on-Insulator Technology and Devices 94-11*, ECS Publishers, NJ, pp.236-244.
49. Nazarov, A.N., Barchuk, I.P. and Kilchitskaya, V.I. (1996) Thermal polarization and depolarization processes in BOX of SOI SIMOX structure, in P.L.F. Hemment et al. (eds.), *Silicon-on-Insulator Technology and Devices 96-3*, ECS Publisher, NJ, pp. 302-308.
50. Gobreht, H. and Hofmann, D. (1966) Spectroscopy of trap by fractional glow technique, *J. Phys. Chem. Sol.* **27**, 509-522.
51. Nazarov, A.N., Colinge, J.P. and Barchuk, I.P. (1997) Research of high-temperature instability processes in buried dielectric of full depleted SOI MOSFETs, *Microelectr. Eng.* **36**, 363-366.
52. Nazarov, A.N., Barchuk, I.P. and Colinge, J.P. (1998) The nature of high-temperature instability in fully depleted SOI IM n-MOSFET, *Proceeding of 4th International High Temperature Electronic Conference (HITEC)*, Albuquerque, NM, pp.226-229.

HYDROGEN AS A DIAGNOSTIC TOOL IN ANALYSING SOI STRUCTURES

A. BOUTRY-FORVEILLE, A. NAZAROV[1] AND D. BALLUTAUD
LPSB-CNRS, 1 place Aristide Briand, 92195 Meudon cedex (France)
[1]*Institute of Semiconductors, Academy of Sciences, Prospekt Nauki 45,*
252650 Kiev, Ukraine.

Abstract.

The interactions of hydrogen and the isotope deuterium used as tracer with buried oxide layers ($Si/SiO_2/Si$) prepared by thermal oxidation or by oxygen implantation have been investigated using Secondary Ion Mass Spectrometry profiling combined with effusion experiments. The amount of deuterium trapping at the silicon/silicon dioxide interfaces and in the buried oxide has been quantified. In SIMOX material the deuterium diffusion profiles show deuterium trapping on implantation induced defects, and deuterium diffusion in the silicon substrate by permeation through the buried oxide layer at temperatures higher than 250°C.

1. Introduction

Oxygen implantation (SIMOX,"separation by implanted oxygen") is one way to produce buried silicon dioxide SiO_2 layers in silicon for the fabrication of silicon-on-insulator (SOI) substrates. Another way to get buried SiO_2 layers (BOX) is thermal oxydation of a silicon substrate followed by chemical vapour deposition of a polycrystalline or amorphous silicon layer, followed by recrystallization by laser zone melting (ZMR) [1].

Hydrogenation, typically using a plasma source, is used to improve the electronic properties of silicon-based electronic devices. In general, it is admitted that passivation occurs when hydrogen is trapped on dangling bonds [2], on point or extended defects such as impurities, silicon grain boundaries or dislocations [3], or at silicon/silicon dioxide interfaces [4]. The plasma hydrogenation leads to a drastic decrease of the concentration of electrically active defects in polycrystalline silicon films and hydrogen annealing of the $Si/SiO_2/Si$ structures is widely used to passivate the electrical activity of the interface states.

Deuterium permeation through the silicon overlayer and trapping in the oxide have been previously studied over a range of high temperatures (500-1000°C) in SIMOX structures, using a deuterium gas source and nuclear-reaction analysis as the characterization technique [5]. The study of the release of hydrogen by bond breaking is also of interest to explain the processes of the generation of electrically active defects

179

P.L.F. Hemment et al. (eds.),
Perspectives, Science and Technologies for Novel Silicon on Insulator Devices, 179–186.
© 2000 *Kluwer Academic Publishers. Printed in the Netherlands.*

in MOS structures involving the injection of hot charge carriers. Some interesting results were obtained recently with deuterium which seems to improve the lifetime of MOS transistors when compared with hydrogen [6].

Hydrogen diffuses rapidly and at the same time may be trapped on defects in both silicon and SiO_2. Furthermore, if deuterium is used as the tracer, this isotope of hydrogen can be analysed with a higher sensitivity. In the present work, the depth distributions of deuterium from a radio-frequency (RF) plasma in $Si/SiO_2/Si$ multilayers prepared by oxygen implantation (SIMOX) or laser processing (ZMR) are reported, using Secondary Ion Mass Spectrometry (SIMS), in order to evidence defects in Si and SiO_2. Effusion experiments combined with isothermal annealings provide some information concerning the stability of deuterium bonding on defects.

2. Experimental

The cross sections of the studied samples are presented on figure 1.

The deuteration of the samples was performed in a deuterium RF plasma reactor (Deuterium plasma pressure: 100 Pa) for 30 min at different temperatures and plasma power densities. The deuterium concentration profiles were obtained by SIMS with a CAMECA IMS4F, with a Cs^+ primary ion beam (14 keV, 5.10^{-3} A/cm^2). The crater depths were measured with a Tencor stylus profilometer and the absolute concentrations were determined by calibration with deuterium implanted Si and SiO_2 standard specimens. Deuterium implanted Si/SiO_2 interfaces specimens were used to

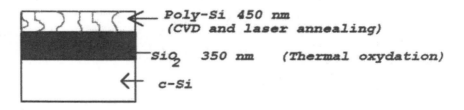

ZMR

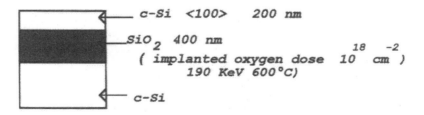

SIMOX

Figure 1: Sample cross sections

check the accumulation of deuterium at the Si/SiO_2 sample interfaces. Effusion spectra were measured by a quadrupole mass spectrometer coupled to an evacuated quartz tube (10^{-5} Pa) which contained the deuterated samples, which were submitted to an isothermal or a linear heating rate (15°C/min).

3. Results

3.1. ZMR SAMPLE

Figure 2 (curves a and b) show the temperature effect on the deuterium distribution in the ZMR SOI structure. The RF plasma treatment at 250 °C leads to deuterium penetration through the silicon layer and its uptake into the buried oxide. Two intensive deuterium peaks respectively at the top silicon/ and at the substrate/BOX interface respectively are observed in the SIMS profiles. No deuterium penetration through the BOX into the silicon substrate is observed at this temperature of deuteration. The deuterium diffusion profile does not depend on the RF plasma power density over the experimental range of parameters studied.

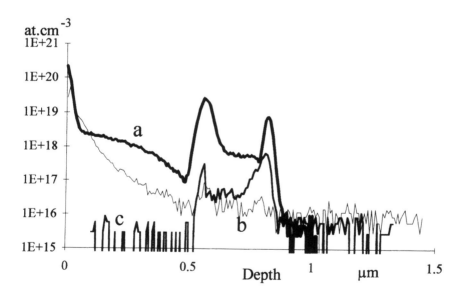

Figure 2: Deuterium concentration profiles in ZMR samples;
(a) After deuteration at 250°C (1W/cm^2 during 30 min).
(b) After deuteration at 150°C (1W/cm^2 during 30 min).
(c) Sample (a) after isothermal annealing at 600°C during 2 hours.

The deuterium diffusion profile in the top polycrystalline silicon layer follows accurately an *erfc function*, with an effective diffusion coefficient of 2×10^{-13} cm^2 s^{-1} at 250°C (Figure 3), although it is not generally the case in polycrystalline silicon, neither in monocrystalline silicon [7][8]. In this particular case, it may be assumed a Fick's law intergranular diffusion of deuterium along grain boundaries perpendicular to the layer surface, probably reduced by trapping effects [9].

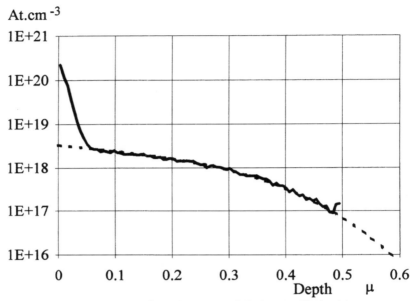

Figure 3: Simulation of the deuterium diffusion profile in the top polycrystalline silicon corresponding to curve (a) figure 2.

The existence of two wide deuterium peaks at the BOX/silicon interfaces suggests the presence of wide transition layers with high concentration of defects, strained and dangling bonds allowing accumulation of deuterium. The wider deuterium peak at the top silicon/BOX interface, compared with the deuterium peak of the BOX/substrate interface, may be related to the roughness of the interface, and/or to the fact that atomic bonds are different at the polycrystalline silicon/BOX interface on one hand, and at the BOX/substrate interface on the other hand.

The deuterium effusion spectra obtained from SIMOX and ZMR samples after deuteration (1W/cm^2 at 250°C during 30 min) are shown in figure 4 (curves a and b). The deuterium effusion spectrum of the ZMR sample on figure 4 (curve a) shows two main effusion peaks at about 550 and 650°C with shoulders at 380°C and 480°C. By comparison with previous results obtained for bulk silicon [10] or microcrystalline silicon [11], the 380 °C effusion feature may be attributed to deuterium accumulated

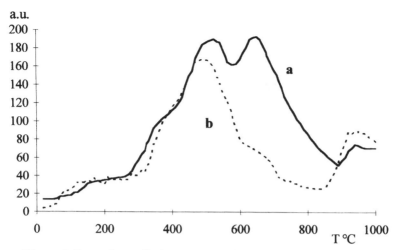

Figure 4: Deuterium effusion spectra of ZMR (curve a) and SIMOX
(curve b) samples (deuterated 30 min at 250°C, 1W/cm^2).

just beneath the surface where defects and traps due to plasma induced damage are in a high concentration, while the breaking of isolated Si-deuterium bonds may be responsible for the effusion peak at 480°C [11]. Then the high temperature peaks at 550 and 650°C would be due to the effusion of deuterium trapped in the buried oxide layer and/or at the Si/SiO$_2$ interfaces.

Figure 2 (curve c) shows the effects of an isothermal annealing (600°C during 2 hours) on a ZMR sample previously deuterated with a RF plasma power density of 1W/cm^2 at 250°C during 30 min. The deuterium peak at the top silicon/BOX interface has strongly decreased, which leads these authors to attribute the effusion peak at 560°C mainly to deuterium detrapping from this interface, while the effusion peak at 650°C is attributed to deuterium coming from the BOX/substrate interface. The deuterium is stable in the BOX, where it can still be detected with a concentration higher than 10^{17} at.cm^{-3} after an isothermal annealing at 600°C during 2 hours (Figure 2, curve c).

3.2. SIMOX SAMPLE

Figure 5 (curves a and b) shows the influence of the temperature on deuterium diffusion profiles in a SIMOX sample. Permeation through the implanted oxide layer occurs at temperatures higher than 250°C (curve a). On the other hand, the deuterium diffusion profile does not appear to depend on the RF plasma power density in the studied experimental parameter range. The deuterium diffusion profiles observed in the top monocrystalline silicon layer are in good agreement with the previous works of D. Mathiot [12]. The deuterium profiles exhibit undulations showing that some deuterium

184

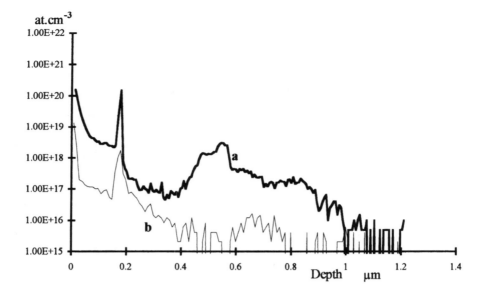

Figure 5: Deuterium concentration profiles in SIMOX samples;
(a) After deuteration at 250°C (1W/cm^2 during 30 min).
(b) After deuteration at 150°C (1W/cm^2 during 30 min).

is trapped on residual reactive sites in the superficial monocrystalline silicon and in the buried oxide layers, which could be related to silicon inclusions or oxide protrusions with different compositions. Fourier Transformed Infra-Red (FTIR) absorption spectroscopy (Figure 6) performed at 7K on hydrogenated SIMOX samples included an absorption peak at about 3650 cm^{-1}, corresponding to the vibration of OH in the Si-O-H stretching mode [13].

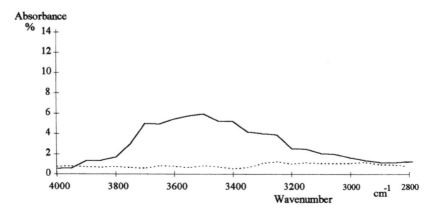

Figure 6: FTIR spectrum of hydrogenated BOX in SIMOX

The SIMOX sample effusion spectrum presents a large peak at 510°C with some shoulders at 380 and 680°C (Figure 4, curve b). The effects of isothermal annealing performed on SIMOX samples after deuteration are shown in figure 7. First the deuterium in-diffuses in the oxide layer, then the deuterium out-diffuses from the oxide layer for temperatures higher than 600°C. After isothermal annealing, the deuterium concentration profile in the BOX becomes constant. By comparison with previous results obtained on monocrystalline silicon [10], the high temperature effusion peak at 680°C would be due to effusion of deuterium trapped in the BOX and/or at the two BOX interfaces.

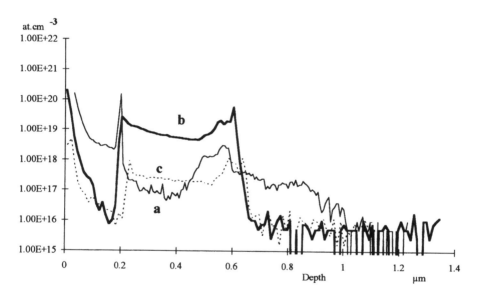

Figure 7: Deuterium concentration profiles in SIMOX samples;
(a) After deuteration at 250°C (1W/cm^2 during 30 min).
(b) Sample (a) after an isothermal annealing at 500°C during 1 hour.
(c) Sample (a) after an isothermal annealing at 600°C during 2 hours.

4. Conclusions

The concentration of deuterium in the buried oxide is governed by two parameters: the diffusivity in the upper silicon layer, which is different in monocrystalline and polycrystalline silicon, and trapping on defects. From this point of view, deuterium diffusion profile analysis is a tool to detect defects in the different layers. In SIMOX samples, trapping offers evidence of reactive sites probably related to silicon inclusions or oxide protrusions with different compositions. However, it must be pointed out that after the isothermal annealing, the deuterium concentration profile in the implanted oxide layer becomes constant. In ZMR samples, the large deuterium peak at the polycrystalline silicon/BOX interface shows pronounced deuterium trapping which may be related to the atomic bond type at this interface.

Permeation through the implanted oxide layer occurs at 250°C, while the deuterium does not diffuse in the silicon substrate through the thermal oxide, over the range of parameters investigated. The deuterium is stable in the oxide layers where it can still be detected with a concentration higher than 10^{17} at.cm^{-3}, after a thermal annealing at 600°C during 2 hours.

5. References

1. Givargizov E. I., Loukin V. A. and Limanov A. B. (1994) Defect engineering in SOI films prepared by zone-melting recrystallization, in J.-P. Colinge, V. S. Lysenko and A. N. Nazarov (eds) *Physical and Technical Problems of SOI Structures and Devices*, NATO ASI series, Kluwer Academic Publishers, Dordrecht., pp.27-38.
2. Ballutaud D., Aucouturier M. and Babonneau F. (1986) Electron spin resonance study of hydrogenation effects in polycrystalline silicon, *Appl. Phys. Lett.* **49**, 1620-1622.
3. Aucouturier M., Rallon O., Mautref M. and Belouet C. (1982) Guerison par l'hydrogène de défauts recombinants dans les couches de silicium polycristallin RAD, *J. Phys. France* **43**, 117-122.
4. Ballutaud D. and Aucouturier M. (1989) Defect microchemistry at the Si/SiO$_2$ interface grown on polycrystalline silicon sheets: hydrogenation effect study, *Solid State Phenomena* **6-7**, 539-546.
5. Myers S. M., Brown G. A., Revesz A. G. and Hughes H. L. (1993) Deuterium interactions with ion-implanted SiO$_2$ layers in silicon, *J.Appl.Phys.* **73** (5) 2196-2206.
6. Devine R. A. B., Autran J.-L., Warren W. L., Vanheusdan K. L. and Rostaing J.-C. (1997) Interfacial hardness enhancement in deuterium annealed 0.25 µm channel metal oxide semiconductor transistors, *Appl. Phys. Lett.* **70**, 2999-3001.
7. Nickel N. H., Jackson W. B. and Walker J. (1996) Hydrogen migration in polycrystalline silicon, *Phys. Rev. B* **53** (12) 7750-7761.
8. Rizk R., de Mierry P., Ballutaud D., Aucouturier M. and Mathiot D. (1991) Hydrogen diffusion and passivation processes in p- and n-type crystalline silicon, *Phys. Rev. B* **44** (12) 6141-6151.
9. Brass A.-M. and Chanfreau A. (1996) Accelerated diffusion of hydrogen along grain boundaries in nickel, *Acta mater.* **44** (9) 3823-3831.
10. Ballutaud D., de Mierry P., Pesant J.-C., Rizk R., Boutry-Forveille A. and Aucouturier M. (1992) Hydrogen effusion from monocrystalline B-doped silicon, *Mat. Science Forum*, **83-87**, 45-50.
11. Lusson L., Elkaim P., Correia A. and Ballutaud D. (1995) Hydrogen diffusion and trapping in micro-nanocrystalline silicon, *J. Phys. III (France)* **5**, 1173-11.
12. Mathiot D. (1989) Hydrogen diffusion in undoped monocrystalline silicon, *Phys. Rev. B* **40**, 5867-5875.
13. Etemadi R., Godet C., Perrin J., Seignac A. and Ballutaud D. (1998) Optical and compositional study of silicon oxide films deposited by plasma-enhanced chemical vapour deposition, *J. Appl. Phys.* **83** (10) 5224-5232.

BACK GATE VOLTAGE INFLUENCE ON THE LDD SOI NMOSFET SERIES RESISTANCE EXTRACTION FROM 150 TO 300 K

A. S. NICOLETT
Faculdade de Tecnologia de São Paulo - FATEC/SP
Brazil (e-mail: nicolett@lsi.usp.br)
J. A. MARTINO
Laboratório de Sistemas Integráveis - LSI/EPUSP
Universidade de São Paulo, Brazil (e-mail: martino@lsi.usp.br)
E. SIMOEN
IMEC
Leuven, Belgium (e-mail: simoen@imec.be)
C. CLAEYS
IMEC
Leuven, Belgium (e-mail: Cor.claeys@imec.be)

Abstract: This work studies the influence of the back gate voltage on the LDD SOI nMOSFETs series resistance from 150 to 300 K. The MEDICI simulated results were used to support the analysis. It was observed that for low temperatures the influence of the back gate bias is higher. However, this influence becomes negligible when the back interface below the LDD region is inverted or if a retrograde profile for the LDD region is used.

1. Introduction

Silicon-On-Insulator (SOI) technology has been intensively investigated due to its many advantages. In this technology, special interest has been paid to fully depleted SOI MOSFETs which have superior behavior compared with partially depleted ones. When the temperature is reduced, the advantages, such as improvement of the mobility and the subthreshold slope, increase [1]. However, series resistance (R_{series}) is an intrinsic problem in SOI devices due to the small silicon film thickness, and this problem increases when the temperature is reduced [2].

The influence of the back gate voltage on the series resistance becomes appreciable mainly at low temperatures and for thin buried oxides [3]. For some applications of SOI devices it is expected that the buried oxide thickness will be significantly reduced. Some extraction methods have used the back interface conditions (depletion or/and accumulation) to extract electrical parameters [4] but the influence of the back gate voltage on the series resistance is normally negligible. The goal of this work is to verify

P.L.F. Hemment et al. (eds.),
Perspectives, Science and Technologies for Novel Silicon on Insulator Devices, 187–193.

the back gate voltage and low temperature influence on the LDD (Lightly Doped Drain) series resistance.

2. Simulation Details

The LDD SOI nMOSFETs studied were simulated by MEDICI [5] with a drawn channel width (W_m) of 20 μm and different drawn channel lengths (L_m) of 0.6, 0.8, 1.0, 1.2, 1.6 and 2.0 μm. A gate oxide thickness (t_{oxf}) of 15 nm, buried oxide thickness (t_{oxb}) of 80 nm, silicon film thickness (t_{si}) of 80 nm, a substrate doping $Na = 1 \times 10^{17}$ cm^{-3} and different LDD concentrations were used in the simulations. The front channel characteristics $(I_{ds} \times V_{gs})$ and $(I_{ds} \times V_{ds})$ were obtained for different back gate voltages (V_{gb}) and different temperatures.

3. Results and Discussions

The series resistance R_{series} was obtained from $I_{ds} \times V_{gs}$ curves of L-array transitors using the technique proposed by G. J. Hu [6] for different overdrive voltages $(V_{gs} - V_{th})$. Figure 1 shows the results of the percentage series resistance change ΔR, defined in Equation (1), as a function of the back gate bias V_{gb} for $(V_{gs} - V_{th}) = 2.45$ V and two different LDD concentrations (N_{LDD}).

$$\Delta R = \frac{\left[R_{series}\left(V_{gb} \neq 0\right) - R_{series}\left(V_{gb} = 0\right) \right]}{\left[R_{series}\left(V_{gb} = 0\right) \right]} x100 \qquad (1)$$

As can be observed, ΔR is significantly influenced by the back gate voltage mainly when a smaller LDD concentration is used. Figure 1 also shows that for $V_{gb} < -15$ V and $V_{gb} < -30$ V, ΔR becomes almost constant for $N_{LDD} = 1 \times 10^{18}$ cm^{-3} and $N_{LDD} = 4 \times 10^{18}$ cm^{-3}, respectively. When the back gate bias becomes more negative, the depletion region of the back interface inside the LDD region increases and the series resistance also increases due to the reduction of the volume of the neutral film available for conduction [7]. When the back interface below the LDD region is inverted, the depletion region reaches its maximum value. This can explain why the series resistance changes hardly for $V_{gb} < -15$ V $(N_{LDD} = 1 \times 10^{18}$ cm^{-3}) and $V_{gb} < -30$ V $(N_{LDD} = 4 \times 10^{18}$ cm^{-3}).

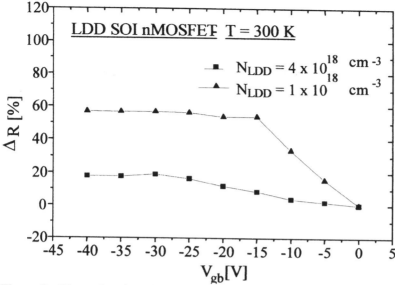

Figure 1: *ΔR* as a function of the back gate bias for two different LDD concentrations obtained from an L-array with t_{oxb} = 80 nm at 300 K.

Figure 2 shows the cross section of an LDD SOI nMOSFET and the analyzed region.

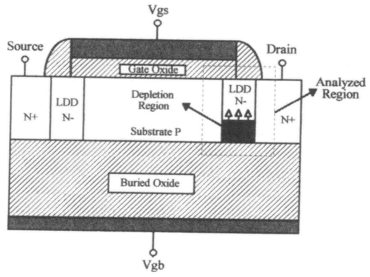

Figure 2: Cross section of a LDD SOI nMOSFET, showing the analyzed region.

Figure 3 shows the variation of the LDD depletion region obtained by bi-dimensional numerical simulation of an SOI nMOSFET with channel length $L_m = 1$ µm , $t_{oxb} = 80$ nm, $N_{LDD} = 1 \times 10^{18}$ cm^{-3}, $V_{ds} = 0.1$ V, $V_{gs} = 2.0$ V, for different V_{gb} values at 300 K. Below each curve the silicon film is depleted and above it is neutral.

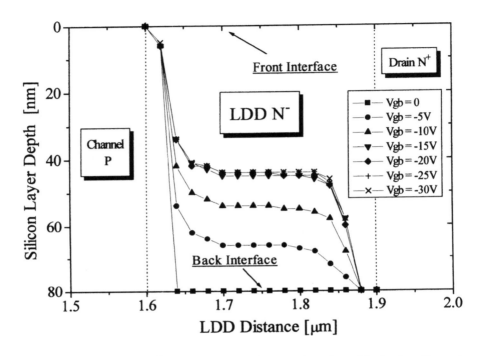

Figure 3: Variation of the LDD depletion region as a function of the back voltage at 300 K, for a LDD SOI nMOSFET with channel length $L_m = 1$ µm.

When the temperature is decreased, the effective LDD concentration is reduced due to incomplete ionization [8]. This effect can be modeled by Equation (2), where N^+_{LDD} is the effective LDD concentration, E_d is the donor energy level, EF_n is the Fermi level, k is the Boltzmann's constant and T is the temperature in Kelvin.

$$N^+_{LDD} = N_{LDD}\left[1 - \frac{1}{1 + 2 \exp\left(\dfrac{EF_n - E_d}{KT}\right)}\right] \qquad (2)$$

Figure 4 shows the influence of the temperature on ΔR for an $N_{LDD} = 4\times10^{18}$ cm^{-3} which was modeled by Eq. (2). It can be seen that the influence of the back gate voltage is higher when the temperature is reduced, since the maximum thickness of the depletion layer is larger due to the reduction of the effective LDD concentration upon cooling.

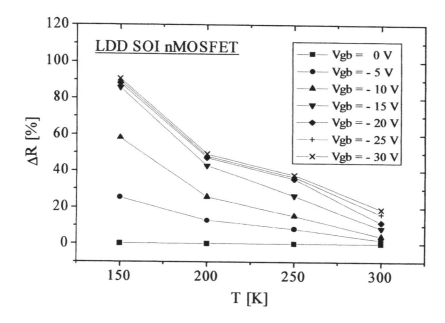

Figure 4: Influence of the temperature on ΔR for different back gate voltages.

The increase of the LDD concentration could reduce the influence of the back gate voltage. However, the LDD region gradually looses its function for higher N_{LDD}. A possible solution would be to provide an LDD region with a retrograde profile. The higher concentration (N_{LDD1}) would be near the back interface to reduce the back gate voltage influence and another one, with smaller concentration (N_{LDD2}), would be near the front interface to reduce the high electrical field near the drain. Figure 5 compares the influence of V_{gb} on ΔR for two different arrays at 150 K.

The first L-array has been simulated with a retrograde profile ($N_{LDD1} = 5\times10^{20}$ cm^{-3}, $N_{LDD2} = 4\times10^{18}$ cm^{-3}), and the second has been simulated with $N_{LDD} = 4\times10^{18}$ cm^{-3}. These initial concentrations (at $T = 300$ K) decrease upon cooling and their values at 150 K can be modeled by Equation (2). It can be seen that the influence of the back gate voltage is reduced when a retrograde profile is used.

Figure 6 compares the $(I_{ds} \times V_{ds})$ curves obtained for two different devices with $L_m = 1$ µm, $W_m = 20$ µm, $V_{gs} = 2.0$ V, two V_{gb} (0 and - 30 V) and $T = 150$ K. The device with a retrograde profile has similar behavior, when compared to the device without a retrograde profile, but with a smaller V_{gb} influence on the series resistance.

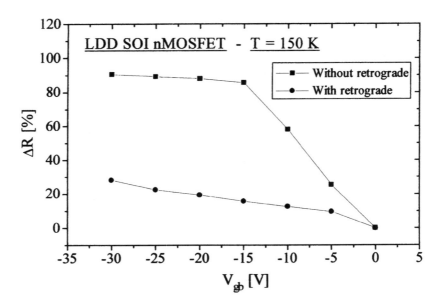

Figure 5: Influence of V_{gb} on ΔR for two different L-array transistors at 150 K.

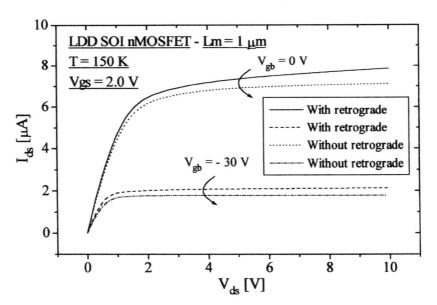

Figure 6: Comparison of $(I_{ds} \times V_{ds})$ curves obtained for two different devices with and without retrograde profiles.

4. Conclusion

The influence of the back gate voltage on the SOI MOSFET series resistance was verified using bi-dimensional numerical simulations. When the back gate bias decreases the series resistance increases until the back interface below the LDD region reaches inversion. In this condition, the depletion region depth is a maximum and the series resistance becomes practically constant. This effect is higher when low temperatures are used, but can be reduced if a retrograde well is incorporated.

References

1. Sanches, F. J. G., Ortiz-Code, A., Nunes, M. G. and Anderson, R. L. (1994) Extracting the Series Resistance and Effective Channel Length of Short-Channel MOSFET's at Liquid Nitrogen Temperature, *Solid State Electronics* 37, 1943- 1948.
2. Balestra, F., Nakabayashi, H., Tsuno, M., Matsumoto, T. and Koyanagi, M. (1994) Performances and Physical Mechanisms in Sub-0.1 μm Gate Length LDD MOSFETs at Low Temperature, *Proceedings of the First European Workshop on Low Temperature Electronics* 4, C6-13 - C6-18.
3. Nicolett, A. S., Martino, J. A., Simoen, E. and Claeys, C. (1998) Back Gate Voltage and Buried-Oxide Thickness Influences on the Series Resistance of Fully Depleted SOI MOSFETs at 77 K, *Proceedings of the Third European Workshop on Low Temperature Electronics* 8, Pr3-25 - Pr3-28.
4. Martino, J. A., Simoen, E. and Claeys, C. (1995), Theoretical and Experimental Study of the Front and Back Interface Trap Density in Accumulation Mode SOI MOSFETs at Low Temperatures, *Proceedings of the 187th Electrochemical Society Meeting* 95-9, 271-277.
5. TMA MEDICI, Version 4.0, 1997.
6. Hu, G. J., Chang, C. and Chia, Y. (1987) Gate-Voltage-dependent Effective Channel Length and Series Resistance of LDD MOSFET's, *IEEE Trans. on Electron Devices* 34, 2469-2475.
7. Reichert, G., Ouisse, T., Pelloie, J. L. and Cristoloveanu, S. (1996) Mobility Modeling of SOI MOSFETs in the High Temperature Range, *Solid-State Electronics* 39, 1347-1352.
8. Selberherr, S. (1989) MOS Device Modeling at 77 K, *IEEE Trans. on Electron Devices* 36, 1464-1474.

CHARACTERIZATION OF POROUS SILICON LAYERS CONTAINING A BURIED OXIDE LAYER

S.I. ROMANOV [1][2], A.V. DVURECHENSKII [1][3], Yu.I. YAKOVLEV[1],
R. GRÖTZSCHEL [4],   U. KREISSIG [4],   V.V. KIRIENKO [1],
V.I. OBODNIKOV [1], A. GUTAKOVSKII [1]
[1] Institute of Semiconductor Physics,
pr. Lavrent'ev 13, 630090 Novosibirsk, Russia
[2] Tomsk State University,
sq.Revolution 1, 634050 Tomsk, Russia
[3] Novosibirsk State University,
Pirogova 2, 630090 Novosibirsk, Russia
[4] Research Center Rossendorf, Incorporated,
P.O.Box 510119, D-01314, Dresden, Germany

Abstract

We have developed new electrolyte solutions for high voltage electrochemical oxidation of porous silicon and have demonstrated their utility by fabricating porous silicon structures containing a buried oxide layer. These structures have high breakdown voltages, over 300V, and are suitable for epitaxial growth of device grade silicon layers. These new structures provide the basis for a versatile SOI technology for the integration high speed and low power devices with high voltage power devices within a monolithic substrate. This paper complements technical details of a novel SOI technology described in the preceding paper in this volume.

1. Introduction

While developing SOPS BOL technology (Silicon-On-Porous Silicon with Buried Oxide Layer) [1-3], we have faced the low performance of electrochemical anodic oxidation (AO) of porous silicon layer (PSL). It was established that the formation potential V_F of BOL was impracticable to be obtained over 30-33V on oxidizing in electronic-grade HCl aqueous solution. This sets limits on industrial applications for fabricating SOI devices. As was apparent after researching dielectric isolation in SOPS BOL structures [1], the breakdown voltage about 100V would be achieved by the use of this electrolyte solution. Such an isolation makes it possible to design all the low power and high speed devices but gives no way of producing high voltage power devices whose use is widespread for automotive electronics, power sources, display drivers, and

195

P.L.F. Hemment et al. (eds.),
Perspectives, Science and Technologies for Novel Silicon on Insulator Devices, 195–204.

so on. It is of value to work out a versatile SOI technology that should provide a means of integrating power devices with the conventional devices. Recently, wafer bonding method has been given special attention because high voltage SOI wafers are readily fabricated [4]. Moreover, it was shown that the high breakdown voltage devices above 500V can be made on SOI structures with thin device layer and 3 μm buried oxide [5].

We believe that SOPS BOL technology can meet the above requirements, once the AO procedure has been improved by the incorporation of the new electrolyte solutions which have been specially developed. We will demonstrate that the new solutions permit BOL with V_F up to 200V to be formed, thereby, far exceeding the values obtained by AO of PSLs in conventional solutions. These electrolyte solutions are now patenting in Russia.

In this paper we describe the electrochemical processing and characterize the advanced PS BOL structures from the viewpoint of their application as the substrates for epitaxial growth and simultaneously the insulator in full-wafer scale SOI. The technical details described here complement details of a novel SOI technology (SOPS BOL) reported in the preceding paper [1].

2. Experiment

The two-layer PS was formed by anodizing 4-inch p^+-type Si(001) wafer with the resistivity of 0.001–0.006 Ω·cm in 20% HF solution as described in details elsewhere [1,6]. This PS consists of top and bottom layers which have different porosity (void fraction) in depth. The porosity of the top-PSL is about 35% or 40% being rather lower than that of the bottom-PSL ranging from 70% to 95%. BOL was fabricated inside this PS by means of the new AO procedure with V_F varying from 50V to 150V under galvanostatic conditions. As we have shown in the preceding paper [1], this BOL contains the compact pore-free oxide, which is formed at an interface of the bottom-PSL with bulk silicon, and the overlying porous oxide including a large number of cavities. An electrolytic cell has a standard two-electrode configuration with a Pt counter electrode for both PSL and BOL formation. In addition, the samples were annealed at 400–950°C for 30 min in a nitrogen ambient at a pressure of 1 atm. These structures and the formation conditions are summarized in Table 1 and are defined in terms of h_1 – the thickness of the top-PSL; P_1 – the porosity of the top-PSL; h_2 - the thickness of the bottom-PSL; P_2 - the porosity of the bottom-PSL; V_F – the formation potential of the anodic oxide film; T_{ann} – the temperature of annealing.

In order to estimate dielectric isolation of a top-PSL from p^+ substrate, current-voltage measurements were made across the PS BOL structure to determine leakage currents and breakdown voltages before and after thermal annealing. For this purpose, before measuring an aluminum 0.1-μm-thick film was deposited to form contact to the PS BOL structure which then was patterned using conventional lithography to produce isolated mesas of width 700 μm.

The compositional and structural characteristics of the PS BOL structures were studied using (i) secondary ion mass spectrometry (SIMS) [7] and elastic recoil detection analysis (ERDA) for measuring the depth distribution of impurities, as well as (ii) Rutherford backscattering spectroscopy (RBS) together with cross-sectional transmission electron microscopy (TEM) for structural disorder estimation and phase location.

The SIMS and TEM experiments were performed with a MIQ-256 Cameca-Riber instrument using a 10-keV Cs^+ primary beam and an electron microscope JEOL-4000EX, respectively. The ERDA and RBS measurements were carried out as described in Refs. 8 and 9.

TABLE 1. Description of the samples used in this work.

Sample	h_1 μm	P_1 %	h_2 μm	P_2 %	V_F V	T_{ann} °C	Solution
1T2	0.1	40	1	60	32.8		HCl aq.
9/0	0.3	40	1	70			
9/10	0.3	40	1	70	100		
17T2	0.3	40	1	80	150	≤950	
18/0	0.3	40	1	80			
18/5	0.3	40	1	80	50		New
18/10	0.3	40	1	80	100		
18/15	0.3	40	1	80	150		
36/0	0.3	40	1	95			
36/10	0.3	40	1	95	100		
22.3	0.3	35	1	70	50		

3. Results

HIGH-VOLTAGE ANODIC OXIDATION. Figure 1 shows typical V_F - *time* curves of the AO process for the two-layer p^+-PS in either the 1M HCl aqueous solution (sample 1T2) or the new electrolyte solution (sample 18/15). The data clearly demonstrate the benefit of using the new solution: the formation potential V_F is increased by a factor of five (from 30V to 150V). This value is no limit. Using this solution we consistently form a BOL with V_F over 200V. As a result the breakdown voltages of the PS BOL structure are changed dramatically, as shown in Figure 2. An initial value of breakdown voltages is about 75V and grows at an accelerated race on annealing up to 300V at 900-950°C. Furthermore, there is no question that the breakdown voltage can overcome this level after annealing at higher temperatures ~ 1300°C necessary to form high quality insulator [10]. The obtained values of V_F and the breakdown voltages far exceed those attainable using the 1M HCl aqueous solution. Consequently, we can fabricate good insulating layers. However, does a certain possibility exist of growing device grade silicon overlayers on these PS BOL substrates after the high-voltage AO? To answer the question the following data were obtained.

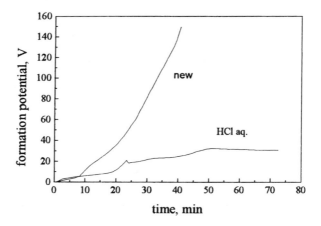

Figure 1. Variation of the formation potential V_F of a buried oxide layer in p^+-PSL with oxidation time at 2 mA/cm^2 in HCl aqueous (sample 1T2) and new (sample 18/15) solutions.

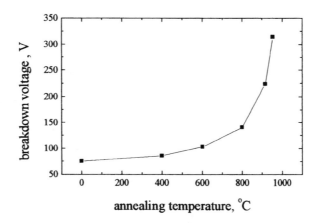

Figure 2. Increase of the breakdown voltage of the PS BOL layer with annealing temperature. Data are from sample 17T2.

PS BOL STRUCTURE. **Figure 3** shows the cross-sectional TEM micrographs of the two-layer PS anodized over a wide range of V_F from 50V to 150V (samples 18/5, 18/15, respectively, without annealing). It is clear that (i) this severe AO does not degrade the PSL and (ii) two dissimilar layers are clearly identified as top layer and bottom layer and are defined by distinct interfaces. Also one can see that the bottom-PSL contains the compact oxide at the interface with the substrate and the porous oxide having the anisotropic columnar structure which is aligned perpendicularly to the wafer surface. It is worth mentioning that the morphology of p^+-type porous layers is well known to be initially highly anisotropic with pores along [001] direction [11]. The observed columns are distorted near the compact oxide layer in the sample oxidized to

150V, Figure 3(b). The distortion of [001]-oriented columns is not revealed in the upper part of this bottom-PSL and in the whole bottom layer of slightly oxidized sample, Figure 3(a). This suggests that the anodic compact oxide located in the interface of the bottom-PSL with the substrate is the main source of internal stress in the PS BOL structure. The bottom-PSL is certain to damp party this stress field and, thereby effectively shield the top-PSL from further strain during oxidizing.

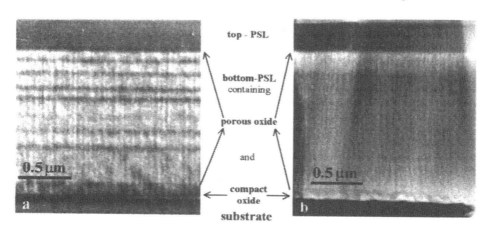

Figure 3. Cross-sectional TEM micrographs of PS BOL structures formed by oxidizing p$^+$-PSL with (a) V_F = 50V (sample 18/5) and (b) V_F = 150V (sample 18/15) without annealing.

OXYGEN CONTENT. To grow epitaxial films it is of importance to know the composition of porous layers and the microscopic state of the top-PSL surface after high-voltage AO.

Figure 4 shows concentration depth distributions of Si, O, C, N, and F in sample 18/10 determined using ERDA [8]. Si and O can be seen to be the most abundant elements. C, N, and F are all minor impurities in the anodically oxidized PSL even after long air exposure (2 weeks). The high concentrations of Si and O are roughly constant throughout the inner part of the PSL with a O/Si ratio of 1.5 to 1.6 whilst the near surface layer of thickness 100nm contains less oxygen with a O/Si ratio of 0.7 to 0.8. These data demonstrate that most Si atoms in the near surface region of the top-PSL have no chemical bonds with O atoms whereas most Si atoms in the bottom-PSL react with O atoms forming silicon oxide during anodic oxidation.

By SIMS measurements, we have studied an oxygen content in different PS BOL structures. Figure 5 shows the $^{32}O_2/^{30}Si$ yield ratio as a function of etching time for samples 22.3, 9/10, 18/5, 18/10, and 18/15. The porosity P_2 and the formation potential V_F were variable parameters. The SIMS data show that O profiles are practically constant when passing the top-PSL (etching time $\leq$ 500s) slightly depending on P_2 and V_F. In contrast, the oxygen concentration in the bottom-PSL is strongly dependent on these parameters. The oxygen content, as shown in Figure 5,

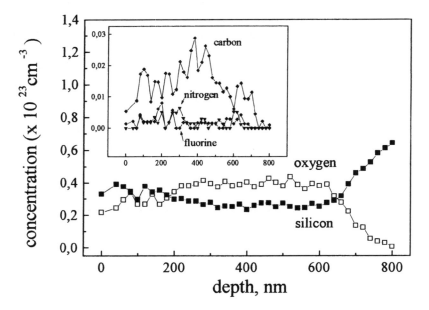

Figure 4. Composition depth profiles of sample 18/10 with $P_2 = 80\%$ after anodic oxidation at $V_F = 100$V.

increases monotonically with V_F. The depth profiles change dramatically with P_2 from uniform distributions at $P_2 = 70\%$ in Figure 5(a) to nonuniform distributions at $P_2 = 80\%$, Figure 5(b). Note that there are oxygen peaks near the interfaces of the bottom-PSL in the first case, Figure 5(a). These results lead to the conclusion that the high-voltage AO of the two-layer PS occurs mainly inside the bottom-PSL rather than the top-PSL, that is, the latter is oxidized to only a small extent as compared to the former.

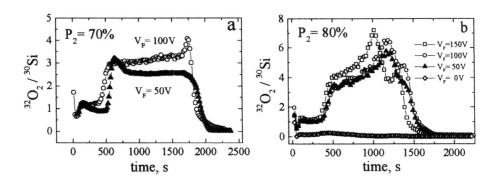

Figure 5. $^{32}O_2/^{30}Si$ yield ratio as a function of etching time for PS BOL structures from samples with (a) a porosity of 70% and (b) a porosity of 80%.

MICROCTRUCTURE OF TOP-PSL. The crystalline structure of top-PSL was analysed using RBS technique [9]. Random and aligned RBS spectra of the two-layer PS before and after AO (samples 9/0 and 9/10, respectively) are presented in Figure 6. This is a typical example. The effect of AO on the structure of the top-PSL proved to be great. When comparing the conforming curves in Figure 6 (a) and (b), one can see the dramatic rise in aligned backscattering yield of He$^+$ ions from the oxidized PSL. A minimum backscattering yield, χ_{min}, calculated as a ratio of the aligned yield to the random yield increases by a factor of ~ 6 after 100V AO. This high χ_{min} may be due to either the presence of silicon oxide film on internal surface of pore walls or lattice distortion of the Si skeleton induced by the anodic film. The latter is not desirable for epitaxial growth.

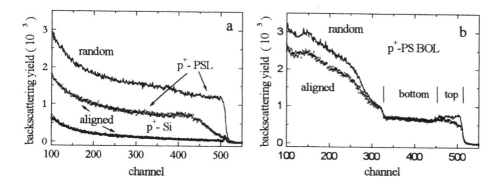

Figure 6. Typical RBS spectra from (a) an as-prepared two-layer p$^+$-PS (sample 9/0), and (b) the final PS BOL structure after anodic oxidation at $V_F = 100$V (sample 9/10).

Further RBS and ERDA analysis have been carried out to discriminate between these supposed effects. The examination was based on the V_F - and P_2 - dependence of χ_{min} (Figure 7, samples 9/0, 18/0, 36/0, 9/10, 18/5, 18/10, 18/15, and 36/10 in Table 1) along with the oxygen concentration-depth relationship of sample 18/10 obtained by ERDA, as shown in Figure 4. As evident from Figure 7(a), χ_{min} increases to ~ 45% for a formation potential V_F of 50V but is then insensitive to V_F up to 150V. The dependence of χ_{min} on porosity in samples anodized to 100V is shown in Figure 7(b). Prior to anodization χ_{min} is typically 10% to 16% but increases to 65% after oxidation of the 70% porosity sample and to about 35% in the sample with a porosity of 95%. According to the ERDA data, the oxygen to silicon atoms ratio in the near surface layer of the top-PSL (sample 18/10 in Fig.4) is 0.65 to 0.70. Such a proportion means that the silicon atoms bonded to the oxygen atoms in anodic amorphous oxide is 32-35% of the total amount of silicon in this surface layer. This is necessary component of χ_{min} which was measured equal to ~ 47%, as seen in the sample with a porosity of 80% in Figure 7(b). The difference of these two terms gives $\Delta\chi_{min}$~12-15%. This value is due to the backscattering yield from the remainder of silicon skeleton in the surface layer of the top-PSL after AO and remarkably coincides with the initial χ_{min} of non-oxidized sample 18/0 that it is about 15.6%. Consequently, the microstructure of the surface monocrystalline layer of the top-PSL is not changed during high-voltage AO. In the case of low-voltage AO described in the preceding paper [1] we have drawn the same conclusion.

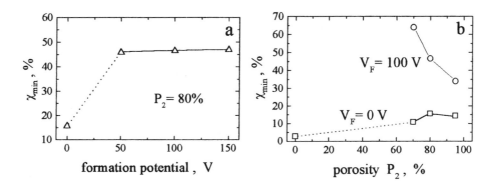

Figure 7. Minimum RBS yield χ_{min} as a function of (a) the formation potential V_F with P_2=80%; and (b) the porosity P_2 for V_F = 0 and 100 V.

This result is important for an understanding of the problem of epitaxial growth on PS BOL substrates. We conclude that the main impediment to growing high quality epitaxial films is associated with the significant decrease in the area of the epitaxy surface of top-PSL as a result of AO. In other words, after AO the 'porosity' of the surface layer is effectively increased due to the Si→SiO$_2$ transition on the pore walls. A conservative estimate of this 'effective porosity' has shown that in the case of samples 9/10, 18/10, and 36/10 the 'porosity' of the top-PSL was increased by 32%, 19%, and 12%, respectively, taking on the final values as 72%, 59%, and 52%. Although growing high quality layers on such porous substrates is difficult to be practical, it does not present an impenetrable barrier to growers and will be a challenging and intriguing task. This work is now in progress.

4. Conclusions

New electrolyte solutions have been developed to fabricate the advanced PS BOL substrates. These structures have high breakdown voltages over 300V after high-voltage AO to 150V and thermal annealing at 900-950°C.

TEM studies have shown that the high-voltage AO does not degrade a two-layer PS. The morphology of the bottom-PSL (high porosity) is anisotropic with porous oxide columns aligned perpendicularly to the wafer surface. There is compact oxide layer at the interface with the substrate. It is suggested that the compact oxide is a source of internal stress in the PS BOL structure whereas the porous oxide damps this stress.

The oxygen content measured by ERDA in one such sample is nonuniformly distributed among the top and bottom layers. The concentration in the bottom-PSL is about two times larger than the oxygen concentration in the top-PSL. Moreover, during anodic oxidation most Si atoms in this bottom-PSL react with O atoms forming silicon oxide whereas most Si atoms in the top-PSL have no chemical bonds with O atoms. As a result the bottom-PSL is converted into SiO_2 phase including Si nanoclusters but the top-PSL is still retained as a monolithic silicon skeleton with oxidized surface.

The high-voltage AO of a two-layer PS occurs mainly inside the bottom-PSL rather than the top-PSL. The concentration of oxygen in the top-PSL is uniformly distributed and remarkably stable during AO. In contrast, its concentration in the bottom-PSL is strongly dependent on the structural and processing parameters. The oxygen concentration increases with the formation potential V_F and changes dramatically with increasing the porosity P_2 from uniform ($P_2 = 70\%$) to nonuniform ($P_2 = 80\%$) distributions.

According to RBS and ERDA measurements, the microstructure of top-PSL is not changed during the high-voltage AO. There is, however, a significant decrease in the area of this substrate which is suitable for epitaxial growth. This needs to be taken into account when depositing high quality layers.

Acknowledgments

This work was supported by the Siberian Branch of the Russian Academy of Sciences (Grant No.15000-421), the State Scientific and Technical Program "Promising Technologies and Devices in Micro-and Nanoelectronics", grant 02.04.1.1.16.□.1, the Russian Foundation for Basic Research (Grants No.96-02-19301, 98-02-17790), the 'Physics of Solid-State Nanostructures' State Program (Grant No.3-011/4), and the Ministry of High School (Grant No.□□-103-98 via Novosibirsk State University).

5. References

1. Romanov, S.I., Dvurechenskii, A.V., Kirienko, V.V., Grötzschel, R., Gutakovskii, A., Sokolov, L.V., and Lamin, M.A. (1999) Homoepitaxy on porous silicon with a buried oxide layer: full-wafer scale SOI, *in this issue the preceding paper.*
2. Romanov, S.I. Russian Federation Patent application 97103165 (1997).
3. Romanov, S.I. Russian Federation Patent application 97103424 (1997).
4. Hunt, C.E., Baumgart, H., Iyer, S.S., Abe, T., and Gosele, U. (eds.) (1995) *Semiconductor Wafer Bonding: Physics and Applications III,* Vol. 95-7, The Electrochemical Society, Inc., Pennington.
5. Nakagawa, A., H. Funaki, H., and I. Omura, I. (1995) High voltage SOI technology (invited paper), in C.E. Hunt, H. Baumgart, S.S. Iyer, T. Abe, and U. Gosele (eds.), *Semiconductor Wafer Bonding: Physics and Applications III,* Vol. 95-7, The Electrochemical Society, Inc., Pennington, pp. 411-419.

204

6. Karanovich, A.A., Romanov, S.I., Kirienko, V.V., Myasnikov, A.M., and Obodnikov, V.I. (1995) A secondary ion mass spectrometry study of p$^+$ porous silicon, *J. Phys. D: Appl. Phys.* **28**, 2345-2348.

7. Canham, L.T., Houlton, M.R., Leong, W.Y., Pickering, C., and Keen, J.M. (1991) Atmospheric impregnation of porous silicon at room temperature, *J. Appl. Phys.* **70**, 422-431.

8. Behrisch, R., Grötzschel, R., Hentschel, E., and Assmann, W. (1992) HIERD analysis of the low-Z deposits on probes from the vessel walls of fusion experiments, *Nucl. Instr. and Meth.* **B68**, 245-248.

9. Grötzschel, R., Hentschel, E., Klabes, R., Kreißig, U., Neelmeijer, C., Assmann, W., and Behrisch, R. (1992) Elemental analysis of thin layers by elastic heavy ion scattering, in E. Wieser (editor). *Institute for Ion Beam Physics and Material Research, Research Programme 1992, Scientific Report 1991*, FZR 92-06, April 1992 Forschugszentrum, Rossendorf e.V., p.53.

10. Oules, C., Halimaoui, A., Regolini, J.L., Perio, A., and Bomchil, G. (1992) Silicon on Insulator Structures Obtained by Epitaxial Growth of Silicon over Porous Silicon, *J. Electrochem. Soc.* **139**, 3595-3599.

11. Grosman, A., Ortega, C., Wang, Y.S., and Gandais, M. (1997) Morphology and Structure of p-type Porous Silicon by Transmission Electron Microscopy, in M.O. Manasreh (series editor), G. Amato, C. Delerue, and H.-J. von Bardeleben (eds.), *OPTOELECTRONIC PROPERTIES of SEMICONDUCTORS and SUPERLATTICES, Volume 5: STRUCTURAL AND OPTICAL PROPERTIES OF POROUS SILICON NANOSTRUCTURES*, Gordon and Breach Science Publishers, pp. 317-331.

TOTAL-DOSE RADIATION RESPONSE OF MULTILAYER BURIED INSULATORS

A.N. RUDENKO, V.S. LYSENKO, A.N. NAZAROV, I.P. BARCHUK,
V.I. KILCHYTSKA, T.E. RUDENKO, S.V. DJURENKO, Ya.N. VOVK
Institute of Semiconductor Physics, NAS of Ukraine
Prospect Nauki 45, 252028 Kyiv, Ukraine

1. Introduction

The high radiation tolerance of SOI CMOS ICs to transient radiation effects and single-event upset is well known [1], however, in contrast SOI CMOS devices are known to be rather susceptible to total-dose radiation effects. This sensitivity to total-dose exposure is associated with radiation-induced positive charge trapping in the thick buried oxides, which results in parasitic back-channel conduction in n-channel MOSFETs [2]. The most widely used method to suppress the radiation-induced back-channel conduction is additional doping of the Si film near the silicon film - buried insulator interface [3], however, at high irradiation doses this method is not always adequate.

2. Device Fabrication

In this work the aim is to improve the total-dose radiation hardness of SOI structures consisting of buried SiO_2-Si_3N_4-SiO_2 (200, 150 and 15 nm-thick, respectively) layers fabricated by the ZMR technique [4]. The lower 200-nm thick SiO_2 layer was grown by high-pressure thermal oxidation of the Si substrate, whilst the Si_3N_4 layer was deposited by the LPCVD process. Several processes were investigated for the formation of the thin SiO_2 spacer between the Si film and the Si_3N_4 layer, where the spacer served to avoid formation of a Si/Si_3N_4-interface. Experiments showed the best suited SiO_2 spacer was obtained by thermal oxidation of deposeted Si_3N_4. MOS test devices were then fabricated using a standard CMOS process. The thickness of the Si film in the fabricated devices was 300 nm, the front gate oxide thickness was 80 nm, and the film doping concentration was 2×10^{16} cm^{-3}. For comparison, test devices were fabricated on ZMR SOI structures with a 300-nm buried SiO_2 layer. Comparative studies were also performed on the devices made on SIMOX substrates fabricated by implantation of 200 keV O^+ to a dose of 1.8×10^{18} $O^+ cm^{-2}$ at a substrate temperature of 600°C followed by anneal at 1300°C for 5 h in argon with 0.5% oxygen.

Fabricated n- and p-channel SOI MOSFETs were irradiated with ^{60}Co gamma rays. The absorbed dose ranged from 10^4 to 5×10^6 rad (Si). The irradiations were

205

P.L.F. Hemment et al. (eds.),
Perspectives, Science and Technologies for Novel Silicon on Insulator Devices, 205–212.
© 2000 *Kluwer Academic Publishers. Printed in the Netherlands.*

carried out at a dose rate of 160 rad (Si)/s, whilst a bias (from -10 to +10 V) was applied to the substrate with other terminals being grounded.

3. Experimental Results

Electrical characterization of the initial and irradiated structures was performed using dynamic transconductance and current-voltage measurements of back-channel n- and p-type SOI MOSFETs and capacitance-voltage measurements of SOI capacitors. The radiation-induced charge, ΔQ_d, and its centroid, X_0^{ir} were evaluated from the midgap voltage shifts [5]:

$$\Delta Q_d = C_d (\Delta V_{mg1} - \Delta V_{mg2}), \qquad (1)$$

$$X_0^{ir} = \frac{|\Delta V_{mg1}| \cdot d}{|\Delta V_{mg1} - \Delta V_{mg2}|} \qquad (2)$$

where d is the insulator thickness, ΔV_{mg1} and ΔV_{mg2} are the midgap voltage shifts for top and bottom buried insulator interfaces, respectively. The variation in C-V curves stretchout was used to determine the radiation-induced surface state density [5].

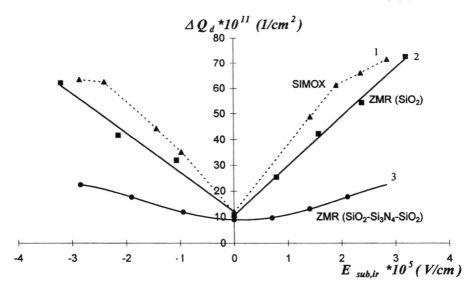

Figure 1. Radiation-induced total charge density vs. applied electric field (D=5x10⁵ rad(Si)).

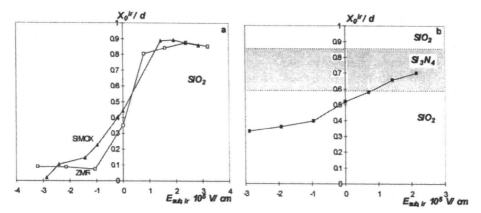

Figure 2. Normalized centroid of the radiation-induced positive charge vs applied electric field during irradiation (a) in buried SiO_2 and (b) SiO_2-Si_3N_4-SiO_2.

The net radiation-induced positive charge density in various buried insulators calculated from experimental C-V curves using equation (1) are shown in Figure 1 as a function of electric field applied to the substrate during irradiation. Figure 2 shows the experimentally determined position of the charge centroid in three-layer (fig. 2b) and single-layer (fig. 2a) buried insulators plotted as a function of the electric field applied to the substrate during irradiation. It can be seen that the replacement of the buried SiO_2 layer by the SiO_2-Si_3N_4-SiO_2 layer results in a significant decrease in the net positive charge buildup (Fig.1) and moves the radiation-induced charge centroid from the Si-SiO_2 interfaces deep into the insulator thickness (Fig.2 a and b).

Together, these effects result in a significant reduction of the back-channel threshold voltage shift when the irradiation is performed with a positive substrate bias ("the worst case" irradiation bias) (Fig.3a).

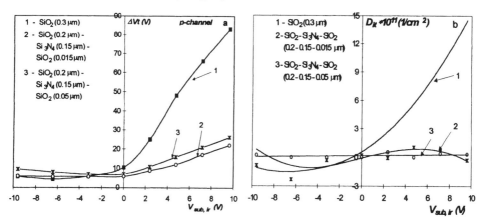

Figure 3. (a) The radiation-induced back-gate threshold voltage shift of the p-channel SOI MOSFET and (b) back interface surface-state density as a function of irradiation bias $(D=5 \times 10^5 \text{ rad(Si)})$.

208

In contrast to structures with buried SiO$_2$ layers, the surface state density at the Si film-buried insulator interface in structures with SiO$_2$-Si$_3$N$_4$-SiO$_2$ layers did not change noticeably with irradiation (curves 2 and 3, Figure 3b). The Si$_3$N$_4$ layer most likely blocks the transport of holes and H-related species produced in the thick underlying SiO$_2$ layer during the irradiation, resulting in a very low radiation-induced surface charge density [6].

4. Model and Discussion

It might be expected that replacement of part of the SiO$_2$ layer by Si$_3$N$_4$ will result in a desirable reduction of the positive charge buildup and the back-channel threshold voltage shift because of a reduced oxide thickness and because the nitride layer introduces electron traps which compensate trapped hole [7]. Indeed, at low doses (D≤5x10^4 rad(Si)) a negative charge buildup in the SiO$_2$-Si$_3$N$_4$-SiO$_2$ structures was observed, as indicated by positive shifts of the C-V characteristics. This could be attributed to the electron trapping in the Si$_3$N$_4$ layer or at the Si$_3$N$_4$-SiO$_2$ interfaces. However, the high-dose radiation response revealed some surprising features. The most stricking result of this work is the symmetrical negative and positive bias radiation responses of the non-symmetrical SiO$_2$-Si$_3$N$_4$-SiO$_2$ structure observed at high-doses (D>5x10^5 rad(Si)) (as shown by curve 3 in Fig.1 and Fig.2b). These results cannot be explained by electron trapping alone, because the negative trapped charge would be greater for irradiation with a negative substrate bias, when electrons are generated and arrive from the thick bottom SiO$_2$ layer. Therefore, it is concluded that the positive charge buildup must be lower for the negative irradiation bias, however, as is evident from figure 1 (curve 3) negative and positive bias radiation responses are in fact symmetrical.

For a better understanding of the cause of the symmetry of the negative and positive bias charge centroid in the non-symmetrical SiO$_2$-Si$_3$N$_4$-SiO$_2$ structure, we performed a simplified analysis to determine the position of the charge centroid in three-layer insulators as a function of the position of the Si$_3$N$_4$ layer and thicknesses of the upper and lower oxide layers. We assumed that there is no net charge generation in the Si$_3$N$_4$ layer. For simplification, we considered the net positive charge trapping at the interfaces only. It was assumed that the net positive charge generated in the top and bottom oxide layers was trapped at the bottom Si$_3$N$_4$-SiO$_2$ and top SiO$_2$-Si interfaces under positive substrate bias and at the top Si$_3$N$_4$-SiO$_2$ and bottom SiO$_2$-Si interfaces under negative substrate bias. The net positive charge trapped at the negative surfaces of the top and bottom oxide layers was presumed to be proportional to their respective thickness. Under these simplifying assumptions one can obtain the following expressions for the net charge centroid for positive irradiation biases:

$$\frac{X_0^{ir}}{d} = 1 - \frac{d_{ox}^b \cdot (d_{ox}^t + d_n)}{(d_{ox}^b + d_{ox}^t) \cdot d} \tag{3a}$$

and for a negative irradiation biases:

$$\frac{X_0^{ir}}{d} = \frac{d_{ox}^t \cdot (d_{ox}^b + d_n)}{(d_{ox}^b + d_{ox}^t) \cdot d} \qquad (3b)$$

where d_{ox}^t and d_{ox}^b are the thickness of the top and bottom oxide layers, respectively, d_n is the thickness of the nitride layer, d is the total insulator thickness normalized to the oxide thickness.

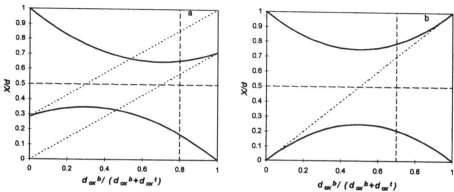

Figure 4. (a) The radiation-induced charge centroid (solid line) as a function of the respective position of the nitride layer (dotted line) and (b) the limiting case of one heavy trapping sheet (dotted line). Top and bottom curves correspond to positive and negative voltage during irradiation, respectively. The vertical line (dashed line) indicates our case.

Fig.4a shows the position of the radiation-induced charge centroid (solid line) calculated from (3a) and (3b) as a function of relations between lower oxide thickness and total oxide thickness, or which is the same, of the respective position of the nitride layer shown by the dotted line. Fig. 4b shows the limiting case of an infinately thin nitride layer ($d_n \rightarrow 0$). It has been found that for all positions of the Si_3N_4 layer, except at the centre of the insulator, the positive and negative bias charge centroid should be non-symmetrical (Fig.4a). This means that in the non-symmetrical three-layer structure the radiation response is also expected to be non-symmetrical. However, analysis shows that the position of the centroid becomes less asymmetrical as the thickness of the nitride layer is reduced. In the limiting case of an infinately thin nitride layer (the case of one heavily trapping sheet), the positive and negative charge centroids are found to be symmetric (Fig.4b) for all locations of nitride layer.

Using SIMS analysis it was found that after recrystallization the nitride layer is heavily enriched with oxygen, and the nitride is no longer stoichiometric (Fig.5). The oxygen concentration distribution was found to be non-uniform with a clearly defined minimum, which was centred upon the nitrogen distribution both before (Fig. 5a) and after recrystallization (Fig. 5b). The trapping properties of the oxynitride layers are known to differ substantially from those of the stoichiometric Si_3N_4. One can assume that most of the charge trapping during irradiation, under both positive and negative bias, occurs in the vicinity of the minimum oxygen concentration, where the composition approximates to Si_3N_4. This, to a first approximation, can explain the symmetricity of the charge centroid for positive and negative irradiation biases.

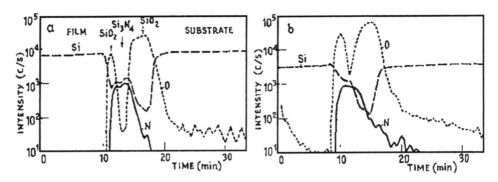

Figure 5. SIMS profiles from SOI structures with a SiO_2-Si_3N_4-SiO_2 buried dielectric (a) before and (b) after recrystallization.

The reduction in the positive charge buildup in the SiO_2-Si_3N_4-SiO_2 layer compared to the single SiO_2 layers, which was observed at high doses ($D > 5 \times 10^5$ rad(Si)), cannot be explained by electron trapping in the Si_3N_4 layer, as the trapped charge (ΔQ_d) had similar values under negative and positive irradiation biases (see Fig. 1). This decrease in trapped charge is essentially larger than can be expected from the decrease in the total oxide thickness.

The high-dose results obtained at relatively low electric fields ($\sim 10^5$ V/cm) can be explained in terms of space charge effects. It is known that in thick buried oxides at high doses (above a few tens of krad) and moderate electric fields the radiation response is controlled both by space-charge effects and low-field recombination processes [7].

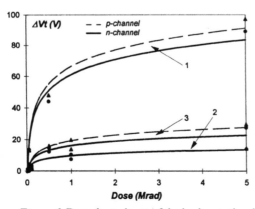

Figure 6. Dose dependence of the back gate threshold voltage shift of SOI MOSFETs with different buried insulators caused by irradiation with a positive substrate bias ($V_{sub,ir}=5V$).

Figure 6 shows the dose dependence of the back-gate threshold voltage shift in p- and n-channel MOSFETs with either SiO_2 or SiO_2-Si_3N_4-SiO_2 insulators. From figure 6 it is evident that at doses, above 5×10^5 rad, for both insulators the threshold voltage shift considerably exceeds the externally applied voltage. On the other hand, space-charge effects are known to become strong even when the threshold voltage shifts are comparable with the applied voltage [7]. The highly non-linear dose dependence shown in figure 6 gives strong evidence for charge buildup limited by space-charge effects. This means that the internal electric field due to trapped charge is compensating the applied electric field. When the net electric field in the insulator approaches zero, there is no longer separation of the generated electrons and holes and under these conditions recombination processes dominate and the charge buildup tends to saturate.

Assuming that all radiation-induced charge is located at the centroid position, the electric field in the charge collection region for the positive and negative irradiation biases it can easily be shown that:

$$E_i = E_0 - \frac{q}{\varepsilon} \cdot \frac{Q_d \cdot \left(d - X_0^{ir}\right)}{d} \tag{4a}$$

$$|E_i| = |E_0| - \frac{q}{\varepsilon} \cdot \frac{Q_d \cdot X_0^{ir}}{d} \tag{4b}$$

where E_0 is the applied electric field, Q_d is the radiation induced charge density and X_0^{ir} is the charge centroid, d - buried insulator thickness.

Using experimentally established values for the charge density and charge centroid in irradiated materials (see Fig.1 and 2: Q_d=5.3x10^{12} cm^{-2} and X_0/d=0.9 for the ZMR buried SiO_2 and Q_d=1.75x10^{12} cm^{-2} and X_0/d=0.7 for the SiO_2-Si_3N_4-SiO_2 layer) one obtains electric field values of 4.2x10^3 and 6.5x10^3 V/cm for single layer and three-layer buried insulators, respectively. Therefore, for both insulators, the experimental radiation–induced charge densities are close to the critical values corresponding to a zero net electric field in the charge collection region, which can be determined for a positive irradiation bias as being equal to:

$$Q_{d\max} = \frac{\varepsilon \cdot E_0}{q} \cdot \frac{1}{1 - \left(\dfrac{X_0^{ir}}{d}\right)} \tag{5}$$

From equation (5) it follows that the variation of the charge centroid from 0.9d to 0.7d results in a decrease of the charge value by a factor of three, which is in good agreement with experimental data.

5. Conclusions

The radiation (^{60}Co) tolerance of SOI structures with buried SiO_2-Si_3N_4-SiO_2 layers fabricated using the ZMR-technique has been investigated. It is shown that introduction of a nitride layer in the buried oxide shifts the radiation-induced charge centroid from the Si-SiO_2 interfaces deep into the insulator and essentially decreases the net radiation-induced positive charge density.

The radiation-induced charge buildup observed at low doses (D≤5x10^{4}rad (Si)) in the SiO_2-Si_3N_4-SiO_2 layers is explained by strong electron trapping in the Si_3N_4 layer which compensates hole trapping in the oxide.

The reduction in the charge buildup observed at high doses in SiO_2-Si_3N_4-SiO_2 layers, which has the same value for negative and positive irradiation biases, is explained in terms of space charge effects. It is shown that the introduction of a heavy trapping localized region into the insulator volume can alter the high dose radiation response of the material by modifying the charge saturation conditions. This provides an explanation for the observed different high dose radiation responses of SiO_2 and SiO_2-Si_3N_4-SiO_2 dielectrics without additional assumptions as to the trapping properties of the oxide layers.

Acknowledgments

The authors would like to express special thanks to Prof. E.I.Givargizov and Dr. A.B.Limanov from the Institute of Crystallography RAS, Moscow (Russia) for help in realizing the ZMR process and to Dr. D. Ballutaud for carrying out the SIMS analysis.

The work has been supported by the INTAS fund (INTAS project No INTAS–93–2075–EXT), NATO Linkage Grant (HTECH.LG 951189) and CNRS collaboration project (No 3061).

References

1. Colinge, J.-P. (1991) *Silicon-on-Insulator Technology: materials to VLSI*, Kluwer Academic Publishers, Dordrecht.
2. Auberton–Herve, A.J. (1990) SIMOX – SOI technologies for high–speed and radiation–hard technologies: status and trends in VLSI and ULSI application, in *Silicon–on–Insulator Technology and Devices*, Montreal, pp.437 – 454.
3. Tsaur, B.Y., Sterino, V.J., Choy, H.K., Chen, C.K., Montanain, R.W., Shot, J.T., Shedd, W.M., La Pierre, D.C. and Blundchard, R. (1986) Radiation hardened JFET devices and CMOS circuits fabricated in SOI film, *IEEE Trans. Nucl. Sci.* NS–33, 1372–1380.
4. Limanov, A.B. and Givargizov, E.I. (1983) Control of structure in zone-melted silicon film on amorphous substrates, *Mater. Lett.* 2, 93–95.
5. Winokur, P.S., Schwank, J.R., McWhorter, P.J., Dressendorfer, P.V. and Turpin, D.S. (1984) Correlating the radiation response of MOS capacitors and transistors, *IEEE Trans. Nucl. Sci.* NS-31, 1453-1458.
6. Ma, T.P. and Dressendorfer, P.V. eds. (1989) *Ionizing Radiation Effects in MOS Devices and Circuits*, Wiley.
7. Saks, N.S. (1978) Response of NMOS capacitors to ionizing radiation at 80K, *IEEE Trans. Nucl. Sci.* 25, 1226-1232.
8. Boesch, H.E., Jr., Taylor, T.L. and Brown, G.A. (1991) Charge buildup at high dose and low fields in SIMOX buried oxides, *IEEE Trans. Nucl. Sci.* NS–38, 1234-1240.

RECOMBINATION CURRENT IN FULLY-DEPLETED SOI DIODES: COMPACT MODEL AND LIFETIME EXTRACTION

T. ERNST, [1,4] A. VANDOOREN, [2] S. CRISTOLOVEANU, [1]
T. E. RUDENKO [3], and J. P. COLINGE [2].

(1) LPCS, ENSERG, BP 257, 38016 Grenoble Cedex, France.
(2) Dept. of Electrical Eng., Univ. of California, Davis, CA95616, USA.
(3) Inst. of Semic. Physics, NASU, 25650, Kiev-28, Prospect Nauki, 45, Ukraine.
(4) STMicroelectronics, 38920 Crolles, France.

Abstract

Gate-controlled carrier recombination is investigated in fully-depleted SOI devices. It is shown how volume and surface recombination in fully depleted gated diodes can be measured and modeled. The compact model is appropriate for including the carrier recombination in SOI/MOS circuit simulations.

1. Introduction

Recombination lifetime is a key parameter for SOI devices [1,2,3] but until recently there has been no direct method for extracting this parameter from fully depleted SOI MOSFETs. The approach proposed in this paper is based on the large peak of forward current observed in P^+PN^+ dual-gate diodes when the front gate voltage is varied [4]. This current peak, identified as a gate-controlled recombination current, is simulated and analytically modeled.

213

P.L.F. Hemment et al. (eds.),
Perspectives, Science and Technologies for Novel Silicon on Insulator Devices, 213–216.
© 2000 *Kluwer Academic Publishers. Printed in the Netherlands.*

2. Experiment

The experimental configuration consists in slightly forward biasing a P^+PN^+ SOI diode. A current peak is observed at various V_D values when V_{G1} is scanned from accumulation to inversion. The low $I_D(V_D)$ slope value in the peak region ($\Delta V_D/\Delta I_D \approx 110$mV/decade at $V_{G1}=V_{G1peak} \approx 0.5$V and $V_{G2}=V_{G2peak} \approx 5$V) unambiguously demonstrates that the current peak is due to excess carrier recombination. When the film is accumulated or inverted, the current drops by one order of magnitude and the diffusion current becomes dominant, with a slope of 80 mV/decade (Fig. 1).

Current shoulders are observed when the front gate is in accumulation ($V_{G1}<-0.5$V) or in inversion ($V_{G1}>1$V). According to $I_D(V_D)$ measurements, these currents are still due to carrier recombination. When the front interface potential is fixed (accumulation or inversion), the potential in the film is no longer controlled by V_{G1} and carrier recombination becomes governed by V_{G2} (Fig. 2).

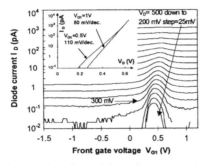 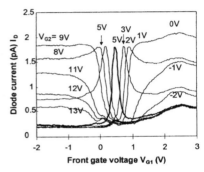

Figure 1. Experimental forward current versus gate voltage for various V_D in a P^+PN^+ diode ($V_{G2}=5$V, W=L=3μm, SIMOX). The insert shows the experimental forward $I_D(V_D)$ characteristics.

Figure 2. Experimental forward current versus front gate voltage for various back gate voltages ($V_D=0.3$V, W=L=3μm).

3. Simulations

The current peaks are reproduced by 2D ATLAS simulation of the gated diode. Firstly, only volume recombination is considered. For low doping or thin film, the lateral and vertical profiles of potential and recombination rate are quasi-flat (Fig. 3). For higher doping or thickness, the potential parabolicity in the vertical direction has to be taken into account.

The amplitude of the current shoulders is much lower that the current peak (as opposed to Fig. 2). This is because the conditions for maximum volume recombination cannot be fulfilled: depletion and quasi-flat 2D potential in the films. In order to reproduce accurately the experimental curves of Fig. 2, we had to include in the simulation a very high recombination rate at the back interface (film-buried oxide). However, in devices such as diodes we have shown that volume recombination may prevail [5]. These two situations, namely volume or surface dominated recombination, will be discussed next.

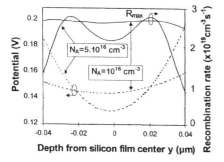

Figure 3. Vertical distributions of electric potential (---) and recombination rate (——) in gated diodes in case of volume recombination ($V_{G1}=V_{G1,MAX}$, $V_{G2}=4V$).

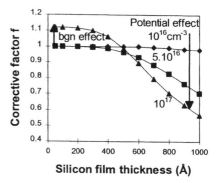

Silicon film thickness (Å)

Figure 4. Calculated correction factor versus silicon film thickness for various base doping.

4. Model and discussion

The model for volume recombination assumes low carrier injection, i.e. nearly flat longitudinal potential. As shown in the simulations of Fig. 6, the peak of maximum amplitude occurs when the whole film is depleted and the potential is symmetrical with respect to the film mid-depth.

The recombination rate R is indeed maximum for $\tau_p\, n=\tau_n\, p=0.5\tau_v\, n_i\, \exp(qV_D/2kT)$ with $n,p\gg n_i$, where $\tau_v=2(\tau_n\tau_p)^{0.5}$ is the effective carriers lifetime. The recombination rate for single-level traps with mid gap energy is integrated over the whole film volume to find the maximum recombination current:

$$R=\frac{pn-n_i^2}{\tau_p(n+n_i)+\tau_n(p+p_i)} \quad \Rightarrow \quad I_{R,max}=t_{si}WL_{eff}f\cdot q\frac{n_i}{\tau_v}\exp\frac{qV_D}{2kT} \tag{1}$$

where f is a calculated corrective coefficient taking into account the potential parabolicity and band gap narrowing (bgn) that depends only on film doping and thickness (Fig. 4). For most practical cases, f≈1 which implies that recombination occurs with constant rate in the device body. This is the basis of the "recombination box" model which offers the possibility to easily extract the recombination lifetime from the current peak. The good fit between modeled and simulated current peaks fully validates our analytical model. The peak width difference between model and experiment can be explained by potential inhomogenities and/or interface states. These parameters can be included in a compact model (Fig 5):

$$I(V_{G1})=\frac{qW_{eff}L_{eff}t_{si}}{\tau_v}f\cdot\frac{n_i\exp(\frac{qV_D}{2kT})}{\cosh(q\frac{\alpha(V_{G1}-V_0)}{kT})} \tag{2}$$

where V_0 is the peak position and α is a parameter taking into account potential inhomogenities and interface states.

216

The compact model is also valid for surface recombination by replacing $f\,t_{si}/\tau_v$ by an effective surface recombination $s=0.5(s_n s_p)^{0.5}$ where s_n and s_p are minority carrier recombination velocities. In this case, α is a coupling factor equal to $d\psi_s/dV_{G1}$ and can be calculated from the Lim and Fossum model [6].

The extracted value of the surface velocity from the experimental peak of Fig. 2 is $s=30\text{cm/s}$.

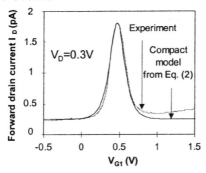

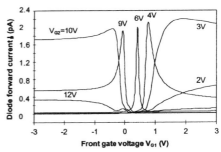

Figure 5. Measured and modeled forward current peak (s=30cm/s, α=0.3, V_0=0.3V, t_{si}=81nm, L_{eff}=2.5μm, W=3μm, N_A=10^{17} cm^{-3}, V_{G2}= 5V).

Figure 6. Simulated forward current peak with back interface recombination (s=30cm/s, V_0=0.3V, t_{si}=81nm, L_{eff}=2.5μm, W=3μm, N_A=10^{17} cm^{-3}, V_{G2}= 5V).

5. Conclusion

The simple recombination-box model offers a universal basis for the rapid extraction of the carrier recombination rate. The model also serves for addressing the junction leakage current in fully depleted MOS-SOI transistors, where a gated diode naturally exists between the N^+ drain, the P^- body and the P^+ body contacts.

6. References

1. Colinge, J.P. (1997) *Silicon-On-Insulator Technology: Materials to VLSI*, Kluwer Academic Publishers, Boston.
2. Cristoloveanu, S. and Li, S.S. (1995) *Electrical Characterization of Silicon-On-Insulator Materials and Devices*, Kluwer Academic Publishers, Boston.
3. Munteanu, D. Weiser, D.A. Cristoloveanu, S. Faynot, O. J. Pelloie, J.L and Fossum, J.G. (1998) Generation/recombination transient effects in SOI transistors: systematic experiment and simulations, *IEEE Trans. Electron. Devices* 45, pp. 1678-1683.
4. Rudenko, T.E. Rudenko, A.N. and Lysenko, V.S. (1995) Electrical properties of ZMR SOI structures: characterization techniques and experimental results, in J.P. Colinge, V.S. Lysenko and A.N Nazarov (eds.), *Physical and Technical Problems of SOI Structures an Devices*, Kluwer Academic Publishers, Dordrecht, pp. 169-179.
5. Ernst, T. Cristoloveanu, S. Vandooren, A. Colinge, J.P. Flandre, D. (1998), ESSDERC Proc., pp. 272-275.
6. Lim, H.-K. Fossum, J.G. (1983) Threshold voltage of thin-film silicon-on-insulator (SOI) MOSFET's, IEEE Transactions on Electron Devices 30, pp. 1244-1251.

INVESTIGATION OF THE STRUCTURAL AND CHEMICAL PROPERTIES OF SOI MATERIALS BY ELLIPSOMETRY

L.A.ZABASHTA, O.I.ZABASHTA, V.E.STORIZHKO,
E.G.BORTCHAGOVSKY[*)], B.N.ROMANYUK[*)], V.P.MELNIK[*)]
[*)]Institute of Applied Physics, Petropavlovskaja St. 58, 244030 Sumy,
[*)]Institute of Semiconductor Physics, NASU, Nauka Av. 45, 252650 Kiev

Abstract

The multiangle ellipsometry is employed to investigate the composition and geometrical structure of SOI materials synthesized by a combined implantation of oxygen and carbon ions. The ellipsometry data are interpreted within the scope of a multilayer model based on both the numerical simulations and the Auger electron spectroscopy data. The optical constants and layer thicknesses in the model are determined by solving the ellipsometry inverse problem with the numerical algorithm developed earlier using Tikhonov's regularization technique. In the framework of the effective medium theory the space factor for the SiO_2 phase filling the effective dielectric layer is estimated for samples prepared differently.

1. Introduction

It is necessary to study the mechanisms underlying the synthesis by high dose ion implantation of buried dielectric and semiconducting phases in semiconductor matrices since these technologies are promising for the production of substrates and novel materials for fast radiation-resistant opto- and microelectronic devices. In order to produce such structures a number of fundamental and practical problems have to be solved to obtained the required implanted impurity profiles and develop nondestructive diagnostics techniques [1,2].

In this paper we report the use of non-destructive ellipsometry analysis techniques to determine both the composition and geometrical structure of silicon-on-insulator (SOI) samples and interpret of the experimental data by means of our methodology developed earlier for the solution of the corresponding inverse problem in order to construct models of the structures under study.

Special attention is paid to thorough investigations of SOI structures synthesized by high-dose ion implantation of both O^+ and C^+ into single crystal silicon with subsequent annealing which permits thin (150 nm) buried SiO_2 dielectric layers of high quality to be formed [1]. Studies of the formation and physical properties of the SOI structures produced [2,3] have been performed earlier using several experimental techniques. The aim of this paper is to show that null or fixed wavelength ellipsometry

P.L.F. Hemment et al. (eds.),
Perspectives, Science and Technologies for Novel Silicon on Insulator Devices, 217–223.
© 2000 *Kluwer Academic Publishers. Printed in the Netherlands.*

can be used in a similar way to automatic scanning ellipsometry for the analysis of multilayered (SOI) structures with more than one interface.

2. Experimental details

Wafers were cut from the central part of a <100> Si crystal with 10 Ω·cm resistivity grown by the Chokhralsky technique. After thermal oxidation (1 hour) in dry O_2 at 850° C until an oxide layer 40 nm thick was formed, the wafers were subjected to two-stage ion implantation in the commercial apparatus Vesuvi-1 with a modified ion source and electrostatic scanning. Three sample series were prepared (Table 1). At the first stage all the samples were implanted with O^+ to the dose of $1.1*10^{17}$ cm^{-2} at 150 keV, the ion current density being 3 mA/cm^2. During the implantation the samples were heated to 100°C. Some of the wafers were additionally implanted (without substrate heating) with C^+ ions to the dose of $2.1*10^{16}$ cm^{-2} at 120 keV (BT and BTAT series) and at 180 keV (CT and CTAT series). The post - implantation annealing was performed in an Ar + 1% O_2 atmosphere for 10 hours as the temperature was increased steadily from 1150 to 1250°C. At the second stage another implantation of O^+ ions to the dose of $1.2•10^{17}$ cm^{-2} was carried out into the substrate heated to 550°C. The post-implantation annealing was performed in the same atmosphere for 6 hours at constant temperature of 1250°C.

The dielectric layer composition and impurity profiles were obtained using Ar^+ sputter depth profiling with an Auger spectrometer.

The ellipsometry angles were measured by a LEF-3M ellipsometer with a He-Ne laser (λ=632.8 nm) in the range of incidence angles φ=45-85° using a two-zone technique. The ellipsometry data were processed by means of the ELLA program package developed earlier [4]. The measurements were taking after removing the oxide in a chemical etchant (HF) and subsequent careful washing with distilled water.

3. Results

As is well known [4] that the interpretation of the ellipsometry data is only possible in the framework of certain theoretical models. The choice of model is based on the a priori information on the structure and properties of the samples or on the data provided by other analytical techniques. In this work the theoretical model was based upon both numerical simulations and impurity depth profiles obtained by Auger electron spectroscopy. Monte-Carlo calculations of the implanted impurity profiles including mixing effects and recoil atoms (TRIM-97) show that for the BTAT samples the O^+ (150 keV) and C^+ (120 keV) ions distribution peaks are close (400 nm), while for CTAT samples C^+ (180 keV) ions are shifted deeper (550 nm) into the sample (fig.1a). Sputter depth profiling using Auger electron spectroscopy (AES) reveals considerable redistribution of the implanted impurities under annealing (detailed studies of the formation of this SOI structures using AES have been performed earlier in [5]). The oxygen distribution peak for ATAT and BTAT samples is shifted towards the surface (300 nm), whereas for CTAT samples two peaks are seen in the distribution of the implanted oxygen (near the vacancy and oxygen distribution peak) [5]. The highest oxygen accumulation is observed for BTAT samples. The second stage oxygen

TABLE 1. Implantation and anneal details

Sample	1st implantation			Anneal		2nd implantation			Anneal	
	Ion	Energy (keV)	D*10^17 (cm^-2)	Temperature (°C)	Time (hr)	Ion	Energy (keV)	D*10^17 (cm^-2)	Temperature (°C)	Time (hr)
AT	O^+	150	1.1	1150 - 1250	10	-	-	-	-	-
ATAT	O^+	150	1.1	1150 - 1250	10	O^+	150	1.2	1250	6
BT	O^+ + C+	150 / 120	1.1 / 0.21	1150 - 1250	10	-	-	-	-	-
BTAT	O^+ + C+	150 / 120	1.1 / 0.21	1150 - 1250	10	O^+	150	1.2	1250	6
CT	O^+ + C+	150 / 180	1.1 / 0.21	1150 - 1250	10	-	-	-	-	-
CTAT	O^+ + C+	150 / 180	1.1 / 0.21	1150 - 1250	10	O^+	150	1.2	1250	6

220

implantation and annealing at 1250^0C results in the formation of an ultrathin SiO_2 buried layer (67% at. of oxygen). The layer thickness is about 15 nm (fig. 1b, curve 3).

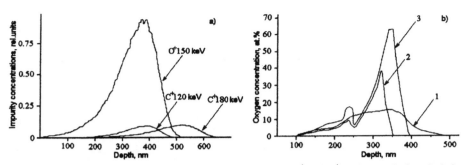

Figure 1. Implanted impurity depth profiles in SOI structures for C^+ and O^+: a) Monte Carlo calculations (TRIM97) and b) Auger spectroscopy data for different stages of the SOI structure formation: 1 – O^+ and C^+ implantation; 2 – annealing, 3 – repeated oxygen implantation and annealing (wafers are identified in Table 1).

The principal ellipsometry parameters obtained are represented in fig.2. The measurements were taken after the passivating film had been removed. As can be seen in the figure, the ellipsometry parameters of the doped structures differ drastically from those of monocrystalline silicon, being dependent on the implanted ion species and the implantation mode ($|\delta\Delta|_{max}=135^0$ at $\varphi = 70^0$ and $|\delta\Psi|_{max} = 8^0$ at $\varphi = 75^0$ for BTAT series samples, $\delta\Delta = \Delta_{Si-SiO_2} - \Delta_{Si(implanted)}$, $\delta\Psi = \Psi_{Si-SiO_2} - \Psi_{Si(implanted)}$).

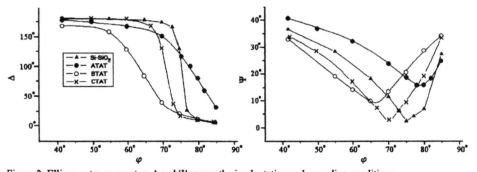

Figure 2. Ellipsometry parameters Δ and Ψ versus the implantation and annealing conditions

The optical pseudoconstants calculated for the implanted Si layer also give qualitative evidence for the formation of systems with differing structure and optical properties (Table 2).

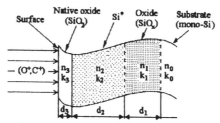

Figure 3. Schematic representation of the samples, deduced from AES profiles.

Taking into account the above considerations, a multilayer optical model was constructed for the samples under study. The model is shown in Figure 3 included: a single-crystal silicon substrate (n_0, k_0), an effective dielectric layer (n_1, k_1, d_1), a partially damaged amorphous Si layer (n_2, k_2, d_2) and a native oxide layer (n_3, k_3, d_3). The thickness of the native SiO_2 oxide in this model was determined independently in a priory investigation of unirradiated single crystal Si plates (because of high correlation coefficients n_{ox} cannot be determined from the ellipsometry data; $d_{ox}=2.5$ nm corresponds to the range $n_{ox} = 1.452 - 1.512$). For the multilayer optic model constructed the unknown optical constants and layer thicknesses were obtained by step-by-step solution of the ellipsometry inverse problem [4]. At each step two or three model parameters were determined in accordance with the degree of their intercorrelation (strongest correlative relation is between the pairs: d_2-d_1, d_2-n_1, n_2-d_1). At the last stage the optical constants of the substrate were verified. The calculated values are given in Table 2.

Table 2. Calculated values of the optical constants

Series	n_{eff}	k_{eff}	n_0	k_0	n_1	k_1	n_2	k_2
AT	3.93	0.30	3.865	0.02	3.24	0.07	3.865	0.02
ATAT	5.40	0.27	3.865	0.02	2.48	0.02	3.868	0.02
BT	3.90	1.04	3.869	0.03	3.99	1.69	3.868	0.02
BTAT	1.94	0.31	3.863	0.02	2.31	0.217	4.269	0.02
CT	4.60	0.29	3.871	0.02	3.26	0.98	3.865	0.02
CTAT	2.91	0.13	3.874	0.04	2.78	0.01	4.170	0.02

(Thicknesses of the 1st and the 2nd layer are fixed for all the models, being $d_1 = 100$ nm and $d_2 = 300$ nm, respectively. The third layer is a natural oxide film with parameters $n_3 = 1.475$, $k_3 = 0$, $d_3 = 2.5$ nm).

4. Discussion

As has been shown previously that carbon atoms additionally introduced by ion implantation into the region of maximum oxygen atom concentration, contribute to the formation of the SiO_2 phase [5]. It is not known, however, how the carbon atoms modify the optical properties of the dielectric layer formed. The ellipsometry data indicate that the additional implantation of carbon atoms and subsequent annealing (BT and CT series samples) result in an increased absorption coefficient in the region where a buried dielectric layer is produced (Table 2). After a second O^+ ion implantation and annealing the absorption coefficient of the dielectric layer remains higher as compared to that of the ATAT samples. In this case optical constants of the recrystallized silicon layer are consistant with the literature data for amorphous silicon [6]. The reason is that in the upper silicon layer a dielectric phase is partially formed. The Auger spectroscopy data (fig.1) shows that the first anneal step also introduces in the seed layer a small additional peak with about 5% at. oxygen, located at a depth of 300 nm (fig.1, curve 2). This new peak further grows (up to about 12% at.) during the second oxygen implantation and the anneal cycle (fig.1, curve 3).

However, in the current study the analysis of the ellipsometry data has been changed in order to obtain composition rather than optical constants. We used effective medium approximation (EMA)[7] to determine the space factor for the SiO_2 phase filling the effective dielectric layer (1) in Fig 3. The latter is a mixture of the SiO_2 phase and amorphous silicon with SiCO inclusions. The dielectric constants of the effective dielectric layer were calculated using the optical constants of the similar layer in the BT series samples (Table 1). The space factor for the SiO_2 phase calculated by this model is 68% at. for BTAT samples and 42% at. for ATAT samples.

The dielectric layer thickness (d_1=94nm) determined from the ellipsometry data is smaller in this case than that for the ATAT samples (d_1=107.1nm) indicating a narrower oxygen depth profile and corresponding to the Auger spectroscopy data revealing variations in the oxygen depth profiles with the implantation parameters [5].

In the future, we intend to combine the algorithms and the corresponding models in a single computational scheme to be used for the consistent reconstruction of structural, chemical and physical properties of subsurface layers and interfaces in multi-layered systems determined from the data of independent experiments. This interdisciplinary approach could improve the reliability of the interpretation and provide new information which was previously inaccessible.

Depending on the problem under study, we also plan to use a variety of other techniques, such as secondary ion mass spectrometry (SIMS), scanning electron microscopy (SEM) and Auger electron spectroscopy (AES) to determine the electrophysical properties of the samples.

5. Conclusions

In this paper it is shown that the multiangle ellipsometry can be applied successfully to the nondestructive analysis and estimation of the composition and geometric structure of such complicated structures as SOI. As was found in the investigations, carbon atoms introduced into the region where the barrier dielectric layer is formed, modify the optical properties of the layer (the absorption coefficient is increased). Using the effective medium theory the space factor for the SiO_2 phase filling the dielectric layer was calculated for samples differing in the preparation procedure.

References

1. Litovchenko, V.G., Romanyuk, B.N., Popov, V.G., Lisovski, I.P., Shkraban, B.N., Melnik, V.P. Lozinsky, V.B., Khokhotva, G.I., Novosiadly, S.P., (1991) Oxygen precipitation mechanisms in silicon-insulator structures, *Ukr.Phys. Journal* **36**, No9, 1424-1429.

2. Romanyuk, B.N., Popov, V.G., Prokofjev A.Yu., Lisovskii, I.P., Lozinskii, V.B., Romanyuk, B.N., Melnik, V.P., Bogdanov E.I., (1995) Process of formation of buried insulating layer in Si at implantation of N^+ and O^+ ions, *Ukr.Phys. Journal* **37**, No 3, 389-393.

3. Romanyuk, B.N., Popov, V.G., Marchenko, R.I., Kljui, N.I., Melnik, V.P., Prokofjev A.Yu., Evtukh A.A., Goltvanskii, Yu.V., Andrijevskii, V.V., Moscal, D.N., Frolov, S.I., (1995) Study of silicon - insulator structures obtained by implantation of nitrogen ions in Si and subsequent thermal oxidation, *Ukr.Phys. Journal* **38**, No 4, 589-595.

4. Dmitruk, N.L., Zabashta, L.A., Zabashta, O.I., Storizhko, V.E. (1997) A stable method for solving the inverse problem of many-angle ellipsometry, *Surface Investigation*, **12**, 1411-1418.

5. Litovchenko, V.G., Efremov, A.A., Romanyuk, B.N., Melnik, V.P. (1998) Processes in ultrathin buried oxide synthesis stimulated by low dose ion implantation, *J.Electrochem. Soc.* **145**, No.8, 2964-2969.

6. Lohner, T., Kotai, E., Khanh ,N.Q, Toth, Z., Fried, M., Vedam, K., Nguyen, N.V., Hanekamp, L.J., van Silfhout, A. (1994) Ion-implantation induced anomalous surface amorphization in silicon, *Nucl. Instr. and Meth.* **B 85**, 335-339.

7. Bergman, D.I. (1980) Exactly solvable microscopic geometries and rigorous bounds for the complex dielectric constant of a two-component composite material, *Phys. Rev. Lett.* **44**, 1285-1287.

EXPERIMENTAL INVESTIGATION AND MODELING OF COPLANAR TRANSMISSION LINES ON SOI TECHNOLOGIES FOR RF APPLICATIONS

J. LESCOT[(1)], O. ROZEAU[(1,2,3)], J. JOMAAH[(2)],
J. BOUSSEY[(2)], F. NDAGIJIMANA[(1)]

(1) Laboratoire d'Electromagnétisme, MicroOndes et Optoelectronique (LEMO-UMR 5530)
(2) Laboratoire de Physique des Composants à Semiconducteurs (UMR 5531)
LEMO-LPCS, BP 257, 38016 Grenoble Cedex1- France
(3) ST Microelectronics, 38920 Crolles, France
E-mail: lescot@enserg.fr

Abstract

This paper provides a modeling technique for coplanar transmission lines implemented on SOI technologies with substrates of various conductivity values especially dedicated to RF design. The model is verified by S-parameter measurements from 45 MHz up to 10 GHz, which clearly proves that such SOI structures can be accurately modelled using a standard analytical approach.

1. Introduction

Compared to GaAs-based industry, key advantages of Si-based monolithic microwave integrated circuits (MMIC's) are low wafer costs and compatibility with low-cost and mature processing technology. However, the major drawback which affects Si high-frequency performance is the poor insulation properties of the common Si substrates which lead to excessive losses of the passive components. Such substrate losses can be significantly reduced by using high-resistivity Si substrates allowed by SOI technology. The motivation of this paper is to provide a modeling technique for coplanar transmission lines on different SOI technologies using substrates of various conductivity values. A comparison with standard Bulk structures is shown.

2. Experimental Setup

2.1. SAMPLES

Cross-section views of the test structures are given in Fig. 1-a and Fig. 1-b for SOI and Bulk materials, respectively. Coplanar transmission lines with a longitudinal length of 20 mm were designed on five substrates with varying conductivities values given in

225

P.L.F. Hemment et al. (eds.),
Perspectives, Science and Technologies for Novel Silicon on Insulator Devices, 225–231.
© 2000 *Kluwer Academic Publishers. Printed in the Netherlands.*

226

Table 1. The results presented in this paper are obtained from a metal thickness (t) of 1 μm, a separation distance (S) of 49 μm and a width (W) of 86 μm. Notice that all structures use the same line geometry and metal properties for a rigorous comparison between Bulk and SOI technologies.

TABLE 1. Technological Parameters

Substrate Type	ρ_{Si} (Ωcm)	e_{ox1}	e_{ox2} (μm)	e_{Si}
Bulk Silicon	10	.5		
Microx[+]	1000			
Unibond[*]	14	0.5	0.4	0.2
SIMOX*	22			
Quartz[*]	≈∞			

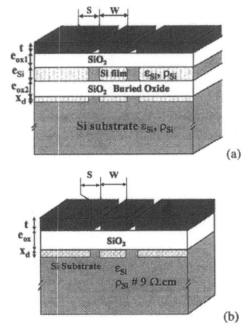

Figure 1. Cross-section view of (a) the SOI and (b) Bulk coplanar structures.

2.2. EXPERIMENTS

Using a Cascade Microtech Probe Station and an HP8510 Network Analyser, coplanar transmission lines were characterized from 45 MHz to 10 GHz. The measured data were

---

[+] Microwave SIMOX [2]
[*] Standard SOI substrates
[*] Reference substrate.

transmitted to a PC computer for automatic parameter extraction. The solution of the standard form of the Telegrapher's equation leads to the complex propagation constant γ and complex characteristic impedance Z_c of the lines of length l directly derived from S-parameters S_{ij} [3]. The transmission line model per unit length is illustrated in Fig. 2-a. A more accurate model for SOI substrates is proposed in Fig. 2-b, where C_D corresponds to the capacitance of the oxide layer while shunt parameters C_{Sub} and G_{Sub} represent ohmic losses and associated capacitance in Si substrate. The values of L, R, G and C can be derived from the complex characteristic impedance and the complex propagation coefficient by the following:

$$R + jL\omega = \gamma Z_c \; , \; G + jC\omega = \frac{\gamma}{Z_c} \; , \; Z_c = 50\sqrt{\frac{(1+S_{11})(1+S_{22}) - S_{21}S_{12}}{(1-S_{11})(1-S_{22}) - S_{12}S_{21}}} \quad (1)$$

And

$$\gamma = \alpha + j\beta = \frac{1}{l}\ln(A \pm \sqrt{A^2 - 1}) \text{ with } A = \frac{(1-S_{11})(1+S_{22}) + S_{21}S_{12}}{2S_{12}} \quad (2)$$

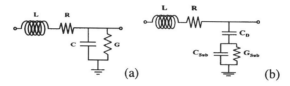

(a)   (b)

Figure 2. Models per unit length for coplanar transmission lines on (a) standard and (b) silicon-based substrates.

3. Quasi-Static Analysis

3.1. SERIES INDUCTANCE L

The inductance per unit length is generally expressed as [6]:

$$L = \frac{1}{4c^2\varepsilon_0 F} \quad (3)$$

where F is a geometrical factor empirically given by

$$k = \frac{W}{W + 2S}$$

$$k' = \sqrt{1 - k^2}$$

$$F = \begin{cases} \dfrac{\ln(\dfrac{2(1+\sqrt{k})}{(1-\sqrt{k})})}{\pi} & \text{for } 0.707 \le k \le 1 \\[3em] \dfrac{\pi}{\ln(\dfrac{2(1+\sqrt{k'})}{(1-\sqrt{k'})})} & \text{for } 0 \le k \le 0.707 \end{cases} \quad (4)$$

3.2. SERIES RESISTANCE R

For an infinite thickness conductor with conductivity σ_M, the sheet resistance is

$$R = \frac{1}{\sigma_M \delta} \text{ Ohms/square where } \delta = \frac{1}{\sqrt{\pi f \mu \sigma_M}} \qquad (5)$$

Here, δ is the skin depth for an infinite conductor and is defined so that the resistance for a rectangular conductor of thickness δ is the same that the infinite conductor in which the current density exponentially decreases. An effective skin depth δ_x for our finite conductor is adopted as follows [3]:

$$j_{Total} = j(o) \int_0^W e^{-\frac{x}{\delta}} dx = j(o)(1 - e^{-\frac{W}{\delta}})\delta = j(o)\delta_x \text{ thus } \delta_x = (1 - e^{-\frac{W}{\delta}})\delta \qquad (6)$$

In the coplanar transmission line case, the skin effect current will be on the conductor sides with field lines extending to the ground planes. An approximation for the addition current is given by

$$j_{Total} = j(o)(\delta_x + \delta_y) \text{ with } \delta_y = \delta_x \frac{W}{t} \qquad (7)$$

This approach can be explained by the fact that W/t is the ratio of the conductor horizontal periphery to that of the vertical one. As a result, the corresponding series resistance value R for the coplanar structure using two ground planes, can be expressed as follows

$$R = \frac{1}{2\sigma_M t(\delta_x + \delta_y)} = \frac{1}{2\sigma_M t(1 - e^{-\frac{w}{\delta}})(1 + \frac{W}{t})\delta} \qquad (8)$$

3.3. CAPACITANCE C_D

In contrast with previous works, the depletion capacitance is taken into account since our structures exhibit comparable SiO_2 and depletion layer thicknesses. By neglecting the interfacial charges, the depletion region thickness x_d is expressed as following [8]:

$$x_d = \sqrt{\frac{2\varepsilon_0 \varepsilon_{Si} |\Phi_{MS}|}{q N_a}} \qquad (9)$$

where N_a is the substrate doping concentration and Φ_{MS} is the work-function difference depending on N_a [8]. Notice that such an approach has been verified by studying the evolution of capacitance C_D with substrate bias. Finally, the dielectric capacitance including fringing effects due to the finite thickness t of metal is given by [7]:

$$C_D = \varepsilon_{SiO_2}\varepsilon_0 \left\{ \frac{\dfrac{S}{h_{eff}} - \dfrac{t}{2h_{eff}} + \dfrac{2\pi}{\ln\left[1 + \dfrac{2h_{eff}}{t}\left(1 + \sqrt{1 + \dfrac{t}{h_{eff}}}\right)\right]}}{} \right\} f_c \tag{10}$$

$$With\ h_{eff} = \begin{cases} \varepsilon_{SiO_2}\left(\dfrac{e_{ox}}{\varepsilon_{SiO_2}} + \dfrac{xd}{\varepsilon_{Si}}\right) & \text{(Bulk substrate)} \\[3mm] \varepsilon_{SiO_2}\left(\dfrac{e_{ox1}}{\varepsilon_{SiO_2}} + \dfrac{e_{Si}}{\varepsilon_{Si}} + \dfrac{e_{ox}}{\varepsilon_{SiO_2}} + \dfrac{xd}{\varepsilon_{Si}}\right) & \text{(SOI substrate)} \end{cases} \tag{11}$$

3.4. CONDUCTANCE C_{Sub} AND CAPACITANCE G_{Sub}

In the model illustrated on Fig. 2-b, G_{Sub} and C_{Sub} are approximated by:

$$G_{Sub} = 2\sigma_{Si}F \text{ and } C_{Sub} = \varepsilon_{Si}\varepsilon_0 2F \tag{12}$$

4. Simulation/Measurement Results

The effective dielectric constant given in Fig. 3 is much larger than the dielectric constant of any layer, which clearly proves that the propagating mode in our structures is a slow-wave one [4]. In the high frequency region, the fundamental mode closely resembles to a dielectric TEM mode, where the substrate acts like a dielectric. As a result, the effective dielectric constant approaches $(\varepsilon_{Si}+1)/2=6.3$.

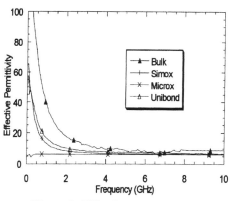

Figure 3. Effective permittivity.

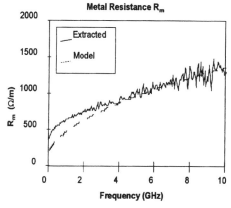

Figure 4. Metal Resistance for Simox Structures.

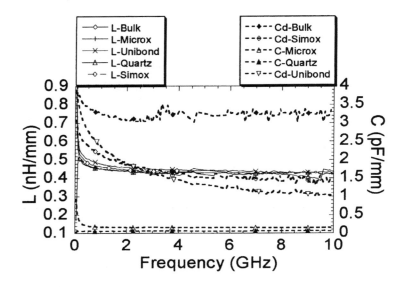

Figure 5. Inductance L and capacitance Cd per unit length.

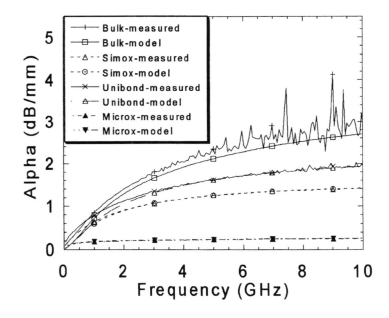

Figure 6. Attenuation constant from model and measurements.

The slow-wave coupling mechanisms can be modeled by the parallel RC network of Fig. 2-b. Using equations (5)-(12), the circuit elements (R, L, C_D, C_{Sub}, and G_{Sub}) have been computed as described in Section 3 and then compared with the measured values.

Figures 4-5 show experimental R, C and L variations versus frequency which agree well with the calculated ones (L=418 pH/mm for all substrates, C_D=1.7 pF/mm, 1.8 pF/mm and 3.1 pF/mm for SIMOX, Unibond and Bulk substrates, respectively).

Due to high resistivity substrate, SOI coplanar transmission lines do achieve better performance than Bulk structures, as shown in Fig. 6 where the measured attenuation constant per unit length is given for the four studied substrates and is successfully compared to values computed using the model.

5. Conclusion

We have first verified that SOI technology is a promising alternative to standard bulk structures for RF applications since it allows the use of substrates with low conductivity. In fact, the measured coplanar transmission lines achieve higher performances than bulk structures using the same transmission line geometry and identical metal. Analytic expressions for interconnect line parameter extraction have been examined. The new simplified equivalent circuit model for coplanar transmission lines on SOI substrates takes into account the depletion capacitance and the skin effect for the frequency dependant metallic losses. Good agreement is observed between experimental data and the proposed model on the full frequency range from 45 MHz to 10 GHz. Our analysis stands for a computationally attractive approach for circuit design and fast optimisation since only a few minutes are required to provide the designer with an accurate transmission line model.

6. References

1. Eggert, D., Huebler, P., Huerrich, A., Kueck, H., Budde, W. and Vorwerk, M. (1997) A SOI-RF-CMOS Technology on High Resistivity SIMOX Substrates for Microwave Applications to 5 GHz, *IEEE Trans. Electronic Devices* **11**, 1981-1989.
2. Agarwal, A. K. et al. (1992) Microx-An all silicon microwave technology, *Proc. IEEE Int. SOI Conf.*, **44**, 144.
3. Eo, Y. and Eisenstadt, W. R. (1993) High-speed VLSI interconnect modeling based on S-parameter measurements, *IEEE Trans. Comp. Hybrids, Manuf. Technol.*, **16**, 555-562.
4. Hasegawa, H., Furukawa, M. and Yanai, H. (1971) Properties of microstrip line on Si-SiO$_2$ system, *IEEE Trans. Microwave Theory Tech.*, **19**, 869-881.
5. Hasegawa, H. and Seki, S. (1984) Analysis of interconnection delay on very high-speed LSI/VLSI chips using an MIS microstrip line model, *IEEE Trans. Microwave Theory Tech.*, **32**, 1721-1727.
6. Gupta, K. C., Garg, R. and Bahl, I. J. (1979) *Microstrip Lines and Slotlines*, Artech House, Dedham.
7. Ko., J. S., Kim, B. K. and Lee, K. (1997) Simple modeling of coplanar waveguide on thick dielectric over lossy substrates, *IEEE Trans. Electronic Devices*, **5**, 856-860.
8. Sze, S.(1981) M.*Physics of Semiconductor Devices*, John Wiley & Sons, 2nd edition.

PERSPECTIVES OF SILICON-ON-INSULATOR TECHNOLOGIES FOR CRYOGENIC ELECTRONICS

C. CLAEYS [1,2] and E. SIMOEN [1]

[1] IMEC, Kapeldreef 75, B-3001 Leuven. Belgium

[2] KU Leuven, ESAT-INSYS, Kard. Mercierlaan 94,B-3001 Leuven, Belgium

1. Introduction

The first experimental data on the use of silicon at low temperatures are nearly 5 decades old and were merely intended to support the physical models under development [1]. About a decade ago, the idea emerged that cryogenic electronics could be considered as the technology of the future. However, economical considerations and difficulties at the system level such as cooling units and packaging, have then reduced the world-wide interest. More recently, due to the advent of nanotechnologies, multiple-level-wiring interconnections, hybrid superconductor-semiconductor systems, the rapid evolution of mini- and micro-coolers, and the appearance of physical limitations for further downscaling, the research activities in the field of cryogenic electronics is showing a revival [2]. An updated review of the different aspects of low temperature electronics is under preparation [3].

The major advantages of operating CMOS at cryogenic temperatures are associated with an increased carrier mobility, steeper subthreshold slope, higher thermal and electrical conductivity and reliability aspects such as electromigration and latch-up suppression [4-5]. When entering the submicron feature area, some of the typical advantages have to be reconsidered in view of the underlying physical limitations [6-7]. However, based on economical reflections as shown in Fig. 1 [2], the cryogenic operation may delay the introduction of a downscaled technology. For standard room temperature operation, there is an exponential cost increase for each new deep submicron generation. However, by cooling down the devices the same speed performance can be obtained for larger gate lengths, shifting the cost-channel length curve to the left for cryogenic operation. From a system cost viewpoint, one also has to take into account the cost of the required cryocooler. The latter determines the critical gatelength for an equal cost performance between cryogenic and room temperature operation. For Fig. 1, the calculations are based on a $L_{crit} = 0.5 \ \mu m$. During the last years, much attention has therefore been given to optimise the cooling systems, in order to come to a cost effective and reliable total system solution.

SOI offers many advantages over bulk CMOS technology, especially from the viewpoints of performance, low power consumption and low voltage operation [8]. Additional niche markets are related to space applications (SOI has an excellent radiation

233

P.L.F. Hemment et al. (eds.),
Perspectives, Science and Technologies for Novel Silicon on Insulator Devices, 233–247.
© 2000 *Kluwer Academic Publishers. Printed in the Netherlands.*

234

hardness) and microsystems. The cryogenic operation of SOI leads to a superposition of the best of both worlds, namely cryogenic operation and SOI. Some years ago, the authors have reviewed in detail the perspectives of SOI technologies for cryogenic applications [9]. Attention was then given to both digital and analog applications. For the latter, considerations of the temperature dependence of the noise performance are of crucial importance [10]. Not only the thermal noise, but also low frequency noise components such as flicker, generation-recombination and Random Telegraph Signal noise have to be taken into account. Recently, Balestra and Ghibaudo have compared the cryogenic behaviour of deep submicron bulk and SOI CMOS devices [11], and they also concluded that SOI technologies have a strong potential.

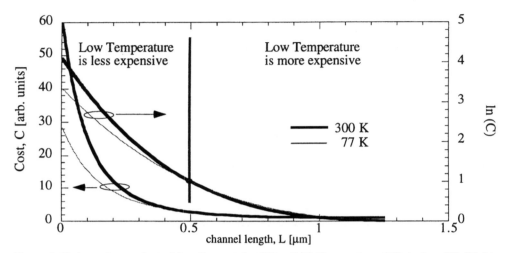

Figure 1. Estimated cost-channel length curve for 300 and 77 K operation of Si devices [2]. Right axis is a log representation of the cost C.

The temperature dependence of some SOI specific physical effects, such as floating body effect, latch and self-heating, will not be reviewed as this has extensively been addressed by the authors in the past [9]. In general, the theoretical predictions compare very well with the experimental observations. Not only technological aspects, but also design optimisation is required to fully benefit from the cryogenic operation. The latter will of course be influenced by the envisaged application. Therefore design aspects such as body contacts, and the use of alternative types of devices, like gate-all-around, accumulation type and twin-gate concepts, have been proposed. Another review paper by the authors discussed the optimisation of the parameter extraction of cryogenic devices [6]. The present paper will mainly focus on the behaviour of short-channel effects at reduced temperatures. As will be pointed out quantum mechanical effects are becoming pronounced for ultimate scaled SOI devices fabricated on ultrathin silicon films. Some future applications entering the field of nanoelectronics will also be covered.

2. Scaling Considerations

Historically DRAM circuits have been the technology drivers forcing the semiconductor industry to follow a rate of growth described by Moore's scaling laws, leading to a shrinkage of the minimum feature size of about 30% every three years and an annual increase of 59% bits/year. This growth has resulted in a reduction of the cost/function by 25-30%/year. Although microprocessor technologies and ASIC products were lagging behind, the required introduction rate of new technologies is now more and more closing the technology gap compared to DRAMs. The scaling trend for the different product families is illustrated in Table 1, taken from the 1997 SIA roadmap [12]. There are strong indications that in the 1999 SIA roadmap update, the 100 nm technology barrier will be pulled forward by a few years.

TABLE 1. Technology generation according to 1997 SIA roadmap [12]

Year	1999	2001	2003	2006	2009	2012
DRAM generation[*1]	180	150	130	100	70	50
MPU Gates[*2]	140	120	100	70	50	35
DRAM production	256M	1G	1G	4G	16G	64G
Logic MPU[1]	6.2M	10M	18M	39M	84M	180M
Logic-ASIC[2]	14M	16M	24M	40M	64M	100M

[*1] Dense lines (DRAM half pitch) (in nm)

[*2] Isolated lines (MPU gates) (in nm)

[1] Logic transistors/cm^2 (packed, including on-chip SRAM)

[2] Usable transistors/cm^2

The scope of this paper is not to review the different scaling effects in MOS devices, but only to address some particular scaling phenomena in SOI technologies which may impact upon their future low temperature applications. Attention is given to quantum mechanical effects impacting the threshold voltage behaviour and transport characteristics, and to series resistance and gatelength extraction. The future applications themselves are discussed in Section 3.

2.1. SCALING IN SOI

It is obvious that SOI is offering the potential for improved device performance, soft error immunity and low power applications. From a processing viewpoint the fact that for a SOI technology a lower number of process steps is needed, has a beneficial impact on the overall cost, although the present day starting material cost may still provide an obstacle for a large-scale breakthrough, especially for the strongly cost driven DRAMs. Recent announcements by leading US and Japanese companies strongly trigger the industrial interest in both

236

partially (PD) and fully depleted (FD) SOI technologies. SOI may postpone the time requirement to meet the 100 nm technology challenge whlist still meeting the system requirements defined by the market.

Scaling the minimum feature size has a direct impact on the short-channel effects (SCE) such as the impact of charge sharing near source/drain carrier extension, drain-induced barrier lowering and punch-through phenomena. For SOI technologies, scaling also requires a reduction of both the silicon film and the buried oxide (BOX) thickness in order to reduce the SCE [13]. This will have a strong influence on the threshold voltage [14], which according to the classical theory is expected to be lowered for reduced gatelengths and film thicknesses. However, as will be discussed below, quantum mechanical effects may have to be taken into account. The latter will be strongly influenced by the operation temperature.

2.2. THRESHOLD VOLTAGE BEHAVIOUR

Classical theory predicts that the threshold voltage V_T decreases monotonically with decreasing film thickness t_s due to the decrease of the charge in the depletion layer, as given by the following formula for buried-channel MOSFETs [15]

$$V_T = V_{FB} + (\frac{qN_At_s}{C_{ox}})(1 + \frac{C_s}{C_s+C_{BOX}}) + \phi_{sth}(1 + \frac{C_sC_{BOX}}{C_{ox}(C_s+C_{BOX})}) + \frac{C_sC_{BOX}V_{sub}}{C_{ox}(C_s+C_{BOX})} \tag{1}$$

with V_{FB} the flat-band voltage, ϕ_{sth} the surface potential at threshold, N_A the doping concentration in the film, C_{ox} the gate oxide capacitance, C_s the film capacitance, C_{BOX} the buried oxide capacitance, and V_{sub} the substrate voltage. However, when t_s decreases below a critical value, quantization effects have to be taken into account resulting in the occurrence of two-dimensional subbands [14]. This is schematically illustrated in Fig. 2a, giving simulated results of the impact of the film thickness on the minimum channel length leading to punch-through. The strong quantization of the SCE occurs for a film thickness below 5 nm, due to an increase of the effective bandgap energy and the decrease of the effective density of states resulting in an increase of the majority carrier extension. The same figure also gives simulated results for surface channel devices, pointing out that in SC-MOSFETs quantization effects are more severe as an increase in the energy bandgap leads to an increase of the depletion layer width with increasing built-in potential.

The quantization effect becomes more pronounced for low temperature operation. A representative parameter to study short-channel effects is the drain-voltage derivative of the threshold voltage, $\gamma = (dV_T/dV_D)$. The classical and quantum mechanical simulated values of this parameter are shown in Fig. 2b and compared with experimental results for 0.1 µm buried channel pMOSFETs on SIMOX with an 8 nm film thickness. No quantization effects are noticed at 300 K. The γ increase for lower temperatures is caused by a decrease of the effective density of states, leading to a decrease of the Debye length. Classically the Debye length would increase with decreasing operating temperatures [16].

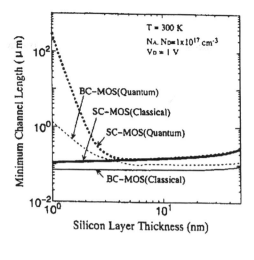

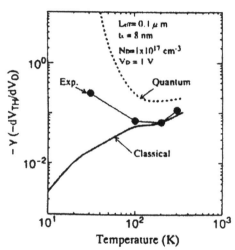

Figure 2. Simulated minimum channel length L_{min} dependence on the silicon film thickness t_s for surface-channel (SC) and buried-channel (BC) MOSFETs/SOI at room temperature, according to the classical and quantum mechanical approach [15] (a), and the temperature dependence of the drain-voltage derivative of the threshold voltage (γ) for 0.1 μm pMOSFETs on SIMOX with t_s= 8 nm [16] (b). The experimental points are the average values for 5 measured devices.

Due to the quantization effects, the threshold voltage no longer decreases for decreasing film thickness, but rather increases for a film thickness below a critical value. The latter depends on technological (geometrical features, doping density) and operational parameters (voltages, temperature).

2.3. TRANSPORT CHARACTERISTICS

In the case of thin film devices, quantization effects may also influence the transport properties in the silicon film thereby modifying the transconductance. The transconductance curve can exhibit oscillations and a step-like behaviour [15,17-18]. Although oscillations can be caused by the scattering processes governing the carrier mobility, it has been postulated that local deviations in the film thickness is the underlying cause. These are related to the morphology roughness of the film-buried oxide interface. It should be mentioned, however, that local fluctuations of the oxide charge at the top BOX layer can not be excluded. The anomalous transconductance behaviour has been observed both at room and cryogenic temperature. To model the mobility variations, one has to take into account that the surface inversion layer can be larger than the silicon film thickness.

To study the impact of quantum mechanical effects on the transport properties experimentally, silicon quantum wire structures have been fabricated [19-21]. The carrier confinement in the two-dimensional electron gas (2DEG) in the silicon film results in a splitting of the conduction band into subbands. The filling of these subbands with electrons shows up as an oscillation in the current transport and associated transconductance. The potential changes $\Delta\Phi$ in the film can be modelled by the equation [19]

$$\Delta\Phi(V_G) = \frac{h^2}{8m^*t_{si}^2} [n^2 - (n-1)^2] \tag{2}$$

with h the Planck's constant, m* the effective electron mass in silicon and n the quantum number for the different subbands. A typical transconductance curve for a 50 nm wide gate-all-around quantum wire, fabricated on a SIMOX substrate is shown in Fig. 3a for operation at 1.5 K. The quantization of the drain current can be clearly observed as local kinks. Figure 3b illustrates the conductance fluctuations in the channel of a 200 nm long and 100 nm wide channel of a SOI MOSFET operating at 300 mK [22]. The spatial potential variations in the channel lead to the formation of islands or quantum dots, the size of which determines the oscillation period of the conductance modulation. The spikes observed in Fig. 3b, for example at $V_G=1.3V$, are caused by RTSs in the drain current with typical time constants from hundreds of microseconds to tens of seconds [23].

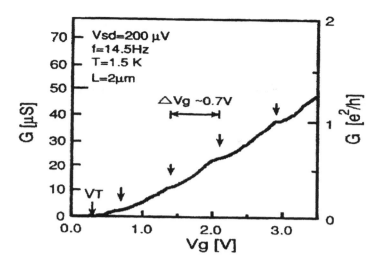

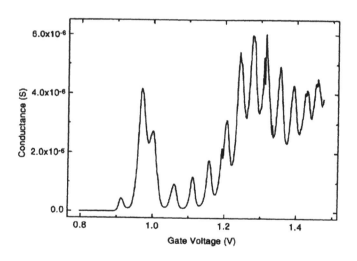

Figure 3. Transconductance as a function of the gate voltage for (a) a 50 nm wide silicon quantum wire gate-all-around structure operating at 1.5 K [21], and (b) a 200 nm long and 100 nm wide SOI MOSFET at 300 mK [22].

2.4. SERIES RESISTANCE AND CHANNEL LENGTH (ΔL) EXTRACTION

Scaling has a strong impact on the parameter extraction techniques, especially on the device resistance and the electrical active gatelength, in devices fabricated in thin silicon films. This is even aggravated for low temperature operation [6]. The total device resistance consists of the series resistance R_S and the channel resistance R_{ch}. Optimised techniques for the determination of the series resistance have recently been proposed and are based on a comparison of the experimental device characteristics with MEDICI simulations, taking into account second-order effects impacting on the carrier mobility [24-26]. It is essential to remark that the resistance is also gate voltage dependent, as illustrated in Fig. 4 for the channel and series resistance of PD MOSFETs operating at 77 K. A simple technique, based on the combination of a long reference device and a shorter L-array transistor is used to determine the series resistance [27].

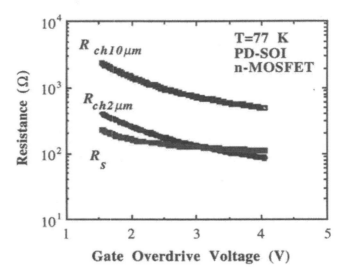

Figure 4. Channel and series resistance versus gate overdrive voltage ($V_G^* = V_G - V_T$) for 10 μm and 2 μm SOI PD n-MOSFETs operating at 77 K.

To derive the effective gatelength or ΔL, one has to take into account the series resistance [28]. A crucial question is whether or not the effective gatelength is temperature dependent due to the occurrence of carrier freeze-out in hte LDD's. Figure 5b illustrates that in the appropriate voltage range, the same ΔL value is found at 77 K and room temperature.

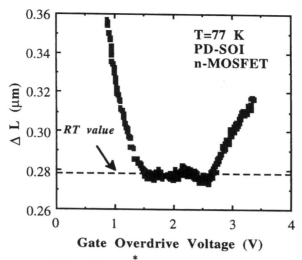

Figure 5. ΔL versus gate overdrive voltage ($V_G^* = V_G-V_T$) for 10 μm and 2 μm SOI PD n-MOSFETs operating at 77 K.

3. Low Temperature SOI Circuits

3.1. TECHNOLOGY AND DESIGN FEATURES

Different approaches such as partially depleted, fully depleted and accumulation-mode device architectures have been reported in the literature. Depending on the application, the partially and fully depleted structures are most often used. For partially depleted devices one has to take into account that the floating body may cause undesired effects such as i) a kink in the drain current characteristics, ii) the observation of threshold voltage instabilities, iii) pronounced transient phenomena, iv) and bipolar action [9]. This can be overcome by using fully depleted devices or by implementing design approaches such as twin gate structures [29-31], gate-all around devices [32] or body ties [33]. However, it has to be remarked that the latter is not so effective at 4.2 K due to the high resistance of the body tie. In cases where analogue applications are aimed at, it is of crucial importance to control the 1/f noise, which increases upon cooling. Due to its importance, the low frequency noise behaviour of SOI technologies has been reviewed in-depth by the authors [34]. Compared to bulk MOSFETs, SOI devices may suffer from additional generation-recombination noise related to the drain current kink and defects in the silicon film. The temperature dependence of the LF noise is illustrated in Fig. 6 for PD and FD devices. In the latter case the excess kink noise is suppressed. Beside an appropriate technology choice, low noise devices can be obtained by using either the twin gate structure or the dual gate concept. The well known room temperature observation that p-channels devices have less noise than their n-channel counterparts remains valid at cryogenic temperatures.

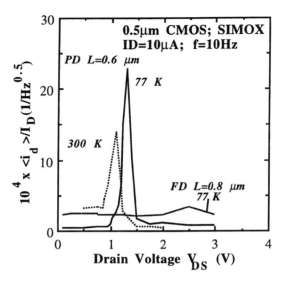

Figure 6. Illustration of the temperature dependence of the LF noise performance of PD and FD SOI devices

3.2. CIRCUIT PERFORMANCE

Submicron cryogenic SOI devices have been reported for a long time. As early as 1991, Wang et al. [35] measured a 0.3 µm technology NAND gate delay time smaller than 100 ps at 85 K. More recently, sub 100 nm cryogenic devices have been fabricated [36]. As illustrated in Fig. 7, unloaded inverter delays are of the order of 6 ps for 75 K operation of 62 nm n- and 85 nm p-channel devices at 1.8 V supply voltage.

4. Future Applications

Scaling down the device geometries will be limited by physical constraints, so that phenomena such as ballistic transport, tunnel effects and quantum mechanical effects may control the device operation. Of course, the technological aspects will have a direct impact on the overall cost. Although 70 nm MOSFETs have already been fabricated, the SIA roadmap is predicting that most likely around 2015 it will be required to go over to 30 nm feature size. This implies that research into so-called nanoelectronics will become of key importance. Whereas for a long time this research domain was reserved for III-V compounds, silicon based devices are now just around the corner. A recent review on the prospects of silicon nanoelectronics can be found in [37]. A particular class of devices gaining more and more interest are the single electron transistors, based on the Coulomb blockade. Although presently processed mainly on bulk material, SOI substrates may give some benefits.

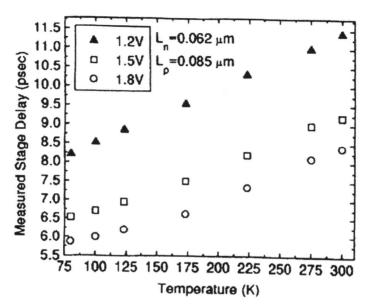

Figure 7. Measured unloaded inverter delay versus temperature for 62 nm n- and 85 nm p-channel devices, for supply voltages fo 1.2V, 1.5V and 1.8V [36].

The operation of single electron transistors (SETs), whereby a small silicon island is separated from the source and drain electron reservoirs by tunnel barriers and capacitively coupled to the gate electrode, is strongly associated with the operating temperature. In general, the minimum operating temperature is given by

$$T \ll q^2 / 2 \, kC \tag{3}$$

with q the electron charge, k the Boltzmann constant and C the capacitance. The Coulomb charging energy of a single electron must be larger than the energy of the thermal fluctuations. Therefore, the original SETs are operating at liquid helium temperature or below. For 77 K operation, it is required to strongly reduce the capacitance (to a few aF) by for example using extremely small geometries, which is imposing tremendous technological challenges. However, SETs successfully operating at 20 K have been reported [38] for devices processed by using the so-called step edge cut off (SECO) fabrication scheme, allowing less stringent lithographic techniques to be used. A further optimisation of these SECO SETs recently resulted in 77 K operation of SETs fabricated in a 0.1 μm technology [39]. Another fabrication technique is based on the pattern-dependent oxidation method, referred to as PADOX [40]. The method, schematically illustrated in Fig. 8, relies on the fact that for narrow silicon lines the oxidation is more pronounced at the end of the wire than in the middle. However, the thinning of the silicon film may result in undesired parasitic effects causing complicated Coulomb blockade oscillations. Therefore, an improvement has been proposed by using a Si_3N_4 capping layer [41] as shown in Fig. 6. The development of single transistor memory cells creates a great potential for SETs to be used to design non-volatile RAMs foe example for mobile computer and communication applications [42]. The real breakthrough will of course depend on the required operating temperature.

Another group of devices, which has already been mentioned in Section 2, are the quantum wires. So far the devices are mainly fabricated to validate theoretical models, including quantum interference phenomena, although future industrial applications should not be excluded. As mentioned before, for SOI based devices these structures are fabricated on very thin silicon films [19].

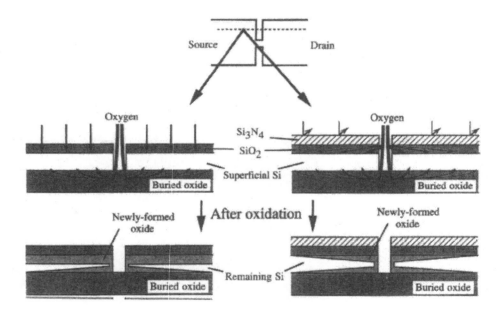

Figure 8. Comparison between the conventional PADOX and the PADOX with Si_3N_4 mask. The latter prevents unnecessary thinning of the silicon film [41].

245

5. Conclusion

Recently cryogenic electronics has seen a revival. In view of this trend there is also a good potential for SOI-based devices, especially as this may lead to some new silicon based devices such as single electron transistors and quantum dots. The real breakthrough of cryogenic devices will depend on the overall system cost, whereby cryocoolers, hybridization aspects (e.g. combination with high T_C superconductors) and packaging are important issues.

6. References

[1] Pearson, G.L. and Bardeen, J. (1949) Electrical properties of pure silicon and silicon alloys containing boron and phosphorus, *Physical Review* **5**, 865-883

[2] Gutiérrez-D., E.A., Claeys, C., Simoen, E., and Koshevaya, S.V. (1998) Perspectives of the cryo-electronics for the year 2000, in L. Brogiato, D.V. Camin and G. Pessina (eds), *Proc. 3rd European Workshop on Low Temperature Electronics (WOLTE-3)*, EDP Sciences, Les Ulis Cedex, pp. 315-320

[3] Gutiérrez-D., E.A., Claeys, C., and Deen, M.J. (2000) Low-Temperature Electronics: Physics, Devices, Circuits & Applications", IEEE Press.

[4] Gaensslen, F.H., Rideout, V.L., Walker, E.J., and Walker, J.J. (1987) Very small MOSFETs for low temperature operation, *IEEE Trans. Electron Dev.* **24**, 218-229

[5] Sun, J.Y., Taur, Y., Dennard, R.H., and Klepner, S.P. (1987) Submicrometer channel CMOS for low temperature operation, *IEEE Trans. Electron Dev.* **34**, 19-27

[6] Simoen, E., Claeys, C., and Martino, A. (1996) Parameter extraction of MOSFETs operated at low temperatures, in C. Claeys and E. Simoen (eds), *Proc. 2nd European Workshop on Low Temperature Electronics (WOLTE-2)*, Les Editions de Physique, Les Ulis Cedex, pp. 29-42

[7] Alawneh, I., Simoen, E., Biesemans, S., De Meyer, K., and Claeys, C. (1998) Comparison of the freeze-out effect in In and B doped n-MOSFETs in the range 4.2-300 K, in L. Brogiato, D.V. Camin and G. Pessina (eds), *Proc. 3rd European Workshop on Low Temperature Electronics (WOLTE-3)*, EDP Sciences, Les Ulis Cedex, pp. 3-8

[8] Colinge, J.P. (1991) Silicon-on-Insulator Technology: Materials to VLSI, Kluwer Academic Publishers, Dordrecht

[9] Claeys, C. and Simoen, E. (1994) The perspectives of Silicon-on-Insulator technologies for cryogenic applications, *J. Electrochem. Soc.* **141**, 2522-2532

[10] Simoen, E. and Claeys, C. (1997) Low frequency noise of cryogenic silicon electronics, in C. Claeys and E. Simoen (eds), *Proc. 14th Int. Conference on Noise in Physical Systems and 1/f Fluctuations*, World Scientific, Singapore, pp. 527-533

[11] Balestra, F. and Ghibaudo, G. (1997) Comparison between the cryogenic behaviour of deep submicron bulk Si and SOI CMOS devices, in C. Claeys, S.I. Raider, M.J. Deen, W.D. Brown and R.K. Kirschman (eds), *Proc. Low Temperatures Electronics and High Temperature Superconductivity IV*, The Electrochem. Soc. Proc. **97-2**, Pennington, New Jersey, pp. 171-186

[12] SIA 1997 Roadmap, Semiconductor Industry Association, 181, Metro Drive, Suite 450, San Jose, CA 95110

[13] Yoshimi, M., Hazama, H., Kambayahi, S., Wada, T., Kata, T., and Tango, H. (1989) Two-dimensional simulation and measurement of high-performance MOSFETs made on a very thin SOI film, *IEEE Trans. Electron Dev.* **36**, 493-503

246

[14]   Omura, Y., Horiguchi, S., Tabe, M., and Kishii, T. (1993) Quantum mechanical effects on the threshold voltage of ultrathin SOI nMOSFETs, *IEEE Electron Dev. Lett.* **14**, 569-571

[15]   Omura, Y., Ishiyama, T., Shoji, M., and Izumi, K. (1996) Quantum mechanical transient characteristics in ultimately miniaturized MOSFETs/SIMOX, in P. Hemment, S. Cristoloveanu, K. Izumi, T. Houston and S. Wilson (eds), *Proc. 7th Int. Symp. on Silicon-on-Insulator Technology and Devices*, The Electrochem. Soc. Proc. **96-3**, Pennington, pp. 199-211

[16]   Omura, Y. and Izumi, K. (1996) Quantum mechanical influences on short-channel effects in ultra-thin MOSFET/SIMOX devices, *IEEE Electron Dev. Lett.* **17**, 300-302

[17]   Butcher, P., March, N.H., and Toshi, M.P. (1993) Physics of Low Dimensional Semiconductor Structures, Plenum, New York

[18]   Omura, Y. and Nagase, M. (1995) Low-Temperature drain current characteristics in sub-10 nm thick SOI nMOSFETs on SIMOX (Separation by Implanted Oxygen) substrates, *Jpn. J. Appl. Phys.* **34**, 812-816

[19]   Colinge, J.P., Baie, X., and Bayot, V. (1994) Evidence of two-dimensional carrier confinement in thin n-channel SOI gate-all-around (GAA) devices, *IEEE Electron Dev. Lett.* **15**, 193-195

[20]   Colinge, J.P., Baie, X., Bayot, V., and Grivet, E. (1996) A Silicon-on-Insulator quantum wire, *Solid-State Electron.* **39**, 49-51

[21]   Morimoto, K., Hirai, Y., Yuki, K., and Morita, K. (1996) Fabrication and transport properties of silicon quantum wire gate-all-around transistor, *Jpn. J. Appl. Phys.* **35**, 853-857

[22]   Peters, M.G., Shi, Y., Dijkhuis, J.I., De Jong, M.J.M., and Molenkamp, L.W. (1996) Random telegraph signals in the Coulomb blockade regime, in M. Scheffler and R. Zimmerman, *Proc. 23rd Int. Conf. Physics of Semiconductors* **3**, World Scientific, Singapore, pp. 2375-2378

[23]   Peters, M.G. (1998) Noise Measurements on silicon quantum dots, Ph.D. thesis, University of Utrecht, The Netherlands

[24]   Nicolett, A.S., Martino, J.A., Simoen, E., and Claeys, C. (1997) Improved channel length and series resistance extraction for short channel MOSFETs suffering from mobility degradation, *J. Solid-State Dev. and Circ.* **5**, 1-4

[25]   Nicolett, A.S., Martino, J.A., Simoen, E. and Claeys, C (1998) Back gate voltage and buried oxide thickness influence on the series resistane of fully depleted SOI MOSFETs at 77 K, in L. Brogiato, D.V. Camin and G. Pessina (eds), *Proc. 3rd European Workshop on Low Temperature Electronics (WOLTE-3)*, EDP Sciences, Les Ulis Cedex, pp. 25-28

[26]   Nicolett, A.S., Martino, J.A., Simoen, E. and Claeys, C. (1998) Back gate voltage influence on the LDD SOI NMOSFET series resistance extraction from 150 to 300 K , this conference

[27]   Simoen, E. and Claeys, C. (1997) Impact of the series resistance on the parameter extraction of submicron MOSFETs operated at 77 K, *Solid-State Electron.* **41**, 659-661

[28]   Schreutelkamp, R., Martino, J.A., Simoen, E., Deferm, L. and Claeys, C. (1995) Combined ΔL and series resistance extraction of LDD MOSFETs at 77 K, in C.L. Claeys, S.I. Raider, R. Kirschman and W.D. Brown (eds), *Proc. Low Temperatures Electronics and High Temperature Superconductivity IV*, The Electrochem. Soc. Proc. **95-9**, Pennington, New Jersey, pp. 290-296

[29]   Gao, M.-H., Colinge, J.P., Lauwers, L., Wu, S., and Claeys, C. (1992) Twin MOSFETs structure for suppression of kink and parasitic bipolar effects in SOI MOSFETs at room and liquid helium temperatures, *Solid-State Electron.* **35**, 505-512

[30] Simoen, E. and Claeys, C. (1994) Low temperature operation of silicon-on-insulator inverters, *J. de Phys. III* **4**, 63-69

[31] Simoen, E. and Claeys, C. (1995) The cryogenic operation of partially depleted silicon-on-insulator invertors, *IEEE Trans. Electron Dev.* **42**, 1100-1105

[32] Simoen, E. and Claeys, C. (1995) Static characteristics of gate-all-around SOI MOSFETs at cryogenic temperatures, *Phys. Stat. Sol. (a)* **148**, 635-642

[33] Simoen, E. and Claeys, C. (1994) The use of body ties in partially depleted SOI MOSTs operating at cryogenic temperatures, *Solid-State Electron.* **37**, 1933-1936

[34] Simoen, E. and Claeys, C. (1996) The low-frequency noise behaviour of silicon-on-insulator technologies, *Solid-State Electron.* **39**, 949-960

[35] Wang, L.K., Seliskar, J., Bucelot, T., Edenfeld, A., and Haddad, N. (1991) Enhanced performance of accumulation mode 0.5 μm CMOS/SOI operated at 300 K and 85 K, in *IEDM Techn. Digest*, pp. 679-682

[36] Assaderaghi, F., Rausch, W., Ajmera A., Leobandung, E., Schepis, S., Wagner, L., Wann, H.-J., Bolam, R., Yee, D., Davari, B., and Shahidi, G. (1997) A 7.9/5.5 psec room/low temperature SOI CMOS, in *IEDM Technical Digest*, pp. 415-418

[37] Van Rossum, M. (1997) Prospects of silicon nanoelectronics, in H. Grünbacher (ed), *Proc. ESSDERC '97*, Editions Frontières, pp. 28-33

[38] Altmeyer, S., Kühnel, F., Spangenberg, B., and Kurz, H. (1996) A possible road to 77 K single-electron devices, *Semicond. Sci. Technol.* **11**, 1502-1505

[39] Altmeyer, S., Hamidi, A., Spangenberg, B., and Kurz, H. (1997) 77 K Single electron transistors fabricated with 0.1 μm technology, *J. Appl. Phys.* **81**, 8118-8120

[40] Takahashi, Y., Namatsu, H., Kurihara, K., Iwadate, K., Nagase, M., and Murase, K. (1996) Size dependence of the characteristics of Si single-electron transistors on SIMOX substrates, *IEEE Trans. Electron Dev.* **43**, 1213-1217

[41] Fujiwara, A., Takahashi, Y., Namatsu, H., Kurihara, K., and Murase, K. (1998) Suppression of effects of parasitic metal-oxide-semiconductor field-effect transistors on Si single-electron transistors, *Jpn. J. Appl. Phys.* **37**, 3257-3263

[42] Yano, K., Ishii, T., Hashimoto, T., Kobayashi T., Murai, F., and Seki, K. (1994) Room temperature single electron memory, *IEEE Trans. Electron Dev.* **41**, 1628-1638

SOI CMOS for High-Temperature Applications

J.P. COLINGE
Dept. of Electrical and Computer Engineering
University of California, Davis, CA 95616, USA

1. Introduction:

Several markets such as oil and gas drilling, automotive and aerospace have an increasingly large demand for electronic circuits capable of operating at high temperatures (Table 1).

Table 1: Applications for high-temperature electronics. [1]

Application	Temperature Range
1. Well Logging Instrumentation	
Oil and Gas Wells	75 - 225°C
Steam injection	200 - 300°C
Geothermal Wells	200 - 600°C
2. Aerospace	
Electronic braking systems	up to 250°C
Rack mounted avionics	up to 250°C
Engine control/monitoring	up to 300°C
'Smart Skins'	up to 350°C
3. Consumer	
TV Distribution	up to 200°C
Microwave Ovens	up to 500°C
Domestic Appliances	up to 200°C
4. Automotive	
Engine Compartment	-40 to 165°C
On-engine and on-transmission	-40 to 165°C
Wheel mounted components	-40 to 250°C
5. Others	
Military	up to 250°C
Communications	up to 250°C
Heavy Industrial Equipment	up to 300°C
Space systems	up to 600°C
Air Pollution Control	up to 550°C
Nuclear reactor monitoring	up to 550°C

High-temperature operation of regular bulk CMOS integrated circuits is usually limited to approximately 200°C because of the increase of the p-n junction leakage currents, the drift of the threshold voltage, the degradation of the mobility in the transistors, and thermally-induced latchup. In addition to these strictly device-related parameter

249

P.L.F. Hemment et al. (eds.),
Perspectives, Science and Technologies for Novel Silicon on Insulator Devices, 249–256.
© 2000 *Kluwer Academic Publishers. Printed in the Netherlands.*

variations, other degradation mechanisms are raising reliability issues when high-temperature operation is to be considered. These include increased electromigration phenomena in aluminum lines, stress and corrosion in the package. While the latter problems can be solved by using tungsten as an interconnect metal and by using appropriate packaging materials, the drift of device parameters with temperature is a problem having no solution as long as classical bulk silicon MOS technology is employed. Silicon carbide and diamond devices can operate at even higher temperatures, but the level of maturity of these technologies is still very low. An exhaustive list of high-temperature circuit and applications can be found in the HITEN database.[2]

SOI MOSFETs presents interesting features as far as high-temperature operation is concerned. Indeed, thin-film SOI MOSFETs present very small junction areas, and their high-temperature leakage current can be 3 to 4 orders of magnitude lower than those of regular MOS devices. In addition, the threshold voltage variation with temperature is much smaller in thin-film SOI MOSFETs than in bulk devices. Functionality of 16k and 256k SRAMs at high temperatures (up to 300 °C) and of SOI CMOS ring oscillators up to 500°C has been demonstrated in the past. [3,4,5] In this Paper the properties of the SOI MOSFET and SOI circuits under high-temperature operation will be reviewed.

2. The SOI MOSFET

The two interesting features of SOI MOSFETs for high-temperature applications are the reduced leakage current and the small variation of threshold voltage with temperature. As far as analog circuits are concerned, the temperature behavior of the output conductance is important as well. We shall examine these parameters in some detail.

2.1. LEAKAGE CURRENT

Classical P-N$^+$ junction theory tells us that the leakage current of a reverse-biased diode is given by the following expression:

$$I_{leak} = q\,A \left(\frac{D_n}{\tau_n}\right)^{1/2} \frac{n_i^{\,2}}{N_a} + q\,A\,\frac{n_i\,W}{\tau_e}$$

where q is the electron charge, A is the junction area, D_n is the electron diffusion coefficient, τ_n is the electron lifetime in p-type neutral silicon, n_i is the intrinsic carrier concentration, N_a is the doping concentration in the p-type material, W is the depletion width, and $\tau_e=(\tau_n+\tau_p)/2$ is the effective lifetime related to the thermal generation process in the depletion region. This current contains two components: a diffusion current, the amplitude of which is proportional to the square of the intrinsic carrier concentration ($I_{diff} \propto n_i^2(T)$), and a generation current, which shows a linear dependence upon the intrinsic carrier concentration ($I_{gen} \propto n_i(T)$). In bulk silicon junctions at room temperature the generation current is usually larger than the diffusion current, but this situation changes at high temperatures because of the stronger temperature dependence of the diffusion current [6]. In thin-film SOI devices it has been observed that the junction leakage current is proportional to $n_i(T)$ if the body of the device is fully depleted when the transistor is turned off. This is always the case in accumulation-mode SOI MOSFETs (Figure 1). This is also the case in enhancement-mode, fully depleted devices up to a critical temperature where the increase of the intrinsic carrier concentration is such that the device is no longer fully depleted.[7] Only above that critical temperature does an $n_i^2(T)$ dependence of the leakage current on temperature become dominant (Figure 2). As can be seen in Figures 1 and 2, the leakage current at 300°C is 1,000 times smaller in SOI than in bulk. This is due to both the smaller junction area in SOI and to the fact that the leakage current is mostly due thermal generation, and not to diffusion, unlike bulk devices.

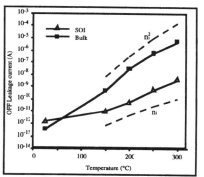

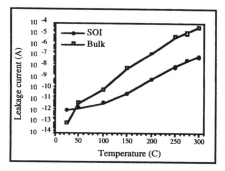

Figure 1: Leakage current dependence on temperature of an SOI accumulation-mode pMOS transistor and pMOS bulk transistor.

Figure 2: Leakage current in bulk and SOI n-channel transistors, of same geometries, as a function of temperature.

2.2. THRESHOLD VOLTAGE

The threshold voltage of SOI fully-depleted transistors is known to be 2 to 3 times less sensitive to temperature than that of bulk devices [6]. When the temperature is raised, the intrinsic carrier concentration increases and the Fermi potential decreases. As a result, the work function difference between the polysilicon gate and the active silicon, Φ_{MS}, changes and the maximum depletion width in the silicon, X_{dmax}, decreases. In bulk devices, the temperature dependence of both of Φ_{MS} and X_{dmax} contributes to a decrease of the threshold voltage as temperature is increased. In enhancement-mode fully-depleted and accumulation-mode SOI devices, there is no variation of X_{dmax} with temperature. There exists a critical temperature, however, (220°C in [8]) beyond which the devices are no longer fully depleted, which causes the threshold voltage variation with temperature to be similar to that in a bulk device (Figure 3). In some SOI devices, such as transistors with top and bottom gates [9], where depletion arises from both the top and the bottom of the silicon film, this critical temperature is increased substantially, such that minimal temperature dependence of the threshold on temperature is observed up to 320°C. Some issues, such as the high-temperature reliability and stability of SOI devices, are under investigation.[10,11,12]. Indeed, instabilities can be caused by the drift of positive ions and the trapping/detrapping or compensation of charges within the buried oxide.

2.3. OUTPUT CONDUCTANCE

The output conductance of a transistor is an important parameter limiting the performances of analog CMOS ICs. Because the body of the SOI devices is electrically floating, impact ionization-related effects (kink effect, parasitic bipolar action,...) tend to degrade the output conductance of SOI MOSFETs. This degradation is minimized, but nevertheless present, if fully depleted devices are used. It is observed that the output conductance of SOI MOSFETs actually improves when temperature is increased (Figure 4) [8]. This is explained by several mechanisms: high temperatures reduce impact ionization near the drain, excess minority carrier concentration in the device body is reduced through increased recombination, and the body potential variations are reduced owing to an increase of the saturation current of the source junction.

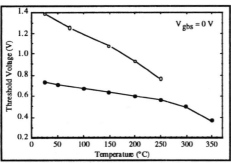

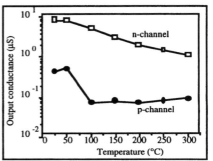

Fig 3: Threshold voltage vs. in N-channel devices (O :bulk; ● FDSOI)

Figure 4: Output conductance dependence on temperature of SOI transistors.

3. SOI Circuits

An interesting property of SOI for high-temperature applications is found at the circuit level: in bulk circuits the leakage currents of all reverse-biased junctions of a circuit branch (e.g. a NAND gate shown in Figure 5) add up. This quickly yields large total leakage currents. An important part of that leakage arises from the large well diffusions. In an SOI circuit, no current can flow to the substrate; therefore, the leakage currents flow in the branches of the circuit, and the total current is fixed by the least leaky device (Figure 5). If one considers analog circuits, the increase of leakage current in bulk devices leads to the loss of the operating point, unless external bias currents are increased with temperature. In SOI, since current never flows to the substrate, there is no need to change bias currents with temperature. Figure 6 shows the gain of an operational transconductance amplifier vs. temperature. The reduction of transconductance caused by the reduction of mobility is almost perfectly compensated by the improvement of MOSFET output conductance.[9] The use of ZTC (zero temperature coefficient) bias conditions allow analog devices to operate at temperatures as high as 400°C (Figure 7).[13,14] The decrease of dc gain with temperature is solely due to the decrease of carrier mobility with temperature.

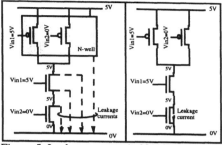

Figure 5: Leakage current in NAND gate (left: bulk, right: SOI)

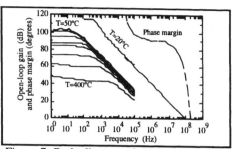

Figure 7: Bode diagram of a SOI Operational Amplifier at different temperatures.

Table 2: Measured performances of a FDSOI $\Sigma-\Delta$ A/D converter. OSR: oversampling ratio; SNR: signal-to-noise ratio; SS: signal swing

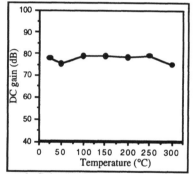

Figure 6: Gain (in dB) vs. temperature in an OTA SOI operational amplifier

Parameter	Temp. 1	Temp. 2	Unit
Temperature	27	350	°C
V_{dd}	±1	±2.5	V
Max. SS	1.5	3.5	V
Accuracy	9	7	bits
F_s	100	500	kHz
Peak SNR	58	30	dB
OSR	128	128	-
Power	60	500	μW
MOS L_{min}	3	3	μm

Several high-temperature SOI circuits and sensors have been reported in the literature. Some of them are presented in Table 3. The performances of a $\Sigma-\Delta$ analog-to digital modulator are presented in Table 2. The accuracy, which is 9 bits at room temperature, is still 7 bits at 350°C.

Table 3: Some high-temperature SOI circuits

Temp.	Circuit	Institution	Ref
300°C	Solenoid driver	Allied Signal	15
300°C	Various Analog & Digital Circuits	Honeywell	16
400°C	Op Amp, Magnetic Sensor	UCL	17
300°C	Voltage Reference	RUB, UCL	18,19
350°C	ΣΔ A/D converter	UCL	20
200°C	Continuous-time filter	UCL	21

Another example is a high-temperature magnetic sensor which could be used in automotive applications. Its performances are presented in Figures 8 and 9. Its sensitivity is larger than that of bulk devices and its production should be significantly less costly than that of III-V semiconductor-based sensors.

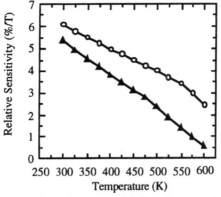

Figure 8. Relative sensitivity of magnetic sensors: SOI (O) and bulk (▲)

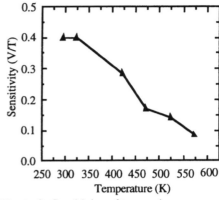

Figure 9: Sensitivity of magnetic sensor vs. temperature (K).

Among the commercially available high-temperature products one can mention the HTMOS™ devices from Honeywell and the HMTS ASICs from Allied-signal.

◊ The HTMOS™ products are designed to meet the data sheet parameters over a -55°C to 225°C with a minimum of 5 years of operation. Currently available products include a Quad Operational Amplifier, Quad Analog Switch, 83C51 microcontroller, mask programmable gate array, and a 32K x 8 SRAM. Additional standard products in various stages of development include a Voltage Reference, Linear Regulator, 12 bit A/D Converter, 16:1/8:2 Multiplexer, 1553 Protocol, and a Clock Oscillator. The HTMOS™ Line is targeted at sensor signal conditioning, data acquisition, and control applications in hostile high temperature environments up to 300°C.[16]

◊ The HMTC's High Temperature ASIC's series include: Torque motor drivers, 32K x 8 read-only memories, Solenoid drivers, LVDT exciters, Operational amplifiers and 90V power FETs. These devices are mainly targeted at the automotive industry.[15]

The stability of sensitive analog SOI circuits with both time and temperature has been studied by the Ruhr-University Bochum. It was found, for instance, that a voltage reference fabricated using SOI MOSFETs could have a stability better than 0.7% over the temperature range -50 to 300°C (Figure 10). Long-term high-temperature experiments show a drift of gate voltages, for a constant drain current bias, of less than

0.5% at 300°C over a period of time of 10^4 minutes.[10] The stability of the input offset voltage of an operational amplifier operating at 250°C is presented in Figure 11. A total drift of less than 25 mV is observed over a period of time of 1000 hours.[18]

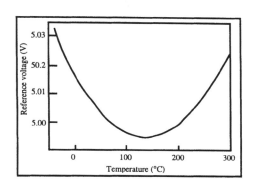

Figure 10: Variation of the output of an SOI reference voltage source with temperature.

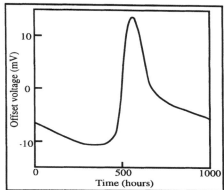

Figure 11: Variation of the input offset voltage of an SOI op amp with time. T=250°C.

4. Conclusion

Owing to full dielectric isolation dielectric, which prevents leakage flow current from junctions into the substrate, SOI circuits can operate at elevated temperatures. Threshold voltage variation with temperature is smaller than in bulk devices, and output conductance actually improves with increased temperature. The performance of analog, digital, and sensor circuits is presented. Commercial SOI high-temperature circuits are available, and experimental circuits operating at up to 400°C have been demonstrated.

5. References

1   http://www.hiten.com/guests/analysis/appl-over.html

2   http://www.hiten.com

3   W.A. Krull and J.C. Lee (1988) Demonstration of the benefits of SOI for high-temperature operation, Proceedings IEEE SOS/SOI Technology Workshop, 69

4   H. Gotou, A. Sekiyama, T. Seki, S. Nagai, N. Suzuki, M. Hayasaka, Y. Matsukawa, M. Miyazima, Y. Kobayashi, S. Enomoto, and. K. Imaoka (1989) A 256 kbit SOI full CMOS SRAM, Technical digest of IEDM, 912-915

5   W. P. Maszara (1990) SOI by wafer bonding: a review, Proceedings 4th International Symposium on Silicon-on-Insulator Technology and Devices, Ed. by D. Schmidt, the Electrochemical Society, vol. 90-6, 199

6   D.P. Vu, M.J. Boden, W.R. Henderson, N.K. Cheong, P.M. Zavaracky, D.A. Adams, and M.M. Austin (1989) High-temperature operation of ISE devices and circuits, Proceedings of the SOS/SOI Technology Conference, 165-166

7   T. Rudenko, A. Rudenko, and V. Kilchitskaya (1997) Modeling and measurements of high-temperature off-state currents in SOI MOSFETs, Electrochemical Society Proceedings **97-23**, 295-300

8   G. Groeseneken, J.P. Colinge, H.E. Maes, J. C. Alderman and S. Holt (1990) Temperature Dependence of Threshold Voltage in Thin-Film SOI MOSFET's, IEEE Electron Dev. Letters **11**, 329-331

9   P. Francis, A. Terao, B. Gentinne, D. Flandre, JP Colinge (1992) SOI technology for high-temperature applications, Technical Digest of IEDM, 353-356

10   http://www.lems.ruhr-uni-bochum.de/forschung/3rdhitec/ge/albuq_ge.htm

11   A.N. Nazarov, J.P. Colinge, and I.P. Barchuk (1997) Research of high-temperature instability processes in buried dielectric of fully depleted SOI MOSFETs, Microelectronic Engineering, Vol. 36, No. 1-4, 363-366

12   A.N. Nazarov, I.P. Barchuk, and J.P. Colinge (1998) The nature of high-temperature instability in fully depleted SOI IM MOSFETs, Proceedings of the 4th International High-Temperature Electronics Conference (HITEC), 226-229

13   B. Gentinne, J.P. Eggermont and J.P. Colinge (1995) Performances of SOI CMOS OTA combining ZTC and gain-boosting techniques, Electronics Letters **32-24**, 2092-2093

14   B. Gentinne, J.P. Eggermont, D. Flandre, J.P. Colinge (1997) Fully depleted SOI-CMOS technology for high temperature IC applications, Materials Science and Engineering **B46**, 1-7

15   http://www.mtcsemi.com/html/technology.html

16   http://www.ssec.honeywell.com/

17   http://www.dice.ucl.ac.be/SOI/~colinge; J.P. Eggermont, D. Flandre, J.P. Colinge (1996) CMOS SOI magnetic field sensors for applications up to 600K, Trans. 3rd International High Temperature Electronics Conference (HiTEC) **1**, X.3-X.8

18   http://www.lems.ruhr-uni-bochum.de/ forschung/3rdhitec/ei/albuq_ei.htm; C. Eisenhut and W. Klein (1997) SIMOX voltage reference for applications up to 275°C using the threshold voltage difference principle, Proceedings of the IEEE International SOI Conference, 110-111

19   J.P. Eggermont, V. Dessard, A. Vandooren, D. Flandre and J.P. Colinge (1998) SOI current and voltage references for applications up to 300°C, Proceedings of the 4th International High-Temperature Electronics Conference (HITEC) 55-59

20   A. Viviani, D. Flandre and P. Jespers (1996) A SOI-CMOS micro-power first-order Sigma-Delta modulator, Proc. IEEE International SOI Conf., 110-111

21   V. Dessard, D. Baldwin, L. Demeûs, B. Gentinne and D. Flandre (1996) SOI implementation of low-voltage and high-temperature MOSFET-C continuous-time filters, Proc. IEEE International SOI Conf., 24-25

QUANTUM EFFECT DEVICES ON SOI SUBSTRATES WITH AN ULTRATHIN SILICON LAYER

Yasuhisa OMURA

High-Technology Research Center and Faculty of Engineering, Kansai University
3-3-35, Yamate-cho, Suita, Osaka 564-80, Japan

1. Introduction

Since the 1970's, many companies have been investing heavily in semiconductor development and manufacturing. This has been followed by a salient growth in tele-communication industries around the world. For about past twenty years, Si MOSFETs have been the leading semiconductor device. During this time, their feature size has been reduced from 10 to 0.1μ m, resulting in a 100 fold increase in circuit functions. Besides their past and current industrial contributions, Si MOSFETs were, in the 1970's, widely used to study two-dimensional (2-D) transport phenomena. What was noticed in those early studies was that 2-D transport phenomena appear only at very low temperatures. This is mainly due to the fact that the inversion layer is bound inside a triangular potential well. There is still some question, however, as to whether or not the actual features of 2-D and 1-D transport in silicon were correctly observed. Since an ultra-thin SOI structure has the capability to form an ideal rectangular potential well, it offers one the opportunity to resolve this issue.

In this paper, we introduce the low-dimensional transport properties observed in ultra-thin SOI structures, and some actual quantum effect devices are demonstrated. Furthermore, the feasibility of future mesoscopic devices is discussed as well.

2. Impact of the extremely-thin SOI structure

The thinning the superficial silicon layer has been investigated intensively[1-3]. An ultra-thin SOI MOSFET is considered to be the most promising device (compared to conventional thick SOI MOSFET) because the ultra-thin structure makes it possible to suppress short-channel effects (SCE) [1,3]. But is this actually a desirable trend? Thus, we have carried out numerical simulations [4] to clarify how the thinning of the superficial silicon layer affects the threshold voltage (V_{TH}) in nMOSFETs/SIMOX [5].

P.L.F. Hemment et al. (eds.),
Perspectives, Science and Technologies for Novel Silicon on Insulator Devices, 257–268.
© 2000 *Kluwer Academic Publishers. Printed in the Netherlands.*

258

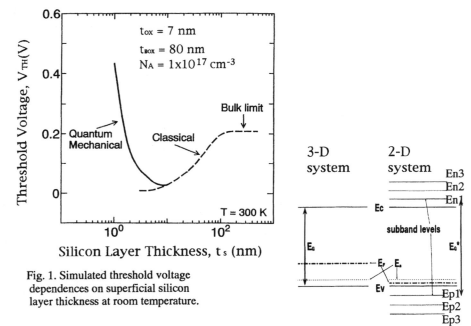

Fig. 1. Simulated threshold voltage dependences on superficial silicon layer thickness at room temperature.

Fermi level definition for 3D and 2D systems

Fig. 2. Schematic band diagrams for ultra-thin SOI nMOSFETs. The 3-D system and 2-D system are compared.
E_C: conduction band bottom, Ev: valence band bottom, E_F: Fermi level, Ea: acceptor level, E_G: forbidden band gap energy in 3-D system, E_G^*: effective forbidden band gap energy, Eni, Epi: subband bottom levels.

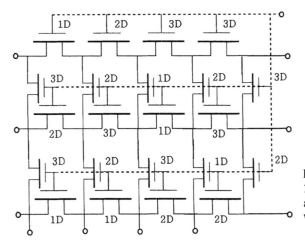

Fig. 3. Network circuit model. 1D, 2D, and 3D SOI MOSFETs are connected by a random-like web.

The V_{TH} dependence on silicon layer thickness (t_s) is shown in Fig. 1. V_{TH} is expressed as [3]

$$V_{TH}=V_{FB} + (qN_At_s/C_{ox})(1+C_s/(C_s+C_{BOX}))$$
$$+ \phi_{sth}(1+C_sC_{BOX}/\{C_{ox}(C_s+C_{BOX})\})$$
$$+C_sC_{BOX}V_{SUB}/\{C_{ox}(C_s+C_{BOX})\}, \qquad (1)$$

where V_{FB} is the flat-band voltage, ϕ_{sth} is the surface potential at the threshold, N_A is the doping concentration, C_{ox} is the gate oxide capacitance, C_s is the body silicon capacitance (ε_s/t_s) and C_{BOX} is the buried oxide capacitance. Classical theory predicts that V_{TH} monotonically decreases as t_s decreases because of the decrease in the depletion charge inside the SOI layer, as indicated by the second term of eq. (1). However, quantum theory predicts that V_{TH} drastically increases as t_s decreases in thickness to less than 10 nm. When t_s becomes smaller than a certain value, the 3-D system changes into a 2-D system, as shown in Fig. 2. In the 2-D quantized system, the clear manifestation of 2-D subband levels results in an increase in the effective forbidden band gap (E_G^*)[6]. E_G^* is defined as

$$E_G^* = E_{p1}-E_{n1}, \qquad (2)$$

where E_{p1} and E_{n1} are the lowest subband bottom levels of holes and electrons, respectively. The Fermi level is expressed as

$$E_F=(E_{p1}+E_a)/2+(kT/2)\ln[N_At_s/4D_{osp}kT] \qquad (3a)$$

for a large value of N_At_s, and

$$E_F=E_{p1}-kT\ln[N_At_s/D_{osp}kT] \qquad (3b)$$

for a small value of N_At_s, where E_a is the acceptor energy level, and D_{osp} is the density of states for holes. Therefore, the increase in E_G^* leads to a relative increase in the Fermi potential needed to induce the inversion layer charge, which results in the increase in V_{TH}. This may be a desirable phenomenon from the view point of V_{TH} design in fully-depleted SOI devices, although it is still a problem from the view point of t_s control.

Noticing that the 1-D SOI MOSFET should have a larger V_{TH} than the 2-D SOI MOSFET, we propose the network circuit shown in Fig. 3 for flexible signal processing. Within the circuit, conventional SOI MOSFETs (3-D MOSFETs), 2-D SOI

MOSFETs and 1-D SOI MOSFETs can be arbitrary distributed. When some terminals are used as input terminals, others work as output terminals. Transfer function of the circuit can be modulated via a gate electrode by means of conductance control.

So how does a reduction of t_s leads to an increase in V_{TH}? When t_s becomes extremely small, the thin SOI system keeps a 2-D nature and E_G^* increases effectively as mentioned before. The influence appears with the increase in the depletion layer width in pn junctions and the decrease of intrinsic carrier concentration (n_i) [6]. When E_F lies inside the band gap, n_i is defined as

$$n_i = \{kT(D_{osn}D_{osp})^{1/2}/t_s\}\exp[-E_G^*/2kT], \qquad (4)$$

where D_{osn} is the density of states for electrons. Possible phenomena, such as enhanced SCE, are discussed with reference to [7,8]. Those phenomena are characteristic of the low dimensionality of the system[8]. Thus, thinning the silicon layer may preclude the down-scaling of SOI MOSFETs through the enhancement of the SCE. Ultimate device structures are discussed later.

3. Transport characteristics characterized by low dimensionality

Thinning the silicon layer also affects some transport characteristics. Reduction of t_s results in the 2-D quantization of the conduction, and the g_m characteristics of SOI MOSFETs show step-like anomalies [9,10]. These anomalies are found not only in majority-carrier devices, but also in minority-carrier devices, especially those exhibiting the minority-carrier injection phenomenon [6]. A schematic diagram of a gated pn junction device with t_s of 8 nm fabricated on a SIMOX substrate is shown in Fig. 4. When both the gate voltage (V_G) and the substrate voltage (V_{SUB}) are swept at the same time at low temperature, we obtain the interesting characteristics shown in Fig. 5, which shows the junction current (I_j) and differential conductance ($g_j \equiv dI_j/dV_C$) dependences on V_C. Vc is the cathode voltage. Step-like and oscillation-like anomalies are found in the g_j curve [6]. The step-like anomalies reflect the density of states of electrons, in the body region, injected from the cathode. Oscillation-like anomalies reflect the change from electrons to holes of the carrier type which are the dominant injected carriers [11]. Our findings from these experiments can be summarized as follows [6,11].
(i) An increase in E_G^* through the 2-D quantization of the system and the suppression of the avalanche phenomenon.
(ii) The suppression of double injection through the reduction of the density of states.
Luckily, these will contribute to the suppression of parasitic bipolar operation, which generally is a serious problem in fully-depleted SOI MOSFET devices.

In the lateral pn-junction device, the negative conductance property based on the band-to-band tunneling (BBT) around the pn junction is also observed at low temperatures[12,13]. The junction current (I_j) dependence on anode voltage (V_A) at 43 K

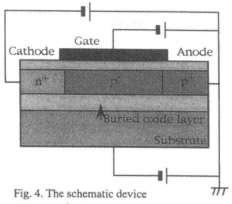

Fig. 4. The schematic device structure of an insulated-gate pn junction device on a SIMOX substrate.

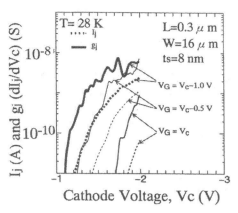

Fig. 5. Experimantally obtained junction current Ij and differential conductance gj dependences on cathode voltage at 28 K.

V_G: gate voltage,

V_C: cathode voltage,

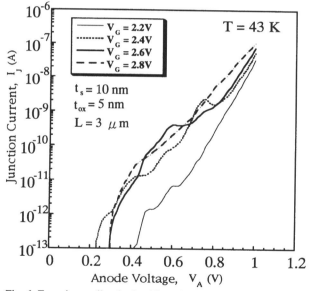

Fig. 6. Experimentally obtained iunction current Ij dependence on anode voltage at 43 K.

for a device with a 10-nm thick silicon layer is shown in Fig. 6. The gate length (L) is 3 μ m and the gate width (W) is 10 μ m. When the gate voltage (V$_G$) ranges from 2.2 V to 2.6 V, clear negative conductance is detected at V$_A$ of around 0.7 V. It should be noted that 2D-to-2D tunneling occurs in this device [12] because the pn junction is adjacent to the edges of the 2D-confined p-type and n-type silicon layers. Negative conductance is not detected at V$_G$ lower than 2.1 V or higher than 2.8 V.

A theoretical consideration [13] indicates that tunneling between the 2-D electron inversion layer and the 2-D holes in the anode is permitted when quantum numbers of 2-D electrons and 2-D holes are identical, i.e. a "resonant" effect exists [13]. It has also been shown, theoretically and experimentally, that tunneling between the 2-D electron inversion layer and 3-D holes is very limited in a thick-silicon layer device [12]. However, we can enhance tunneling in extremely-thin-silicon layer devices.

The thin silicon layer also influences the carrier mobility. Historically, the issue of whether the electron mobility (μ n) in fully-depleted thin SOI MOSFETs is larger than that of bulk MOSFETs has been controversial [14]. Recently, a general concept that the electronic state of an SOI device with t$_s$ larger than the inversion layer thickness (t$_{inv}$) is identical to that of a bulk device when the range of effective electric field corresponds to the predominance of phonon scattering, has been formulated [15]. This means that μ n in both devices must be identical. However, when t$_s$=5 nm, where the inversion layer covers the entire ts, the electronic states in those devices are moderately different, as shown in Fig. 7. In this case, the electron density becomes high over the whole silicon layer, and the conduction electrons are bound inside the rectangular potential well. When electrons interact predominantly with 2-D phonons (surfons), the scattering probability decreases [16]. Therefore, in extremely thin SOI devices, intrinsic mobility enhancement is expected, which will make it easy to observe the velocity overshoot effect (VOE) mentioned below. On the other hand, mobility simulation results from extremely-thin-silicon layer devices have been presented recently; the assumption was that full interaction with 3-D phonons occurred [17,18]. Simulations show that the mobility value should have a local maximum around t$_s$=3nm and that it decreases rapidly with further reduction of the silicon layer thickness. This issue must be studied further.

4. Prospective directed to mesoscopic device applications and reality

The gate length of current SOI MOSFETs ranges from 0.1 to 0.05 μ m [19-21]. Since the energy relaxation length (λ) is identical to the channel length (L$_{eff}$), there must be a strong suppression of the energy relaxation of carriers under high electric fields [22]. It is expected that, in practice, the salient VOE can be found for L$_{eff}$ < 0.1 μ m, as shown in Fig. 8, according to an analysis of the substrate current characteristics in ultra-thin SOI nMOSFETs [23]. The average mean-square velocity of electrons is given empirically by

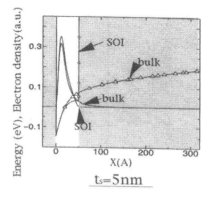

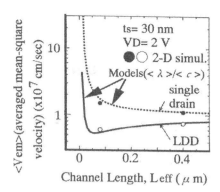

Fig. 7. Simulated electronic states around the surface for bulk Si and SOI Si layer and electron densities for ts = 5 nm. The Poisson equation and Schroedinger equation are self-consistently solved.

Fig. 8. Simulated averaged mean-square velocity dependences on channel length. Empirical models are shown by a dotted line for single-drain structure and a solid line for LDD structure, and closed circles and open circles show data from two-dimensional device simulations.

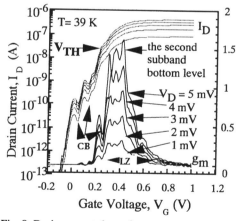

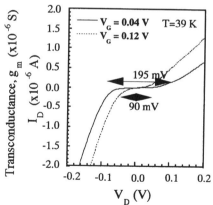

Fig. 9. Drain current dependence on gate voltage at 39 K for five different values of drain voltage in device B.

Fig. 10. Drain current dependence on drain voltage at 39 K. Leff=50nm. Coulomb gaps are found. More clearly defined behaviours are observed at lower temperatures.

$$<V_{em}>=<\lambda>/<\tau>, \tag{5a}$$

$$<\lambda>=9V_D \lambda_0 \exp(2.3V_D \lambda_0/L_{eff}), \tag{5b}$$

where λ_0 is the energy relaxation length at the local-field limit and $<\tau>$ is the averaged energy relaxation time. One reason why the VOE is not clearly observed in state-of-the-art devices probably originates from insufficient 2-D quantization (t_s is still too large). When t_s reaches the de Bloglie wavelength, the fact that there are (i) adequate phase coherence in the depth direction and (ii) enhanced non-stationary energy transport along the channel current, promises an observation of the VOE [19,23,24]. The semi-monochromatic wave nature of electrons is salient because almost all the electrons occupy the lowest energy level. When the VOE is actually utilized, the degradation [25] of the SOI device drivability is superior to that of a bulk device. This would happen with decreasing L_{eff} and must be overcome. This means that the merit of ultra-thin SOI MOSFETs holds even in sub-0.1 μm region. However, it should also be pointed out that the electron-electron scattering process (inelastic scattering) may contribute to the fact that the velocity overshoot effect is not easily observed in thin SOI devices[21].

When mobility enhancement in the 2-D quantized system is actually achieved in ultimately miniaturized SOI MOSFETs, the phase-coherent nature of the mesoscopic system, i.e., ballistic conduction [26] and quantized conductance [27], will be more important than previously found.

Recent experimental results of mesoscopic conduction in 2- or 6-nm-thick SOI nMOSFETs with a 50-nm-channel length fabricated on SIMOX substrates [28] are described next. Since the silicon/buried oxide interface of the SIMOX substrate has a roughness value (R_{ms}) that is less than 1 nm and an undulation period larger than 1.5 μm, most of the device area is formed in an extremely flat region.

Drain current vs. drain voltage characteristics of two types of 50-nm-channel nMOSFETs/SIMOX with a 2-μm channel width at room temperature are reported in [28]; device A has a gate oxide layer thickness (t_{ox}) of 2 nm and a silicon layer thickness (t_s) of 2 nm, and device B has t_{ox} of 3 nm and t_s of 6 nm. The drain current value is smaller than expected because of the parasitic resistances ($\sim$6 KΩ) of the shallow source and drain regions.

High-temperature mesoscopic effects in a 50-nm-channel nMOSFET/SIMOX with a 6-nm-thick silicon layer (device B) are shown in Fig. 9. In device A, the lowest subband bottom level ($E_{n,1}$) is about 100 meV from the conduction band bottom and, in device B, the $E_{n,1}$ is about 20 meV from the conduction band bottom. Figure 9 shows the dependence of drain current on gate voltage (V_G) at 39 K for five different values of drain voltage in device B. At threshold, transconductance (g_m=dI_D/dV_G) shows a clear anomaly corresponding to crossing the lowest subband bottom level ($V_G \sim$0.2 V). It is believed that the g_m oscillation in the subthreshold current region (shown by CB) corresponds to the Coulomb blockade (CB) because of the finite R_{ms} value, which is supported by the existence of the Coulomb gap of 90∼195 mV at V_G of 0.04 and 0.12 V as shown in Fig. 10. It is considered that the combination of the large undulation period and the local roughness results in coupled barriers inside the short channel region. Since the period of the Coulomb oscillation in the subthreshold region is 0.08 V, the effective gate capacitance of the built-in single electron transistor (b-SET) is 2.0 aF and the effective gate area (S_G) is 1.8×10^{-12} cm^{-2}. Furthermore, the asymmetric tunneling barrier capacitance can be estimated as 0.82∼1.8 aF from Coulomb gaps shown in Fig. 10. When the barrier width is assumed to be $\sqrt{S_G}$, the tunneling barrier thickness ranges from 4.5 to 10 nm.

The other g_m oscillation for $V_G > V_{TH}$ (shown by <-LZ->) may be caused by the localization (LZ) effect between the source and drain junctions since the channel conductance becomes infinitesimal small when $V_G > V_{TH}$ at 1.1 K (not shown here). In contrast, in device A, the CB was observed in the subthreshold region only for L_{eff}=50nm. This suggests that the potential fluctuation, which strongly contributes to the carrier scattering, is quite large in device A because the Rms roughness has almost the same value as the silicon layer thickness. It is suggested that the localization effect is stronger in device A. Thus, further reduction of the interface roughness in SIMOX substrates should make it possible to build higher-temperature operating Si quantum effect devices in the near future.

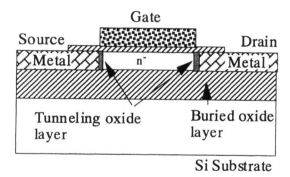

Fig. 11. Ultimately miniaturized device structure: A tunnelin-barrier junction (TBJ) MOS device structure. The device structure can be scaled down from the TBJ MOS device to the single-electron transistor.

Finally, we will discuss the ultimate miniaturized device structure. Some researchers suggest that the double-gate device is the ultimate MOSFET structure. Is it technologically realistic, believable, and/or an optimum target? As long as the device utilizes conventional pn junctions, it will badly suffer from the constraints of the SCE and the band-to-band tunneling between the source and the drain [8]. To eliminate those

266

constraints, the junction structure should be greatly changed. We propose a tunneling-barrier junction (TBJ) MOS device structure using an extremely-thin silicon dioxide layer as shown in Fig. 11. Simulated drain current vs. drain voltage characteristics will be shown in a future publication [29]. The medium-scaled version of this device represents the single-electron transistor [30] and the ultimate miniaturization of the structure yields the resonant-tunneling device [31, 32]. Therefore, this device structure promises very useful and flexible device applications.

Thus, ultra-thin SOI MOS device structures not only offer the promise of many future applications, but also represent a powerful tool for advancing device physics.

5. Conclusion

What happens to an SOI MOSFET when the silicon layer is very thin is discussed in detail. We show that quantum mechanical short-channel effects impose limits on the down-scaling of SOI MOSFET devices. However, the impact of quantum mechanical effects on the threshold voltage can be used in the design of network circuits. Thinning the silicon layer will bring new possibilities, such as tunneling enhancement, mobility enhancement, velocity overshoot enhancement and many kinds of quantum effects that will yield new device concepts. These will contribute to the advancement of device physics and new device applications, and will also be very useful in the analysis of future mesoscopic devices.

Acknowledgement

A part of this study was supported financially by The Japan Securities Scholarship Foundation (No. 974). Major part of this research was financially supported by the Kansai University Research Grants: Grant-in-Aid for Joint Research, 1998.

References

1. Yoshimi, M., Wada, T., Kato, K. and Tango, H.(1987) High performance SOI MOSFET using thin SOI film, Ext. Abstract of 1987 IEEE IEDM, 640-643.
2. Vasudev, P. K., Terrill, K. W. and Seymour, S.(1988) A High Performance Submicrometer CMOS/SOI Technology Using Ultrathin Silicon Films on SIMOX, Tech. Dig. of 1988 Symp. on VLSI Technol.,61-62.
3. Omura, Y., Nakashima, S., Izumi, K. and Ishii, T.(1991) 0.1-μm-Gate, Ultrathin-Film CMOS Devices Using SIMOX Substrate with 80-nm-Thick Buried Oxide Layer,Ext. Abstract of 1991 IEEE IEDM, 675-678.
4. Omura, Y., Horiguchi, S., Tabe, M. and Kishi, K.(1993) Quantum-mechanical effects on the threshold voltage of ultrathin-SOI nMOSFET's, IEEE Electron Devices Lett., **14**, 569-571.
5. Izumi, K., Doken, M. and Ariyoshi, H.(1978) CMOS devices fabricated on buried SiO2 layers formed by oxygen implantation into silicon, Electron. Lett., **14**,593-594.

6. Omura, Y.(1995) Two-Dimensionally Confined Carrier Injection Phenomena in Sub-10-nm-Thick SOI Insulated-Gate pn-Junction Devices, Ext. Abstract of 1995 Int. Conf. on Solid State Devices and Materials, 563-565.

7. Omura, Y. and Izumi, K.(1996) Quantum Mechanical Influences of Short-Channel Effects in Ultra-Thin MOSFET/SIMOX Devices, IEEE Electron Devices Lett., 17, 300-302.

8. Omura, Y., Ishiyama, T., Shoji, M. and Izumi, K.(1996) Quantum mechanical transport characteristics in ultimately miniaturized MOSFETs/SIMOX, Proc. of 7th Int. Symp. on SOI Technol. and Dev.(The Electrochem. Society), Vol. PV96-3, 199-211.

9. Colinge, J. P., Baie, X. and Bayot, V.(1994) Evidence of Two-Dimensional Carrier Confinement in Thin n-Channel SOI Gate-All-Around (GAA) Devices, IEEE Electron Device Lett., 15, 193-195.

10. Omura, Y. and Nagase, M.(1994) Low-Temperature Drain Current Characteristics in Sub-10-nm-Thick SOI nMOSFET's on SIMOX Substrates, Ext. Abstract of 1994 Int. Conf. on Solid State Devices and Materials, 993-995.

11. Omura, Y.(1996) Two-Dimensionally Confined Injection Phenomena at Low Temperatures in Sub-10-nm-Thick SOI Insulated-Gate p-n-Junction Devices, IEEE Trans. on Electron Devices, 43, 436-442.

12. Omura, Y.(1996) Negative Conductance Properties in Extremely Thin Silicon-on-Insulator Insulated-Gate pn-Junction Devices (Silicon-on-Insulator Surface Tunnel Transistor), Jpn. J. Appl. Phys., 11A (Part 2), L1401-L1402.

13. Omura, Y.(1998) Features of Indirect-Band-to-Band Tunneling in an Insulated-Gate Lateral pn Junction Device on a SIMOX Substrate with an Ultrathin 10-nm-Thick Silicon Layer, Proc. of 3rd European Workshop on Low Temperature Electronics(San Miniato, Italy, June, 1998); J. Phys. IV, Vol. 8, Pr3-63 - 66.

14. Yoshimi, M., Hazama, H., Takahashi, M. and Tango, H.(1988) Observation of Mobility Enhancement in Ultrathin SOI MOSFET's, Electron. Lett., 24, 1078- 1079.

15. Shoji, M., Omura, Y. and Tomizawa, M.(1995) Universal Mobility Behavior in Si/SOI Inversion Layers and Mobility Degradation in Extremely Thin Si/SOI, Proc. of 1995 IEEE Int. SOI Conf., 108- 109.

16. Ezawa, H.(1971) Phonons in a Half Space, Ann. Phys., 67, 438-460.

17. Takagi, S., Koga, J. and Toriumi, A.(1997) Mobility Enhancement of SOI MOSFETs Due to Subband Modulation in Ultra-Thin SOI Films, Ext. Abstract of the 1997 Int. Conf. on Solid State Devices and Mat., 154-155.

18. Shoji, M and Horiguchi, S.(1997) Phonon-Limited Electron Mobility in Ultra-Thin SOI MOSFETs, Ext. Abstract of the 1997 Int. Conf. on Solid State Devices and Mat., 156-157.

19. Omura, Y. and Izumi, K.(1994) Physical Background of Substrate Current Characteristics and Hot-Carrier Immunity in Short-Channel Ultrathin-Film MOSFET's/SIMOX, IEEE Trans. on Electron Devices, 41, 352-358.

20. Momose, H. S., Ono, M., Yoshitomi, T., Ohguro, T., Nakamura, S., Saito, M. and Iwai, H.(1994) Tunneling gate oxide approach to ultra-high current drive in small-geometry MOSFETs,Ext. Abstract of 1994 IEEE IEDM, 593-596.

21. Ohba, R. and Mizuno, T.(1997) Non-Stationary Electron/Hole Transport in Sub-0.1 μ m MOS Devices-Degradation Mechanism and Low-Power Applications-, Ext. Abstract of the 1997 Int. Conf. on Solid State Devices and Mat., 158-159.

22. Henrickson, L., Peng, Z., Frey, J. and Goldsman, N.(1990) Enhanced Reliability in Si MOSFET's with Channel Lengths under 0.2 micron, Solid-St. Electron., **33**, 1275.

23. Omura, Y.(1995) A Simple Model for Substrate Current Characteristics in Short-Channel Ultrathin-Film Metal-Oxide-Semiconductor Field-Effect Transistors by Separation by Implanted Oxygen, Jpn. J. Appl. Phys., **34**, 4722-4727.

24. Omura, Y.(1995) An Improved Analytical Solution of Energy Balance Equation for Short-Channel MOSFET/SOI and Transverse-Field Induced Carrier Heating, IEEE Trans. on Electron Devices, **42**, 301-306.

25. Fossum, J. G. and Krishnam, S.(1993) Current-Drive Enhancement Limited by Carrier Velocity Saturation in Deep-Submicrometer Fully Depleted SOI MOSFET's, IEEE Trans. on Electron Devices, **40**, 457.

26. Teilel, S. L. and Wilkins, J. W.(1983) Ballistic Transport and Velocity Overshoot in Semiconductors-I: Uniform Field Effects, IEEE Trans. on Electron Devices, **30**, 150.

27. Nakajima, Y., Takahashi, Y., Horiguchi, S., Iwadate, K., Namatsu, H., Kurihara, K. and Tabe, M.(1994) Quantized Conductance of a Silicon Wire Fabricated Using SIMOX Technology, Ext. Abstract of the 1994 Int. Conf. on Solid State Devices and Materials, 538-540.

28. Omura, Y., Kurihara, K., Takahashi, Y., Ishiyama, T., Nakajima, Y. and Izumi, K.(1997) 50-nm-Channel nMOSFET/SIMOX with an Ultrathin 2- or 6-nm-Thick Silicon Layer and Their Significant Features of Operations, IEEE Electron Device Lett., **18**, 190-192.

29. to be submitted to IEEE Electron Device Lett.

30. Matsumoto, K., Ishii, M., Shirakashi, J., Segawa, K., Oka, Y., Vartanian, B. J. and Harris, J. S.(1995) Comparison of Experimental and Theoretical Results of Room Temperature Operated Single Electron Transistor made by STM/AFM Nano- Oxidation Process, IEEE IEDM(Washington, DC), pp. 363-366.

31. Suematsu, T., Kohno, Y., Suzuki, N., Watanabe, M., and Asada, M.(1993) Metal(CoSi2) /Insulator(CaF2) Resonant Tunneling Transistor, Ext. Abstract of 1993 IEEE IEDM, 553-556.

32. Namatsu, H., Horiguchi, S., Takahashi, Y., Nagase, M. and Kurihara, K.(1997) Fabrication of SiO2/Si/SiO2 Double Barrier Diodes using Two-Dimensional Si Structures, Jpn. J. Appl. Phys., **36**, 3669.

Wafer Bonding for Micro-ElectroMechanical Systems (MEMS)

CYNTHIA A. COLINGE
California State University
Dept. of Electrical and Electronic Engineering
6000 J Street
Sacramento, California 95819-6019

1. Introduction

The wafer bonding technique was originally developed to produce SOI substrates, in which case hydrophilic oxide-to-oxide bonding is involved. Later on it became possible to bond other types of materials including bare (hydrogen terminated) silicon, compound semiconductors and nitride materials. Now wafer bonding is used to produce not only SOI substrates, but high-voltage silicon devices, optical devices, and micro-electro-mechanical systems. In this paper a review of the various processes for microfabrication of micro-electro-mechanical systems (MEMS) devices will reveal that silicon direct wafer (SDB) bonding combined with techniques such as bulk and surface micromachining, has added advantages for three-dimensional applications. We will report results showing that SDB can be used for the construction of complex structures. Furthermore, we will discuss the advantages of nitride bonding and present examples of practical applications utilizing this technique.

2. Review of MEMS Processing Techniques

The successful design and fabrication of microstructures with 3-D features has led to the development of various constructional technologies. The most commonly used techniques for fabricating MEMS are bulk and surface micromachining, and LIGA. Bulk micromachining involves the removal of material from a substrate by wet or dry etching with silicon etchants. The thickness of bulk micromachined devices can range from several microns to 600 μm. Examples of bulk micromachined devices are traditional piezoresistive pressure and acceleration sensors, most microvalves and micropumps, as well as ink jet print heads, light modulator arrays, implantable neural probes and laterally-driven resonant devices [1,2,3]. Although there are a variety of devices that can be made using bulk micromachining, the restrictions on feature shape can be severe. Isotropic etches can produce unwanted undercutting of patterned layers while anisotropic etches are bound by the silicon <111> crystal planes. Therefore, both processes constrain the possible shapes of mechanical structures. Deep reactive ion etching is becoming an excellent replacement for wet etching due to its ability to produce high aspect ratio structures independent of crystal orientation.

P.L.F. Hemment et al. (eds.),
Perspectives, Science and Technologies for Novel Silicon on Insulator Devices, 269–280.
© 2000 *Kluwer Academic Publishers. Printed in the Netherlands.*

Surface micromachining uses deposited or grown thin-films on the surface of the silicon substrate as sacrificial layers which support additional deposited films such as polycrystalline silicon or metals. The thickness of surface micromachined devices is restricted to the thickness of the grown/deposited layer, typically < 10μm. However, there are numerous examples of surface micromachined devices such as micromotors, polysilicon based accelerometers, aluminum micromirror devices, polysilicon resonant structures, and others [4]. Compared to the bulk micromachining process, surface micromachining does not suffer from constraints due to crystal orientation.

In the LIGA process, synchrotron X-rays are used to project a pattern into a thick resist layer. Because of the absence of diffraction, the limit on depth is the available thickness of stress-free resist (~1000 μm). The resist structure can be used directly in some applications, or replicated in different materials (e.g., plastics, metals and ceramics) by electroplating and molding. LIGA has been used to form assembled mechanisms, and has been combined with sacrificial layer processing to form integrated movable structures. The main drawback of the LIGA process is the expense of the exposure source. As a result, lithographically inferior, but much cheaper alternatives, such as excimer lasers are currently being investigated [5].

3. Introduction to Wafer Bonding

"Wafer bonding" refers to the phenomenon where mirror-polished, flat and clean wafers of almost any material, when brought into contact at room temperature, are locally attracted to each other by Van-der-Waals forces or hydrogen bridge. Wafer bonding is alternatively also known as "direct bonding" or "fusion bonding". The bonding at room temperature is usually relatively weak compared to the covalent bonds of a semiconductor. Therefore, for many applications the room temperature bonded wafers must undergo a heat treatment to strengthen the bonds across the interface. Because of the large number of papers published on wafer bonding over the last decade, refer to the proceedings of a series of symposia devoted to wafer bonding [6,7,8,9], recent review articles [10,11,12,13], and a special 1995 issue of the Philips Journal of Research [14].

In 1985 Lasky proposed a mechanism to describe the bonding process between two oxidized wafers at both room temperature and high temperature [15]. This initial work was used to produce SOI by bonding two oxidized wafers together at room temperature, annealing the bond at high temperature in an oxygen ambient, and thinning one of the wafers to a desired SOI thickness. Since then the process has been perfected and a more thorough understanding of the chemical reactions at the bonded interface are known.

Hydrophilic silicon wafer bonding involves the following steps:

    i) The surfaces of two silicon wafers, which may or may not contain structures such as cavities are conditioned chemically to enhance hydrophilicity in preparation for the bonding process. The wafer surfaces are covered by one or two monolayers of water.

    ii) The two wafer surfaces are brought into contact at room temperature in a sufficiently clean environment (either in a conventional cleanroom or in a "microcleanroom", see e.g.[16]) in order to avoid particles between the wafers.

    iii) Directly after room temperature bonding, the adhesion between the two wafers is determined by Van der Waals interactions or hydrogen bridge bonds which is one or two orders of magnitude lower than typical for covalent bonding. For most practical applications a higher bond energy is required which may be accomplished by an appropriate heating step ranging from 200 °C to 1100°C.

Stengl *et al.*, developed the first detailed model for direct wafer bonding for oxidized wafers which describes the bonding chemistry at different temperatures [17]. When silica surfaces are hydrated, water molecules cluster on the oxidized wafer surface as shown in Figure 1.

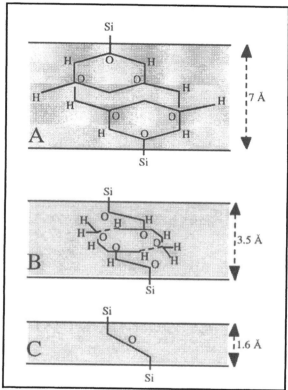

Figure 1: Stengl's et al. proposed model for silicon wafer bonding at different temperatures. A: Room temperature, $SiOH:(OH_2)_2:(OH_2)_2:HOSi$; B: T = 200°C, $SiOH:HOSi + (H_2O)_4$; C: T > 700°C, $SiOSi + H_2O$

When two such surfaces are brought into contact, hydrogen bonding occurs via the adsorbed water. At temperatures above 200°C the adsorbed water separates from the SiOH group and forms a tetramer water cluster. Within Stengl's model for temperatures greater than 700°C, the water clusters decompose and diffuse away leaving Si-O-Si bonds. Maszara's experimental findings appeared to agree with Stengl's model, where reaction bonding proceeds by two different reactions [18]. In Maszara's work, bonded wafers were measured for surface energies (the energy required to open the bonded wafers, measured using the crack propagation method). Surface energies for room temperature contact bonded wafers varied between 60-85 ergs/cm^2 which is consistent with the surface energy of silica bonded through hydrogen bonding. In addition, the surface energy increased at the transition temperature of 300°C. This is the temperature where hydrogen bonds begin to convert to Si-O-Si bonds. Surface energy was constant for bonded wafers annealed in the region of 600°C to 1100°C for all anneal times between 10 seconds to 6 hours. Maszara concluded that the bonding process does not involve mass transport in that temperature range. Rather the surface energy is limited in that regime by the amount of contacted area of the bonded wafers, which is a function of how well the wafers can elastically deform at a specific temperature. The kinetics of the deformation are so fast that the bond strength appeared to be a function of temperature only. For temperatures greater than 1100°C, the surface energy did not increase with time of anneal, but this was attributed to the viscous flow of the oxides at these high temperatures.

In contradiction to the Stengl *et al.* model, it has been shown recently that long time annealing (> 40 hours) at temperatures as low as 120°C can lead to the reaction of SiOH groups across the bonding interface according to

$$Si\text{-}OH + HO\text{-}Si \quad \rightarrow \quad Si\text{-}O\text{-}Si + H_2O \qquad (1)$$

The water molecules resulting from this reaction, as well as the water molecules present on the hydrophilic wafer surface prior to the initial bonding, can react with the surrounding silicon by the reaction

$$2H_2O + Si \quad \rightarrow \quad SiO_2 + 2H_2 \qquad (2)$$

After bonding in air the surface energy can increase up to 1,000 - 1,500 ergs/cm^2. However, when bonding in air the full surface energy of 2,000 - 3,000 ergs/cm^2 can not be reached at low anneal temperatures. The reason appears to be the interference of trapped nitrogen molecules at the bond interface locally preventing reaction (1) to occur. Indeed, if room temperature wafer bonding is performed under vacuum conditions (a few Torr) almost full surface energy and bonding strength may be reached at 120°C [19].

4. Review of Wafer Bonding for MEMS Applications

Wafer bonding can be used with bulk or surface micromachining, to create unique three-dimensional structures. Examples of several types of bonding processes and the applications of each including; bonding of sacrificial layers, bonding of different crystal orientations, and bonding with etch-stops will be examined here. Additionally, the advantages of bonding nitride materials for the realization of microfluidic valves AND piezoresistive sensors will be presented.

4.1. SACRIFICIAL DIRECT WAFER BONDING

In situations where layer patterning is desired for realization of bridges and cantilevers across deep holes or grooves, conventional techniques in photolithography will not be sufficient. By means of low-temperature sacrificial wafer bonding, layer deposition, resist spinning, and layer patterning, the realization of structures across deep groves in a silicon wafer is possible. V.L. Spiering *et al.*, successfully used this process to fabricate thin, flat, and dimensionally precise metal bridges across trenches for the purpose of making electrical contact as shown in Figure 2.

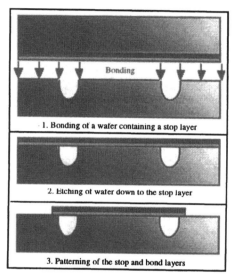

Figure 2: Aluminum bridges across deep silicon groves [20].

4.2. BONDING (100) TO (110) SILICON

Because anisotropic etchants are highly selective to the silicon <111> plane, the ability to fabricate densely packed diaphragms in (100) silicon is limited; windows opened and etched at the backside of the wafer are relatively large compared to the membrane formed on the front-side. However, T. Fujii *et al.*, combined technologies of wafer bonding, polysilicon preferential polishing, and silicon micromachining in a unique way, to develop densely packed microdiaphragm pressure sensors as shown in Figure 3. The new structure can accommodate many membranes in a small area using wafer bonding of (100) to (110) silicon and anisotropic etching. Additionally, this technique enables the use of large scale integration (LSI) to achieve intelligent systems on a single chip.

274

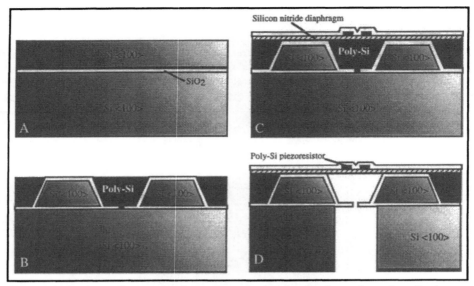

Figure 3: Process flow and schematic cross-sectional view of a microdiaphragm pressure sensor. A: wafer bonding; B: sacrificial poly-Si for cavity formation; C: microdiaphragm formation; D: sensor formed by anisotropic etching [21].

4.3. BONDING AND THINNING

Bonding and thinning techniques can be useful for fabrication of thin single-crystal materials which are important for the integration of both a sensor and electronic circuitry. Y. Matsumoto *et al.*, fabricated a capacitive accelerometer using silicon direct bonding of two oxidized wafers as shown in **Figure 4**. The bottom wafer ("handle wafer") formed one side plate for a capacitive sensor. The middle, silicon dioxide bonded layer, determined the gap of the capacitive sensor (1μm). The top layer (10 μm single-crystal silicon) was fabricated by grinding and polishing the bonded wafers. This layer formed the mass and beams, and also formed another plate of the capacitive sensor. Many holes were etched in the mass plate and subsequently the silicon dioxide was removed through the holes as a sacrificial layer.

5. Nitride Wafer Bonding

Nitride is an ideal material for microfluidic devices such as valves and microchannels due to its inertness in harsh environments; oxygen, water and sodium do not react with nor diffuse through it. In addition, it is the material of choice for diaphragms because thin nitride membranes are robust and inherently dielectrically isolate conductive materials. It has the added benefit of being highly selective as a masking material, important for bulk micromachining, when etching silicon in solutions of tetra-methyl-ammonium-hydroxide (TMAH) or potassium hydroxide (KOH). Although there are many publications on bonding of silicon, oxides, and compound semiconductors, little is known about bonding with nitrides. Therefore experiments were performed to analyze the feasibility of bonding "low-stress" membrane quality nitride. In particular surface pretreatments (cleaning techniques) and "bond strength" (surface energies) data were analyzed. Following the experimental studies, two MEMS test structures were fabricated; a membrane for a microfluidic valve and a pressure sensor including a membrane with piezoresistors.

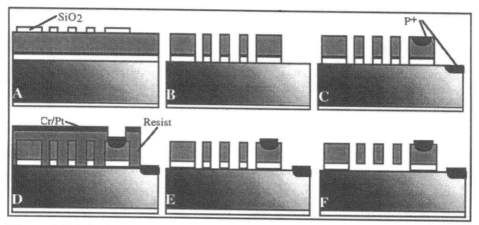

Figure 4: Fabrication process of an SOI accelerometer based on bond and etch/grind technology. A: Oxidation and lithography; B: RIE; C: Boron implantation; D: Lithography and Cr/Pt evaporation; E: Cr/Pt lift-off; F: SiO_2 etching [22].

5.1 CLEANING TECHNIQUES

Nitride, grown by low pressure chemical vapor deposition (LPCVD), was cleaned using a variety of wet and dry (plasma) processes. Following the cleaning steps, samples were bonded and imaged using an infrared system designed for observation of the bonded interface. In all but one case, wafers did not bond. The cleaning step which did bond included an (1) Oxygen plasma exposure using a Branson/IPC barrel asher at 200 Watts, O_2 200 ccm, 1 Torr, ten minutes (2) Immediate DI water rinse, five minutes (3) Standard RCA 1 and RCA 2 clean, 80°C, ten minutes each (4) Spin/rinse dry [23]. To better understand the effect of the cleaning processes on the surface chemistry of the nitride, X-ray photoemission data was found for all the cleaning techniques used. Analysis of the data revealed that the chemical preparation necessary for successful bonding exhibited a hydroxyl (OH) terminated nitride surface (hydrophilic surface), while all others did not show a hydroxyl group. The hydroxyl termination was observed through shifts in the oxygen 1S peak where a value identified as an OH bond was found. This result suggests that the bonding mechanism at room temperature for nitride-nitride is probably due to a similar bonding mechanism for oxides [24] and furthemore, for successful bonding of nitride proper cleaning must be followed.

276

5.2. SURFACE ENERGIES ("BOND STRENGTH")

Once an appropriate pretreatment was found, the surface energies as a function of temperature were determined using the crack propagation method. A matrix of samples were bonded to provide a comparison to the nitride. These samples included bare silicon, oxidized silicon and oxidized nitride (oxynitride). Bonded samples were annealed for 24 hours in N_2, at room temperature, 450, 850, and 1100°C. Results shown in Figure 5 reveal that in all cases lower surface energies were found for nitride-nitride bonded samples. Surface energies were higher when nitride substrates had been oxidized prior to bonding. Even though the measured oxide layer grown on the nitride was only 25Å, it was sufficient to enhance surface energy; bulk fracture of this sample was achieved at 1100°C. This is most likely due to the ability of the hydrophilic reaction by-products (H_2O, O_2, or H_2) to diffuse through this oxide layer, increasing the degree of contact of the two surfaces. Since nitride acts as a good diffusion barrier to O_2 and H_2O these reaction by-products will remain at the bonding interface, inhibiting intimate contact of the two bonding surfaces; three-dimensional bulk fracture was not achieved for nitride-nitride bonded samples.

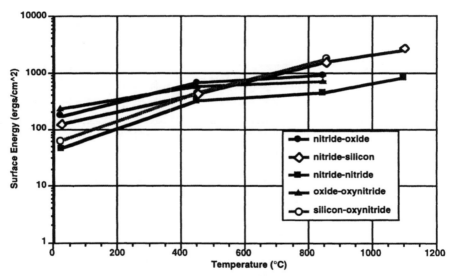

Figure 5 Surface energies plotted as a function of temperature for a matrix of samples. Absence of data points at 1100°C indicates 3-D bulk fracture of bonded samples.

5.3. NITRIDE MEMBRANES FOR MICROVALVES

The ability to bond nitride coated wafers is important for microfluidic applications operating in harsh chemical environments. Additionally, this technique allows the fabrication of membranes useful as microfluidic valves. We showed previously that the fabrication of such a structure is feasible. The substrates used for the study were double-side polished, four inch n-type, 10-20 Ω/cm (100) wafers. The wafers were bonded using the chemical treatment described above, and annealed at 1100°C, 24 hours in nitrogen ambient. Wafers were patterned using an Electronic Vision double-sided alignment tool. Square windows were patterned and opened on the front-side using reactive ion etching. An alignment step was used to align the top square windows to the back-side of the wafer. Square windows were then opened on the back-side. The wafers were etched in 25% by weight tetramethyl ammonium hydroxide (TMAH). These processing steps and a plan view of a membrane are shown in Figure 6. Although surface energies for nitride-nitride bonding are lower compared to oxides, this particular process can provide a unique way of incorporating robust membranes at a buried interface; across a four inch wafer we achieved 100% yield for all the membranes.

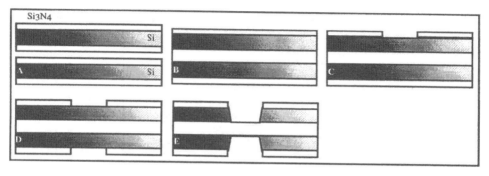

(a)

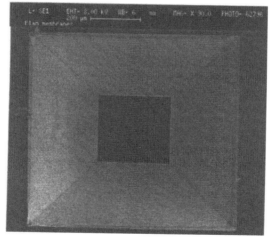

(b)

Figure 6: (a) Process flow and schematic cross-sectional representation of a buried nitride bonded membrane. A: Nitride deposition; B: Bonding; C: Pattern first window; D: Double-side align, pattern second window; E: TMAH etch. (b) SEM plan view of final structure.

278

5.4. ENCLOSED CHANNEL WITH PIEZORESISTIVE ELEMENT

Wafer bonding and etch back have become increasingly important in microelectromechanical systems (MEMS) manufacturing to create heterostructure substrates with single crystal silicon layers on oxide, nitride, or other materials [25]. We have applied the wafer bond and etch back technique to demonstrate the feasibility of fabricating a high-sensitivity single crystal piezoresistive transducer element on silicon nitride (Si_3N_4)membranes. Currently piezoresistive pressure sensors consist of polysilicon piezoresistors deposited onto either Si_3N_4 membranes [26] or p^+ silicon membranes with an intermediate oxide layer for dielectric isolation [27]. Nitride membranes are preferred for two reasons: they inherently dielectrically isolate the piezoresistors, and thin ($\approx2000Å$) membranes are realizable [28]. In addition, single crystal piezoresistors are strongly preferred for their high piezoresistive coefficients, thus making them more sensitive than polysilicon resistors [29].

Growth or deposition of single crystal silicon on Si_3N_4 is not possible using traditional microelectronics processes, however it is readily achieved by wafer bond and etch back techniques. This can be accomplished by bonding a nitride deposited wafer with boron incorporated as an etch stop. The wafer is bonded to a second support wafer with an etched cavity used to define the membrane. Once bonded, the top wafer can be processed (thinned) using the same procedures and chemistry as used for Bond and Etch back Silicon-On-Insulator (BESOI). Processing steps, shown in Figure 7, included selective etching to an SOI layer. Ultimately, the SOI is used to provide the necessary electronic circuitry for the sensor [30].

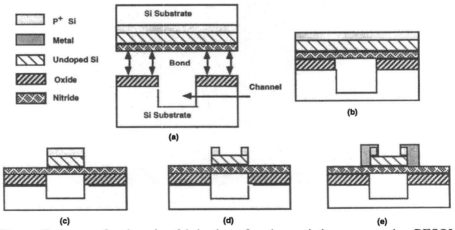

Figure 7: Process for the microfabrication of a piezoresistive sensor using BESOI. (a) Bonding step. (b) Etchback to p+ etchstop. (c) Photlithography defining piezoresistors. (d) Patterning resistors. (e) Metaliztion.

6. Conclusion

The wafer bonding technique has been expanded to bonding nitride materials. A review of various processes for microfabrication of micro-electro-mechanical systems (MEMS) devices reveal that silicon direct wafer (SDB) bonding combined with techniques such as bulk and surface micromachining, has added advantages for three-dimensional applications. The practical applications of structures and devices using SDB was outlined. Focus was placed on structures and devices with nitride membranes, and those fabricated utilizing nitride bonding. Despite the fact that bond strength and yield are typically lower for nitride-nitride compare to Si-Si bonding, we have demonstrated that wafer bonding can be used to fabricate membranes for microfluidic valves and that a modified bond and etch back process can be used to produce single crystal silicon on nitride membranes for use in the fabrication of devices such as pressure sensors.

7. References

1   E. Klaasen, K. Petersen, J.M. Noworolski, J. Logan, N.I. Maluf, J. Brown, S. Storment, W. McCulley and G.T.A. Kovacs (1996) Silicon fusion bonding and deep reactive ion etching: a new technology for microstructures, Sensors and Actuators A, **52**, 132-139

2   Y. Lindén, L. Tenerz, J. Tirén and B Hölk (1989) Fabrication of three-dimensional structures by means of doping-selective etching (DSE), Sensors and Actuators A, **16**, 83-88

3   P.W .Green, R.R.A. Syms and E.M. Yeatman (1995) Demonstration of three-dimensional microstructure self-assembly, J. Microelectromechanical Systems, **4-4**, 170-176

4   Y.B. Gianchandani, and K. Najafi (1992) A bulk silicon dissolved wafer process for micromechanical devices, J. Microelectromechanical Systems, **1-2**, 77-85

5   H. Miyajima and M. Mehregany (1995) High-aspect-ratio photolithography for MEMS applications, J. Microelectromechanical Systems, **4-4**, 220-229

6   U. Gösele, J. Haisma, M. Schmidt and T. Abe, eds. (1992) First Int. Symp. Semiconductor Wafer Bonding: Science, Technology and Applications, Electrochem. Soc. Proc. **92-7**

7   H. Baumgart, Ch. Hunt, M. Schmidt and T. Abe, eds. (1993) Second Int. Symp. Semiconductor Wafer Bonding: Science, Technology and Applications, Electrochem. Soc. Proc. **93-29**

8   H. Baumgart, Ch. Hunt, S. Iyer, U. Gösele and T. Abe, eds. (1995) Third Int. Symp. Semiconductor Wafer Bonding: Science, Technology and Applications, Electrochem. Soc. Proc. **95-7**

9   U. Gösele, H. Baumgart, Ch. Hunt, T. Abe, eds. (1997) Fourth Int. Symp. Semiconductor Wafer Bonding: Science, Technology and Applications, Electrochem. Soc. Proc. **98-36**

10   U. Gösele, H. Stenzel, M. Reiche, T. Martini, H. Steinkirchner and Q.Y. Tong (1996) History and future of semiconductor wafer bonding, Solid State Phenomena **47/48**, 33-44

11   F.S. d'Aragona and L. Ristic (1994) In Sensor Technology and Devices, ed. L Ristic, Boston:Artech House, 157- 201.

280

12   J. Haisma, B.A.C.M. Spierings, U.K.P. Biermann and A.A. van Gorkum (1994) Diversity and feasibility of direct bonding: a survey of a dedicated optical technology, Appl. Optics 33, 1154-1169

13   W.P. Maszara (1991) Silicon-on-insulator by wafer bonding: a review, J. Electrochem. Soc. 138, 341-347

14   Special Issue on Direct Bonding (1995) Philips Journal of Research 49, 1-182

15   J.B. Lasky, S.R. Stiffler, F.R. White, F.R. Abernathery (1985) Silicon-on-Insulator (SOI) by Bonding and Etch-Back, IEDM Tech. Dig. , 684-687

16   G. Cha, W.S. Yang, D. Feijo, W.J. Taylor, R. Stengl, and U. Gösele (1992) Electrochem. Soc. Proc. 92-7, 249

17   R. Stengl, T. Tan, and U. Gösele (1989) A Model for the Silicon Wafer Bonding Process, Jpn. J. of Appl. Physics, 28-10, 1735-1741

18   W. P. Maszara. G. Goetz, A. Cavigilia, and J.B. McKitterick (1988) Bonding of Silicon Wafers for Silicon-on-Insulator, J. Appl. Phys., 64-10, p. 4943-4950

19   Q.Y. Tong, L.J. Kim, T.H. Lee and U Gösele (1998) Low-vacuum wafer bonding, Electrochem Solid-State Letters 1, 52-53

20   V.L. Spiering, J.W. Berenschot, M. Elwenspoek and J.H.J. Fluitman (1995) Sacrificial wafer bonding for planarization after very deep etching, J. Microelectromechanical Systems, 4-3, 151-157

21   T. Fujii, Y. Gotoh and S. Kuroyanagi (1992) Fabrication of microdiaphragm pressure sensor utilizing micromachining, Sensors and Actuators A, 34, 217-224

22   Y. Matsumoto, M. Iwakiri, H. Tanaka, M. Ishida and T. Nakamura (1996) A capacitive accelerometer using SDB-SOI structure, Sensors and Actuators A, 53, 267-272

23   W. Maszara,, G. Goetz, A. Caviglia and J.B. McKitterick (1988) Bonding of silicon wafers for Silicon-on-Insulator, J. Appl. Phys., 64-10, 4943-4950

24   C.A. Desmond, J.J. Olup, P. Abolghasem, J. Folta and G. Jernigan (1997) Analysis of nitride bonding, Electrochemical Society Proceedings 98-36, 171-178

25   D.W. Burns, H. Guckel, "Thin films for micromechanical sensors", J. Vac. Sci. Tech. A, 8 (4), p. 3606, July 1990

26   S. Sugiyama, I. Igarashi, "Micromachining and Its Application for Pressure Sensors", Extended Abstracts of the 21st Conference on Solid State Devices and Materials, p. 189, Tokyo, 1989

27   R. Schellin, G. Hess, "A silicon subminiature microphone based on piezoresistive polysilicon strain gauges", Sensors and Actuators A, 32, p. 555, 1992

28   P. Murphy et al., "Subminiature Silicon Integrated Electret Capacitor Microphone", IEEE Trans. on Electrical Insulation, 24 (3), p. 495, June 1989

29   G. Blasquez et al., "Capabilities and Limits of Silicon Pressure Sensors", Sensors and Actuators A, 17, p. 387, 1989

30   C.A. Desmond, R. Mlcak and D. Franz (1995) Fabrication of a high-sensitivity pressure sensor with nitride membranes and single-crystal piezoresistors using wafer bond and etch back, Electrochemical Society Proceedings 98-36, 509-517

A COMPREHENSIVE ANALYSIS OF THE HIGH-TEMPERATURE OFF-STATE AND SUBTHRESHOLD CHARACTERISTICS OF SOI MOSFETS

T. E. RUDENKO, V. S. LYSENKO, V. I. KILCHYTSKA, and
A. N. RUDENKO
Institute of Semiconductor Physics, NA S of Ukraine
252650, Kiev 28, Prospect Nauki, 45, Ukraine

1. Introduction

Thin-film SOI MOSFETs are known to have superior device properties for high-temperature applications, namely, latch-up free operation, smaller threshold voltage shifts, and much lower off-state leakage currents [1-3]. The aim of this work is a detailed characterization of the high-temperature off-state and subthreshold characteristics of SOI MOSFETs using simulations and measurements. It is shown that at high temperatures (200-300^{0}C) both subthreshold and off-state characteristics of enhancement-mode devices can be explained in terms of diffusion models adjusted for high-temperature conditions.

The measurements presented in this study were performed on 20-μm long n-channel enhancement-mode MOSFETs fabricated on ZMR SOI substrates with various film thickness and doping.

2. Off-State Characteristics

An increase in the off-state leakage current with temperature is the primary factor limiting the upper operating temperature in conventional bulk Si CMOS devices. Much lower off-state leakage currents at elevated temperatures are expected to be in SOI devices due to smaller source-drain junction areas and due to elimination of high well-to-substrate diffusion leakage [1-2,4].

In the general case, the off-state current in a SOI MOSFET consists of two components: generation current in the depleted region of the reverse-biased drain junction I_{gen} and diffusion current from undepleted part of the Si film I_{diff}. Generation current in the drain region in the n-channel device can be given by [4]:

$$I_{gen} = q \cdot d_{Si} \cdot Z \cdot \frac{n_i}{\tau_g} \left\{ \sqrt{\frac{2 \cdot \varepsilon_s \cdot k \cdot T}{q^2 \cdot N_A}} \left[\sqrt{\ln\left(\frac{N_A}{n_i}\right) + \frac{q \cdot V_D}{k \cdot T}} - \sqrt{\ln\left(\frac{N_A}{n_i}\right)} \right] \right\}, \quad (1)$$

281

P.L.F. Hemment et al. (eds.),
Perspectives, Science and Technologies for Novel Silicon on Insulator Devices, 281–293.
© 2000 *Kluwer Academic Publishers. Printed in the Netherlands.*

where d_{Si} is the silicon film thickness, Z is the device width, n_i is the intrinsic carrier concentration, τ_g is the generation lifetime, N_A is the channel doping, the other symbols have their usual meaning. The diffusion current in n-channel enhancement-mode devices and p-channel accumulation-mode devices can be expressed respectively as:

$$I_{diff} = q \cdot Z \cdot \sqrt{\frac{D_n}{\tau_m}} \cdot \left(1 - e^{\frac{-qV_D}{kT}}\right) \cdot \int_0^{d_{Si}} \frac{n_i^2}{N_A} \cdot e^{\frac{q \cdot \varphi(x,T)}{kT}} dx, \qquad (2)$$

$$I_{diff} = q \cdot Z \cdot \sqrt{\frac{D_p}{\tau_{rp}}} \cdot \left(1 - e^{\frac{q \cdot V_D}{kT}}\right) \cdot \int_0^{d_{Si}} N_A \cdot e^{\frac{q \cdot \varphi(x,T)}{kT}} dx, \qquad (3)$$

with D_n, D_p the electron and hole diffusivity, τ_m, τ_{rp} the electron and hole recombination lifetime, respectively. $\varphi(x,T)$ in (2) and (3) represents the potential distribution across the film thickness in undepleted Si film. For the case of an accumulation-mode device, one should also take into account generation in the depleted Si film under the gate In this paper, we will discuss mainly the enhancement-mode devices, except as otherwise indicated.

Notice that in a general case of non-zero gate voltages and/or non-zero flat-band voltages, the total carrier density responsible for the off-state diffusion current cannot be expressed by a commonly used approximation $(n_i^2/N_A)/d_{Si}$, but is determined by the potential distribution at proper conditions (integrals in eqns.(3), (4)).

Below is shown. that at temperatures above 150-200°C (depending on the Si film thickness and drain voltage) the off-state current is caused by the diffusion mechanism. Therefore, the main trends of the high-temperature off-current in SOI MOSFETs can be derived from eqns.(2), (3).

2.1. THE TEMPERATURE DEPENDENCE

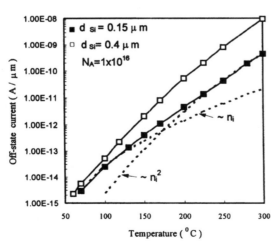

Fig.1 presents the temperature dependence of the off-current in the devices with various film thickness. One can see that in the device with a 0.15 μm film, below a temperature of 150°C, the off-current follows $n_i(T)$ that indicates the predominance of the generation current. At higher temperatures the off-current varies as $n_i^2(T)$, suggesting that the diffusion mechanism is prevailing. In the device with a 0.4 μm-thick film the diffusion current becomes dominant at T=100°C.

Figure 1. The temperature dependence of the off-current in SOI MOSFETs with various film thickness at V_D=3 V.

2.2. THE DRAIN VOLTAGE DEPENDENCE

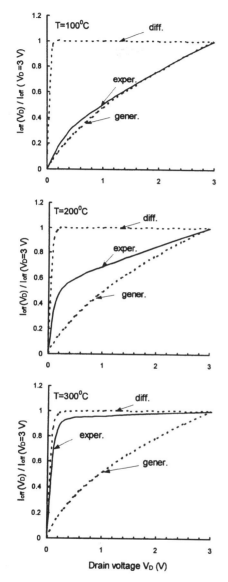

Figure 2 . The drain voltage dependence of the off-state current at various temperatures. The drain current is normalized to the current value measured at V_D=3 V (V_{gf}=-5 V, V_{gb}=0 V).

Identification of the off-current components from the drain voltage dependence is based on the fact that the diffusion current tends to saturate with V_D, when $V_D > kT/q$, whereas the generation current increases with V_D as the generation layer width (see eqns.(1)-(3)). Fig.2, which shows the off-state current as a function of the drain voltage at various temperatures, clearly demonstrates the change of the dominant mechanism of the leakage current with temperature. As can be seen from the figure, at T=100^{0}C the off-current is predominantly caused by the generation in the depleted drain region, whereas at T=300^{0}C, the off-current is entirely determined by diffusion mechanism.

The relation between generation and diffusion components depends on the drain voltage, the lower the drain voltage, the higher is the relative contribution from the diffusion current.

2.3. THE INFLUENCE OF THE SILICON FILM THICKNESS

According to eqn.(2), the diffusion current in a SOI MOSFET is determined by the potential and minority carrier concentration distributions, which vary with the gate biases and the Si film thickness.

Fig.3 shows the minority carrier concentration distributions in the SOI films with various film thickness calculated for the n-channel enhancement-mode device biased in the off-direction at T=300^{0}C. Fig.3a and Fig.3b illustrate the single- and double-gate control cases , respectively. The dashed line shows the minority carrier concentration in quasi-neutral Si with the same doping. It is evident that the carrier concentration distribution in a SOI film is nonuniform across the film thickness and highly dependent on the film thickness. An average minority carrier concentration at given conditions is essentially lower than n_i^2/N_A, even in rather thick films.

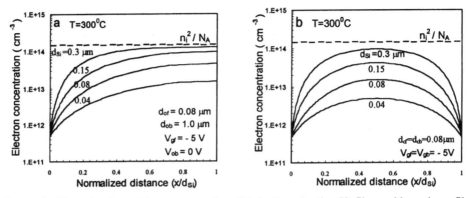

Figure 3. The minority carrier concentration distributions in the Si films with various film thicknesses in the n-channel enhancement-mode device strongly biased in the off-direction in single-gate (a) and double-gate (b) regimes at T=300°C. (N_A=1x10^{16} cm^{-3}).

In Fig.4 the total on- and off-state minority carrier densities (integrated over the film thickness) are plotted as a function of the film thickness (T=300°). Solid and dashed lines illustrate the double- and single-gate control cases, respectively. In contrast to the

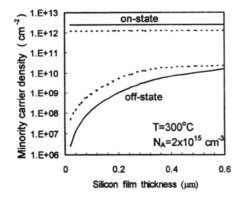

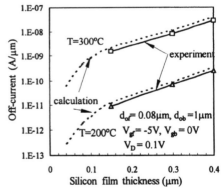

Figure 4. The total on- and off-minority carrier densities vs the film thickness in double-gate (solid line) and single-gate (dashed line) regimes

Figure 5. The experimental and calculated off-current as a function of the Si film thickness.

on-state, in which the drain current flows predominantly in the surface channels and thus is essentially independent on the film thickness, the off-state diffusion current results from the middle part of the Si film, where the minority carrier concentration is extremely sensitive to the film thickness. From Fig.4 it follows that the high-temperature off-state current and on-to-off current ratio will be strongly nonlinearly dependent on the film thickness. The measurements performed on the devices having various film thicknesses support the above conclusion (Fig.5).

2.4. THE HIGH-TEMPERATURE ADVANTAGES OF THE DOUBLE-GATE REGIME

Fig.4 clearly demonstrates the high-temperature advantages of double-gate regime (in addition to the second channel in the on-state). It is evident that for given Si film parameters the minimum achievable off-current and maximum on-to-off current ratio can be obtained for the double-gate regime. For example, for a 0.1 μm-thick film the improvement in on-to-off current ratio at T=300°C due to double-gate regime is expected to be about 10 (a factor of 2 is due to the second channel in the on-state, and a factor of 5 is due to a lowering of the off-state diffusion current).

2.5. THE BACK-GATE BIASING EFFECTS

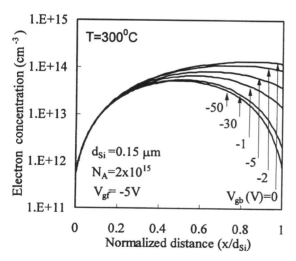

In the single-gate regime the off-state minority carrier density and thus the off-state diffusion current can be reduced by biasing the back-gate in the off-direction. It is illustrated by Fig.6, which shows the minority carrier concentration distributions across the film thickness calculated for various back-gate biases at T=300°C with the front-gate strongly biased in the off-direction. For these conditions, the front Si film interface is strongly accumulated. Increasing the back-gate voltage in the off-direction results in a decrease of the electron concentration in the

Figure 6.. The minority carrier concentration distri-butions in a SOI film with the front interface in strong accumulation and various back-gate biases.

middle and in the lower parts of the film. When the back interface exhibits strong accumulation, further biasing the back-gate in the off-direction does not noticeably change the carrier distribution. These conditions, when both front and back interfaces are in strong accumulation, correspond to the minimum achievable off-state current, as in the off-state in the double-gate regime.

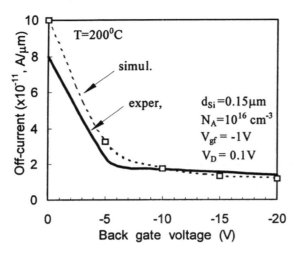

In Fig.7 the experimental off-current and simulated diffusion current are plotted as a function of the back-gate voltage.

As would be expected, increasing the back-gate bias in the off-direction decreases the high-temperature off-current. This effect then tends to saturate, which can be explained by a decrease of the total minority carrier density contributing to the diffusion current.

Figure 7. The experimental and simulated back-gate voltage dependence of the off-current at the high temperature.

2.6. OFF-STATE CHARACTERISTICS IN ACCUMULATION-MODE DEVICES

The major distinction between the off-state in enhancement- and accumulation-mode devices lies in the fact that in the latter ones there is the depleted region under the gate outside of the drain region which gives a contribution to leakage current. This leakage current component is prevailing at room and moderate temperatures as evidenced by the temperature dependence of the off-current [3]. However, at rather high temperatures the off-current in accumulation-mode devices also follows $n_i^2(T)$. This suggests a diffusion mechanism, though this is not as apparent as in the case of enhancement-mode devices, because in accumulation-mode devices the majority carriers are responsible for the off-state diffusion current (see eqn.(3)). The origin of the above temperature dependence is explained by Figs.8 and 9.

Fig.8 shows the potential and minority carrier concentration distributions in an SOI film with both front and back interfaces in strong accumulation calculated for various temperatures. Fig.9 represents the potential and majority carrier concentration distributions at the same temperatures in the same film, but with both interfaces in strong inversion. These conditions represent the limiting case of the off-state in enhancement- and accumulation-mode devices, respectively, which occurs, when both front and back gates are strongly biased in the off-direction. One can see that minority and majority carrier distributions in Figs.8b and 9b are the same. Besides, in both cases the temperature dependence of the carrier concentration and the total carrier density (integrated over the film thickness) follows n_i^2. This interesting situation for majority

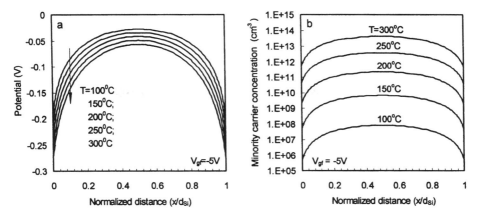

Figure 8. Potential (a) and minority carrier concentration (b) distributions in strongly accumulated SOI film at various temperatures (the off-state in an enhancement-mode device), $d_{Si}=0.15$ μm, $N_A=1\times10^{16}$ cm^{-3}.

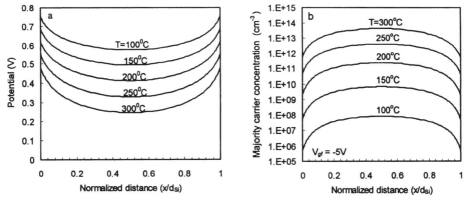

Figure 9. Potential (a) and majority carrier concentration (b) distributions in strongly inverted SOI film at various temperatures (the off-state in a accumulation-mode device), $d_{Si}=0.15$ μm, $N_A=1\times10^{16}$ cm^{-3}.

carriers results from the fact that the Si film is inverted and not quasi-neutral. For this case, the temperature dependence of the majority carrier concentration results from the temperature dependence of strong inversion potential (Fig.9a). Therefore, at sufficiently high temperatures and low drain voltages, one can expect to see the similarities of the off-current behavior in enhancement- and accumulation-mode devices.

2.7. THE INFLUENCE OF THE SILICON FILM QUALITY

The off-state leakage current is considered to be very sensitive to the crystalline quality. Indeed, the generation current is inversely related to the generation lifetime ($1/\tau_g$). On the other hand, the diffusion current is proportional to $(1/\tau_m)^{1/2}$. Therefore, in terms of leakage current, the influence of the Si film quality is expected to be more pronounced at room and moderate temperatures, than at high temperatures, when the diffusion current dominates. Besides, from eqns.(2), (3) follows that in relatively short-channel devices, in which the channel length is less than the minority carrier diffusion length, $L \leq L_n$ $\left(L_n = \sqrt{D_n \tau_m} \right)$, the diffusion current and thus the high-temperature of-state current should not depend on τ_m or L_n. This makes predictable the high-temperature characteristics of short channel devices without knowledge of these parameters depending on the material quality.

3. Subthrehold Characteristics

In this Section, we analyze the physics that is responsible for the temperature degradation of SOI MOSFET characteristics in the subthreshold region.

3.1. A CLASSICAL APPROACH

It is common practice to characterize MOSFET operation in the subthreshold (weak inversion) region by the inverse subthrehold slope factor S (subthreshold swing), which is defined as:

$$S = \frac{dV_g}{d(\log I_D)} = \ln(10) \cdot \frac{dV_g}{d\varphi_s} \cdot \frac{d\varphi_s}{dI_D} \cdot I_D, \tag{4}$$

where V_g is the (front) gate voltage, I_D is the drain current, φ_s is the (front) surface potential. Using a charge-sheet model gives: $I_D \cdot d\varphi_s / dI_D = kT / q$. The term $dV_g / d\varphi_s$, which reflects the coupling between the (front) gate voltage and the (front) surface potential, is expressed as:

$$\frac{dV_g}{d\varphi_s} = \frac{d\left(\varphi_s + \frac{Q_s}{C_o} \right)}{d\varphi_s} = 1 + \frac{C_s}{C_o}, \tag{5}$$

with C_o being the (front) gate oxide capacitance, $C_s = dQ_s / d\varphi_s$ the differential surface capacitance, for the case of SOI, C_s is "the effective" substrate capacitance (the capacitance between the inversion layer and the back gate electrode). Substituting both terms in (4) gives a classical expression [5]:

$$S = \frac{kT}{q} \cdot \ln(10) \cdot \left(1 + \frac{C_s}{C_o} \right). \tag{6}$$

It is generally believed that in the subthreshold region C_s can be obtained using the depletion approximation, so that in a bulk or thick-film (partially depleted) SOI device C_s is usually expressed as $C_s = \varepsilon_{Si} / W_{d\,max}$ with $W_{d\,max}$ the maximum depletion width. In a thin-film (fully depleted) device C_s is taken to be equal to either the capacitance of the depleted silicon film $C_s = \varepsilon_{Si} / d_{Si}$ for the case of accumulation at the back interface, or to the capacitance given by the series connection the silicon film capacitance C_{si} and the back oxide capacitance C_{ob} : $C_s = (C_{Si} \cdot C_{ob}) / (C_{Si} + C_{ob})$ for the case of depletion at the back interface. (For simplicity, in the present analysis we neglect the surface states). Therefore, according to a classical model in a fully depleted device C_s (or $dV_{gf} / d\varphi_{sf}$) is assumed to be temperature independent. As a result, a classical expression (6) foresees a superlinear temperature dependence for S in a partially depleted device (due to narrowing of $W_{d\,max}$ with temperature), and a linear temperature dependence in a thin-film, fully depleted device.

However, there are reasons to expect that at high temperatures the above approach based on the depletion approximation and a charge-sheet model is no longer valid. Using numerical simulations it will be demonstrated below that at T>150°C the classical expression (6) becomes incorrect for both thin-film, fully depleted and partially depleted devices.

3.2. BULK OR THICK-FILM SOI CASE

Fig.10 shows a classical plot of the surface charge density as a function of the surface potential for T=25°C and T=300°C. The dashed line shows the depletion approximation charge $Q_d(\varphi_s) = \sqrt{2\varepsilon_{Si} q N_A \varphi_s}$. As can be seen from Fig.10a, at T=25°C for weak inversion conditions ($\phi_F \leq \varphi_{sf} \leq 2\phi_F$), the total surface charge density is entirely determined by the depleted charge: $Q_s(\varphi_s) = Q_d(\varphi_s)$. However, at T=300°C an electron charge density (Q_n) becomes important and cannot be neglected. As a result, in

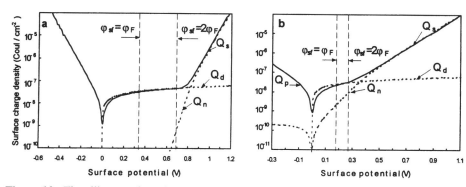

Figure 10. The silicon surface charge density as a function of the (front) surface potential in a bulk (or in a thick-film SOI) device at T=25°C (a) and T=300°C (b).

the subthreshold region the total surface capacitance $C_{stot} = dQ_{stot} / d\varphi_s$ is higher than the depletion capacitance $C_d = dQ_d / d\varphi_s$ due to an increased contribution from Q_n. According to eqn.(6) this would result in an increase in S additional to that expected due to decreasing $W_{d\,max}$.

A careful analysis of the $Q_n(\varphi_s)$-dependence reveals that at T=300°C the relationship $Q_n \sim exp(q\varphi_s/kT)$ is no longer accurate. The reason is in lowering of the surface electric field in the weak inversion region with temperature due to decreasing ϕ_F resulting in the inversion layer broadening, so that the charge-sheet model becomes inaccurate. This can be taken into account by introducing in eqn.(4) the correction factor $n = \left[(kT / q) \cdot (d \ln Q_n / d\varphi_s) \right]^{-1}$. Therefore, an expression for the high-temperature inverse subthreshold slope factor can rewritten as:

$$S = \frac{nkT}{q} \cdot \ln(10) \cdot \left(1 + \frac{C_{stot}}{C_o} \right),\tag{7}$$

where n is the correction factor, which takes into account the effect of lowering of the surface electric field in a weak inversion region with temperature (for the case of a bulk device, at T=300°C n is about 1.3-1.1, depending on the doping concentration), C_{stot} is the total surface differential capacitance, which includes the contribution from the free carriers and is higher than the depletion capacitance.

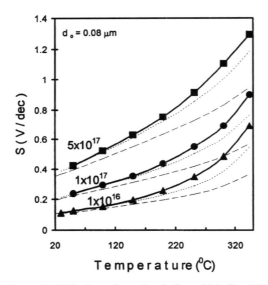

The temperature dependence of S in a bulk or partially depleted (by a conventional definition) SOI device, calculated for various doping concentrations by applying the above corrections, is shown in Fig.11. For comparison, the $S(T)$-dependence expected from the classical expression is shown in the same figure. Therefore, using the classical expression for predicting the subthreshold characteristics in bulk Si and partially depleted SOI devices at high temperatures would result in large errors, even in devices with high channel doping.

Figure 11. S(T) -dependence in a bulk or thick film SOI device for various doping concentrations calculated using the classical expression (dashed lines), with applying the corrections for the increase in C_s due to free carriers (dotted lines) and for lowering the surface electric field (solid lines).

3.3 THIN-FILM SOI CASE

Fig.12 presents the charge density components in a Si film with d_{Si}=0.15μm and N_A=1x10^{16} cm^{-3} as a function of the front surface potential at T=27^{0}C and T=300^{0}C for the case of depletion at the back interface. In the temperature range 20-300^{0}C the above device parameters (d_{Si} and N_A) ensure full depletion (by a conventional definition) of the Si film. As can be seen from Fig.12a, at T=27^{0}C in the weak inversion region the total charge density in the Si film is entirely determined by an ionized doping charge, such that one can use the depletion approximation in deriving $dV_{gf}/d\varphi_{sf}$. However, at T=300^{0}C (Fig.12b) the contributions to Q_{tot} from both electron and hole charge densities Q_n, Q_p become important and cannot be neglected. This would result in an increase in the effective substrate capacitance or. $dV_{gf}/d\varphi_{sf}$.

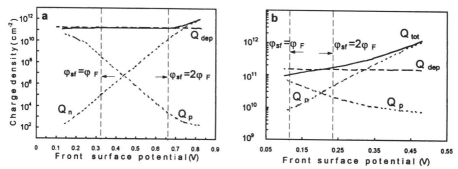

Figure 12. The charge density components in the Si film (d_{Si} =0.15 mm, N_A=1x10^{16}cm^{-3}) of a SOI MOSFET with depleted back inteface vs. the front surface potential calculated for T=27^{0}C (a) and T=300^{0}C (b).

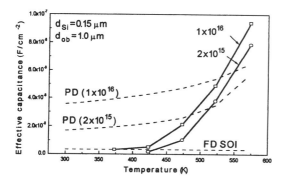

Figure 13. The effective substrate capacitance in a SOI device as a function of temperature calculated at $\varphi_{sf}=(3/2)\varphi_\phi$: Solid line - numerical simulation; dashed lines - depletion capacitance for a thick-film, partially depleted SOI device (PD) for a fully depleted device (FD).

It is illustrated in Fig.13, which shows $C_{seff}=dQ_{tot}/d\varphi_{sf}$, calculated at $\varphi_{sf}=(3/2)\phi_F$, and the capacitance obtained from the depletion approximation for the fully depleted and partially depleted device. It is seen that the capacitance calculated by taking into account the free carriers exceeds the capacitance of a fully depleted device and even can exceed the capacitance expected for "the partially depleted" device with the same channel doping. This must be taken into consideration in calculating $dV_{gf}/d\varphi_{sf}$.

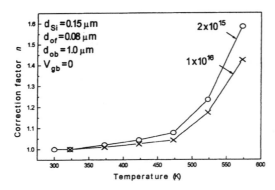

Figure 14. The correction factor *n* accounting for the inversion layer broadening as a function of temperature calculated for a SOI MOSFET with the back interface depleted

The correction factor *n* obtained from the $Q_n(\varphi_{sf})$-dependencies, which accounts for the inversion layer broadening, is plotted in Fig.14 as a function of temperature. It should be noted that for the same temperatures *n* is higher for the case of SOI than for the case of a bulk device with the same channel doping , because of the lower surface electric field. Substituting values for C_{seff} and *n* into eqn.(6) gives the $S(T)$-dependence, which can equally be determined directly from the calculated $Q_n(V_{gf})$-dependence.

Fig.15 shows the experimental and calculated temperature dependence of the subthreshold swing for the case of depletion and for the case of accumulation at the back interface for the device with a channel doping of 2×10^{15} cm^{-3}. The corresponding $S(T)$-

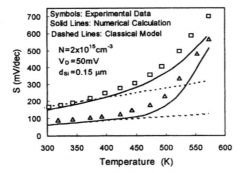

Figure 15 Experimental and calculated temperature dependencies of the subthreshold swing in a SOI MOS FET with accumulated and depleted back interface (the upper and lower curves, respectively).

dependencies expected from the classical model are shown by dashed lines. One can see, that for both cases of accumulation and depletion at the back interface the experimental subthreshold swing fits well the linear temperature dependence only at T≤150°C and rapidly rises with further increase in temperature, in spite of the fact that by the conventional definition the device remains fully depleted. The subthreshold swing obtained by numerical simulation agrees well with experimental data in the whole temperature range being studied. It should be pointed out that in the present simulation the generation leakage current was not taken into account. Therefore, the nonlinear increase of the subthreshold swing with temperature in thin-film SOI devices originates from the temperature dependence of subthreshold (diffusion) current in itself. It is related to the increase of the effective substrate capacitance (or $dV_g/d\varphi_{sf}$) caused by increasing free carrier charge densities, and to lowering the surface electric field in the weak inversion region resulting in an increase of $d\varphi_{sf}/dI_D$.

4. Conclusion

The high-temperature off-state and subthreshold characteristics of SOI MOSFETs have been explained in terms of diffusion models using both simulations and measurements. A strong nonlinear dependence on the film thickness is found to be for off-state diffusion currents. It is shown that in SOI MOFETs biased in the off-direction the carrier density responsible for the off-current is essentially lowered compared to quasi-neutral films. A dramatic decrease in the high-temperature off-state current and improvement in on-to-off current ratio is expected to be for thin films and double-gate regime due to a volume accumulation effect.

The validity of a classical expression for the subthreshold swing in SOI MOSFETs is revised for high temperature conditions. Using numerical simulation, it is demonstrated that at high temperatures (above 150^0C) in the subthreshold region a conventional approach based on the depletion approximation and a charge- sheet is no longer valid. The rise in the free carrier charge densities with temperature results in an increased effective substrate capacitance as compared to that determined from the depletion approximation. Another correction must be introduced into the classical expression to account for the inversion layer broadening caused by lowering the surface electric field in a weak inversion region with temperature. Ignoring these things would introduce large errors in predicting the subthreshold characteristics in thick-film ("partially depleted") SOI devices. Taking into account the above corrections can explain the observed nonlinear temperature dependence of the subthreshold swing in thin-film ("fully depleted") SOI MOSFETs.

5. Acknowledgments

This work has been supported by the INTAS fund (INTAS project N-INTAS-93-2075-EXT) and NATO Linkage Grant HTECH.LG 951189.

6. References

1. Colinge, J.-P. (1991) *Silicon-On-Insulator Technolgy: Materials to VLSI*, Kluwer Academic Publishers, Dordrecht.
2. Francis,P., Terao,A., Gentinne,B., Flandre,D., and Colinge,J.-P. (1992) SOI technology for high-temperature applications in *IEDM Tech. Dig.* p.13.5.1..
3. Flandre,D., Terao, A., Francis, P., Gentinne,B., and Colinge,J.P. (1993) Demonstration of the potential of accumulation-mode MOS transistors on SOI substrates for high-temperature operation (150-300^0 C), *IEEE Electron. Device Lett.*, **14**, 10-12.
4. Jeon,D.-S., Burk, D. (1991) A temperature-dependent SOI MOSFET model for high-temperature applications (27-300^{0}C). *IEEE Trans.El.Dev.***38**, 2101-2111.
5. Wouters, D.J., Colinge,J.P., and Maes,H.E. (1990) Subthreshold slope in thin-film SOI MOSFETs, *IEEE Trans. on Electron Dev.*, **37**, .2022-2033.

Influence of Silicon Film Parameters on C-V Characteristics of Partially Depleted SOI MOSFETs.

**D.Tomaszewski[1), A.Jakubowski[2), J.Gibki[2), T.Dębski[1),
M.Korwin-Pawłowski[3)**

[1)   Institute of Electron Technology, Al.Lotników 32/46,
      02-668 Warsaw, Poland
[2)   Institute of Microlectronics and Optoelectronics, Warsaw
      Technical University, ul.Koszykowa 75, 00-662 Warsaw, Poland
[3)   General Semiconductor Ireland, Macroom Co., Cork, Ireland

Abstract: Discussion of an effect of silicon film thickness, recombination lifetime and generation lifetime on C-V characteristics of partially-depleted SOI MOSFETs in the strong inversion region has been presented. The analysis has been carried out using a physical model of these devices. It has been shown, that the film thickness and recombination lifetime are relevant for accurate simulation of C-V characteristics of PD SOI MOSFETs.

1. Introduction

The fabrication of the SOI (*Silicon-On-Insulator*) CMOS integrated circuits is anticipated to become one of the most promising branches of the electronic industry. These integrated circuits (ICs) offer many advantages as compared to their conventional bulk silicon counterparts. Their packing density is greater, because they do not require large doped wells with p-n junction isolation.. In SOI ICs the isolation between n- and p-channel transistors is mainly dielectric. The power consumption in the SOI ICs is smaller, because the capacitive area of the p-n junctions is much smaller and also the temperature effect on the SOI ICs operation is reduced. Large nonlinear p-n

P.L.F. Hemment et al. (eds.),
Perspectives, Science and Technologies for Novel Silicon on Insulator Devices, 295–305.
© 2000 *Kluwer Academic Publishers. Printed in the Netherlands.*

junctions are replaced by smaller linear capacitances of the buried oxide layer. Hence the delay of signals in the SOI ICs is much smaller. Also thanks to the reduced volume of the active regions of the SOI devices they are less susceptible to a hard radiation. Also the I-V characteristics of the SOI MOSFETs are advantageous as compared to the bulk MOS devices, particularly the subthreshold slope. This, together with smaller power consumption and smaller delay make the SOI CMOS ICs very valuable in the portable apparatus and telecommunication applications. However the SOI technology is also a source of several parasitic effects, which degrade the electrical characteristics of the devices. These effects are characterized in this work. They require investigation for two main purposes: characterization of the SOI devices and computer-aided design (CAD) of the SOI ICs. Reliable models of the SOI MOSFETs can be very helpful in facilitating circuit design.

The paper describes the effect of variations of silicon layer parameters on the capacitance-voltage characteristics of partially depleted MOS SOI transistors (PD SOI MOSFETs). The analysis is based on physical DC and AC analytical models of these devices.

2. Model

The model presented here accounts for the basic phenomena present in the devices [1,2], namely:
- carrier drift below the Si-SiO$_2$ interface,
- diffusion and recombination of minority carriers in the Si film,
- avalanche multiplication of carriers in the drain-"body" junction, carriers generated in this way are injected into the film, so the source-"body" junction becomes forward biased,
- thermal generation and recombination of carriers in space charge regions.

These phenomena are involved in the current continuity equations, so DC and AC models have been derived consistently [3]. Both models result from the solution of the extracted DC and AC components of the current continuity equations. The extraction has been done using the S^{3}A (_Sinusoidal Steady-State Analysis_) approach [4]. It yields a pair of equations (1) and (2) which are solved for the DC and AC components of the body-source voltage: V_{BS}, v^*_{bs}.

$$I_S + I_D = 0 \tag{1}$$

$$\overset{*}{i_s} + \overset{*}{i_{gf}} + \overset{*}{i_d} + \overset{*}{i_{gb}} = 0 \tag{2}$$

In the above equations I_S, I_D denote DC components of the source and drain currents, respectively. The variables with asterisks denote phasors of the currents flowing into the device. The gate currents are purely capacitative, whereas the source and drain currents have both conduction and displacement components.

This method is particularly important in the case of the AC model, where eventual equivalent circuits should be only a consequence of the solution of the continuity equation. Moreover the AC model is fully non-quasistatic, so it accounts for excitation frequency effects on device conductance-voltage and capacitance-voltage characteristics.

The effects of the phenomena mentioned above can be observed on the C-V characteristics of the PD SOI MOSFETs [5]. The capacitances of these devices exhibit unusual effects in the saturation region above the V_{DS} voltage corresponding to the *kink-effect*. One of these effects is the decrease of the C_{gfd} capacitance below zero for high V_{DS} voltages. The other is the increase of the C_{gfs} capacitance beyond its typical value for the saturation region ($2/3C_{ox}$ for the conventional MOSFETs). These effects have been recorded both from the C-V characteristics obtained using numerical simulations and from the C-V curves obtained experimentally. The floating body and the avalanche multiplication are claimed to be responsible for these effects, because they activate the additional high-frequency (HF) signal paths through the floating body between device electrodes. Threfore not only ionization rates but also the parameters of the silicon film will influence I-V and C-V characteristics of the PD SOI MOSFETs.

The C-V data obtained using our model have been compared with experimental data. The parameters of the devices and the physical parameters are given in the Table 1. The results of the comparison of the data are shown in fig.1. Unfortunately the model cannot predict the measured characteristics of the physical devices accurately due to a very simple, one-dimensional model for the *pinch-off* region in the SOI MOSFET operating in the saturation range. It requires further improvement to achieve greater accuracy. Also the C-V characteristics in the subthreshold region require further investigation, however due to the fact that our model accounts for the main physical phenomena it can be used to estimate the effect of these mechanisms on the C-V data.

Table 1. Numerical values of the parameters of the PD SOI MOSFETs used for model verification.

Parameter	Symbol	Value	
channel dimensions	W	100	μm
	L	10	μm
gate and buried oxide layer thicknesses	$t_{ox,f}$	32.8	nm
	$t_{ox,b}$	400	nm
silicon film thickness	t_{Si}	150	nm
doping concentration in the silicon film	N_B	$9 \cdot 10^{16}$	cm^{-3}
carriers saturation velocity in the silicon	υ_{max}	$1 \cdot 10^7$	cm/s
electrons mobility in the inversion layer	$\mu_{c,f}$	320	cm^2/Vs
fixed charge densities at the front and back Si-SiO₂ interfaces	$N_{ss,f}$	$3 \cdot 10^{10}$	cm^{-2}
	$N_{ss,b}$	$1 \cdot 10^{11}$	cm^{-2}
recombination life-time of electrons in the "body" area	$\tau_{n,B}$	$1 \cdot 10^{-7}$	s
generation/recombination life-times of carriers in the space charge areas of the drain-"body" and source-"body" junctions	τ_J	$1 \cdot 10^{-6}$	s
parameters of the avalanche multiplication in the "pinch-off" region	α_{sat}	$1.4 \cdot 10^8$	m^{-1}
	β_{sat}	$1.5 \cdot 10^8$	V/m
parameters of the avalanche multiplication in the depletion region of the drain-"body" junction	α_b	$2 \cdot 10^7$	m^{-1}
	β_b	$1.7 \cdot 10^8$	V/m

3.   Results

In our work we analyze the effect of the following structural and material parameters variations on device C-V charateristics:

- thickness of the Si layer,
- minority carriers lifetime in the body (recombination lifetime),
- carriers lifetime in the space charge regions of source-body and drain-body junctions.

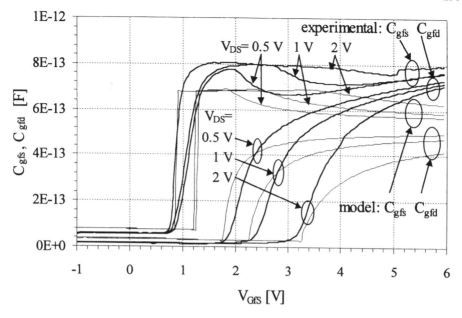

Fig.1. Comparison of the experimental $C_{gfs}(V_{Gfs})$ and $C_{gfd}(V_{Gfs})$ characteristics of the SOI MOSFET and of the model; thicker lines correspond to the experimental data; parameters are listed in Table 1.

3.1 Effect of the silicon layer thickness

We have computed the $C(V_{DS})$ characteristics of the PD SOI MOSFET for three values of the Si film thickness: 150 nm (nominal), 200 nm and 300 nm. The overall results are shown in figures 2 (C_{gfs}, C_{sgf} capacitances) and 4 (C_{gfd}, C_{dgf}) whilst finer details are shown in figures 3 and 5. The data have been obtained for the following transistor bias: $V_{GbS}=0$ V, $V_{Gfs}=2$ V.

The results presented on charts show that an increase of the Si film thickness (i) reduces the effects of the floating body on the C-V characteristics (ii) increases the C_{gfs} capacitance and (iii) decreases of the C_{gfd} capacitance in the range beyond the kink voltage. Then the transistor characteristics approach the ones for a conventional MOSFET. These conclusions have been also drawn from a consideration of the DC transistor characteristics.

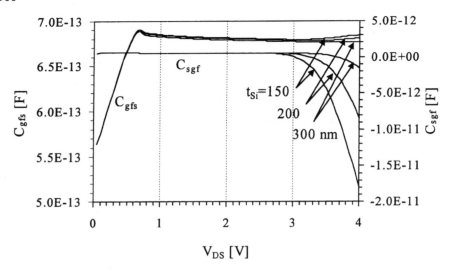

Fig.2. The $C_{gfs}(V_{DS})$, $C_{sgf}(V_{DS})$ characteristics of the model for three Si film thicknesses: 150, 200 and 300 nm (V_{GbS}=0 V, V_{GfS}=2 V). Details are shown in fig.3.

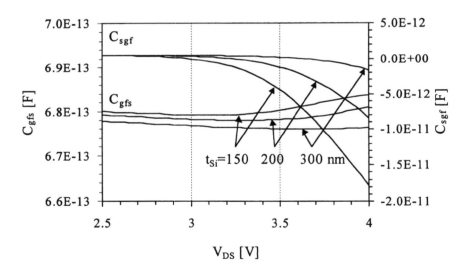

Fig.3. Details of the voltage dependence of C_{gfs} and C_{sgf} calculated for three Si film thicknesses.

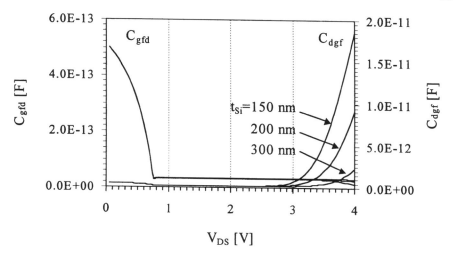

Fig.4. The $C_{gfd}(V_{DS})$, $C_{dgf}(V_{DS})$ characteristics of the model for three Si film thicknesses: 150, 200, 300 nm.

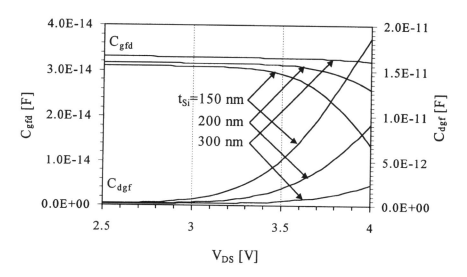

Fig.5. Details of the calculated $C_{gfd}(V_{DS})$, $C_{dgf}(V_{DS})$ characteristics for three Si film thicknesses

3.2 Effect of the recombination lifetime in the body.

We have computed the C-V_{DS} characteristics for PD SOI MOSFETs for three values of the recombination lifetime in the Si film, namely: 10^{-8} s, 10^{-7} s (nominal) and 10^{-6} s. The details of the results are shown in figures 6 and 7 where the voltage (V_{DS}) dependence of C_{gfs}, C_{sgf} and C_{gfd}, C_{dgf} are plotted, respectively.

The results of these computations show that the recombination lifetime in the body strongly influences the C-V curves deep in the saturation region whilst a decrease of the lifetime due to the poor silicon film quality reduces the floating body effects in the PD SOI MOSFETs. In this case the excess holes injected into the body via avalanche multiplication recombine with electrons more efficiently. Thus the changes in the body potential are smaller and the C-V data are more similar to the chatracteristics of the conventional bulk silicon MOSFET. These conclusions are consistent with the deductions from the DC data.

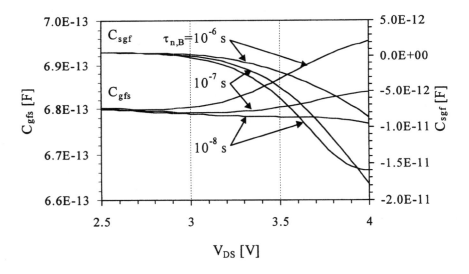

Fig. 6.   Details of the calculated $C_{gfs}(V_{DS})$, $C_{sgf}(V_{DS})$ characteristics for three values of the recombination lifetime in the body

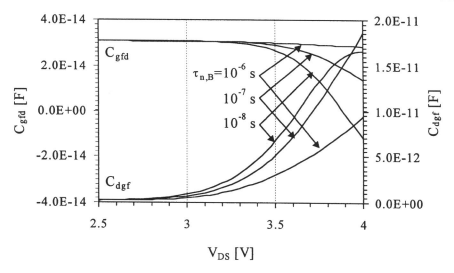

Fig. 7.   Details of the calculated $C_{gfd}(V_{DS})$, $C_{dgf}(V_{DS})$ characteristics for three
values of the recombination lifetime in the body

3.3   Effect of the generation/recombination in the junctions space charge areas.

We have computed the $C(V_{DS})$ characteristics of PD SOI MOSFETs for
three values of the generation/recombination lifetime in the space charge region
of the junctions: 10^{-7} s, 10^{-6} s (nominal) and 10^{-5} s. The details of the
simulations are shown in figure 8.

The results of computations show that the generation/recombination
lifetime in the space charge region does not effect the C-V data. Thus we can
arrive at the conclusion that the AC current components related to the
phenomena in these regions do not influence the body-source voltage phasor.
However the τ_J variable is relevant for the determination of the I-V
characteristics in the low-voltage range and in the subthreshold range [3].
These ranges have not been accounted for in the discussion mentioned above.

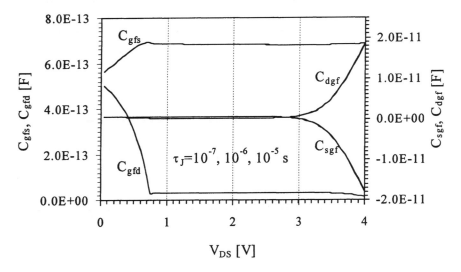

Fig.8. The computed $C_{gfs}(V_{DS})$, $C_{sgf}(V_{DS})$, $C_{gfd}(V_{DS})$, $C_{dgf}(V_{DS})$ characteristics for three generation/recombination lifetimes in the junction depletion region.

4.   Conclusions

The dependence of the C-V characteristics upon silicon layer thickness, recombination lifetime and generation lifetime have been computed using a new mathematical model of the PD SOI MOSFET. The simulation results are generally consistent with experimental data. However our model requires further improvement, mainly in the subthreshold range. Nevertheless, as the model accounts for the main physical phenomena present in the real devices, it can be used to estimate the effect of these mechanisms on the C-V data.

Acknowledgment

The authors wish to thank Prof. P.L.F.Hemment for a critical review and editorial improvements of this paper.

Bibliography

1. Colinge, J.-P. (1991) *Silicon-On-Insulator Technology: Materials to VLSI*, Kluwer Academic Publishers, Boston/Dordrecht/London

2. Cristoloveanu, S., Li, S.S. (1995) *Electrical characterization of Silicon-On-Insulator materials and devices*, Kluwer Academic Publishers, Boston/Dordrecht/London

3. Tomaszewski, D. (1998) *A small-signal model of a SOI MOSFET capacitances*, PhD thesis, ITE, Warsaw

4. Laux, S.E. (1985) Techniques for Small-Signal Analysis of Semiconductor Devices, *IEEE Trans. on Electron Devices*.**32**, 2028-2037

5. Flandre, D. (1993) Analysis of Floating Substrate Effects on the Intrinsic Gate Capacitance of SOI MOSFET's Using Two-Dimensional Device Simulation, *IEEE Trans. on Electron Devices* **40**, 1789-1796

6. Tomaszewski, D., Jakubowski, A., Gibki, J. (1998) A Small-Signal Non-Quasistatic Model of PD SOI MOSFETs, *4th Symposium Diagnostics and Yield, SOI - materials, devices and characterizaton, D&Y'98*, Warsaw, April 22-25

7. van der Ziel, A. (1976) *Solid State Physical Electronics*, 3rd Ed., Prentice-Hall

EFFECT OF SHALLOW OXIDE TRAPS ON THE LOW-TEMPERATURE OPERATION OF SOI TRANSISTORS

V.S.LYSENKO, I.P.TYAGULSKI, I.N.OSIYUK and Y.V.GOMENIUK
Institute of Semiconductor Physics, National Academy of Sciences of Ukraine
Prospect Nauki 45, 252650 Kiev, Ukraine

In this paper investigations of low-temperature operation of n-channel accumulation mode SOI MOSFET are reported. It has been shown that the charge state of shallow traps situated in the thin transition layer between the gate oxide and silicon overlayer affects the transistor characteristics at temperatures below 20 K, giving rise to the peaks in transconductance.

1. Introduction

It is recognized that device operation at low temperatures greatly improves the electrical performance of CMOS circuits [1]. The importance of accumulation mode (AM) SOI MOSFETs with well controlled threshold values, sharp subthreshold slopes and an absence of the floating body effect has now been widely accepted [2]. Colinge [3] has considered conduction mechanisms in thin-film AM SOI transistors and has shown that there exist three main components of the drain current, namely, current in the accumulation layer near the front interface, current in the quasi-neutral layer of the film body (body current component) and the component in the accumulation layer near the back interface. The presence and relative contribution of each component is controlled by the front- and back-gate biases.

In this paper we report, for the first time, experimental evidence for the effect on the low-temperature operation of the n-channel AM SOI transistor of shallow traps located in the thin transition oxide layer near the insulator-semiconductor interface.

2. Experimental

The investigated n^+–n–n^+ transistors were fabricated in SOI substrates prepared by the laser zone-melt recrystallization (ZMR) technique [4]. The nominal thickness of the silicon overlayer and the buried oxide were 300 nm and 1000 nm, respectively. The electron concentration in the film was about 7×10^{15} cm^{-3}. A double-layered dielectric

307

P.L.F. Hemment et al. (eds.),
Perspectives, Science and Technologies for Novel Silicon on Insulator Devices, 307–313.
© 2000 *Kluwer Academic Publishers. Printed in the Netherlands.*

was used as the gate insulator consisting of SiO_2 (40 nm) and Si_3N_4 (40 nm) layers. The width to length ratio of the AM MOSFETs was 20 μm / 500 μm. The input (I_d vs. V_g) characteristics of the transistors were measured in the temperature range from 7 K to 260 K. To avoid potential redistribution effects along the channel, the drain bias was chosen to be low (V_d=0.015 V). In [2] the similar characteristics were measured for p-channel transistors at room and liquid-helium temperatures.

It was shown in [2] that at different combinations of voltages applied to front and back gates different situations may occur giving rise to the onset of the three current components contributing to the drain current. In the I_d vs. V_g curves the onset of the different current components can be seen as "kinks" corresponding to a large change in transconductance g_m=dI_d/dV_g. If a sufficiently high positive voltage is applied to the back gate, all three possible conduction modes eventually appear when the front gate voltage varies from the negative to positive values, which corresponds to the transition from inversion to accumulation at the front interface. However, it is difficult to analyse the mode corresponding to accumulation near the back interface due to the large time constant for the formation of an inversion layer. Because of this constraint the system is essentially in a non-equilibrium state where the non-equilibrium depletion effects superimpose with changes of the static bias. At low temperatures the role of these effects increases. To avoid these complexities we created the depletion layer at the back silicon interface by applying a high negative voltage (–30 V) to the back gate (silicon substrate).

We have shown in our previous investigations of thermally stimulated and field charge release in Si-SiO_2 structures [5–7] that the thin non-stoichiometric insulating layer between the oxide and silicon contains electrically active traps, where charge exchange with the allowed bands in the semiconductor occur via a tunnel-activation mechanism. The activation energy of this process is typically tens meV giving rise to current peaks in the temperature range from 10 to 16 K. Filling of these traps is achieved by applying the appropriate bias to the gate at 20 K with subsequent cooling down to about 6 K. Switching off the filling voltage at low temperature puts the system into the non-equilibrium state when a proportion of the charge is trapped on the interface centres. In such a way we control the charge state of these centres during measurements of the transistor input characteristics.

3. Results and Discussion

Figure 1 shows the input characteristics of a transistor at different temperatures. It can be seen in the figure that there are one or two regions on the voltage axes where large variations of the drain current can be observed. These regions correspond to the onset of the different conduction mechanism. For example, in the curve at 260 K, the reduction of the current with increasing magnitude of the negative gate bias from –1.5 V to –2.5 V is related to the formation of the depletion layer in the silicon film and the increase of the thickness of this layer. However, at this temperature it is not possible to fully deplete the silicon film, since at the negative gate bias below –2 V the current is still rather high and is relatively insensitive to the gate voltage. In this case the formation of an inversion layer screens the electric field and the thickness of the depletion layer does not change with decreasing gate voltage. At a temperature of 160 K and with $V_g=-1$ V the front depletion layer reaches the boundary of the back interface depletion layer and current rapidly reduces to a lower value ($\sim 10^{-13}$ A). At zero gate bias the transistor is in the conducting state, with a value typical of the room temperature current. When temperature is lowered to 60 K the current at zero bias reduces due to both the increase of the resistivity of the quasi-neutral body region and the reduction of its thickness. At the same time the threshold voltage of the body current onset (V_{th1}) shifts towards positive voltages, reflecting the temperature dependence of the maximum thickness of the depletion layer. Figure 2 shows the dependence of the threshold voltage V_{th1} on temperature.

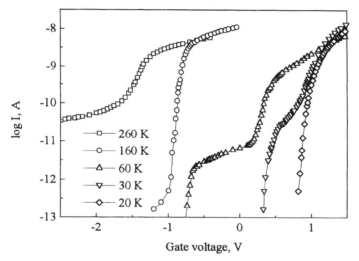

Figure 1. Input characteristics of a SOI transistor at various temperatures.

On the other hand, in the curve measured at 60 K (Figure 1) a second region exists where there is a sharp increase in the current giving rise to another threshold voltage value. This voltage is positive and corresponds to the onset of the conduction channel along the accumulation layer at the front silicon film interface. As the temperature is further reduced the body conduction channel no longer contributes to the total drain current, since the bulk resistance increases due to carrier freezing-out. At temperatures below 20 K we deal only with the component due to conduction in the front accumulation layer.

Measurements of the input characteristics at temperatures below 20 K were carried out with account of the charge state of shallow traps in the transition layer, the existence and some parameters of which are presented in [3]. Filling of the electron traps was achieved by applying at a temperature of 20 K a positive bias (+10 V) to the gate and grounding drain and source.

In figure 3 the $I_d(V_g)$ dependencies measured at temperatures 9 and 13 K with

310

empty and filled traps are shown. It can be seen that filling the centres causes a shift of the characteristics towards positive voltages, which corresponds to filling the traps with electrons. Specific features (steps) can be seen in the curves, the physical nature of which is discussed later. It should be noted that these features are not related to the effects of carrier localization at low temperatures since the current levels in the range of their manifestation are rather high. These features are better seen in the dependencies of transconductance vs. gate voltage shown in the figures 4 and 5 for empty and filled traps, respectively.

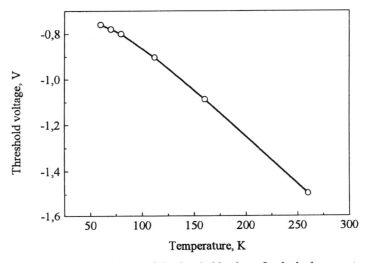

Figure 2. Temperature dependence of the threshold voltage for the body current component

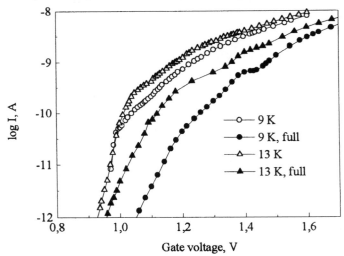

Figure 3. Input characteristics measured at low temperatures for full (full) and empty (open) interface traps.

It is seen in figure 4 that for the case of empty centres the steps in the transconductance appear at temperatures above 11 K. It is known that the shallow interface centres are not filled if the filling temperature is lower than a critical value, i.e. the filling process is thermally activated. On the other hand, the thermally stimulated charge release (TSCR) measurements have shown that the centres are fully emptied at temperatures ranging from 16 to 18 K. Small humps in the curves shown in figure 4 are explained by the fact that a positive bias applied to the gate results in the filling of shallow electron traps. This process is more pronounced at a temperature of 15 K.

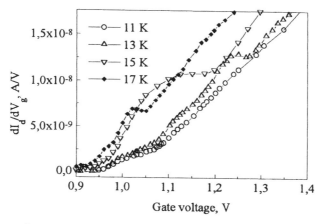

Figure 4. Transconductance measured at low temperatures when the shallow interface traps are empty.

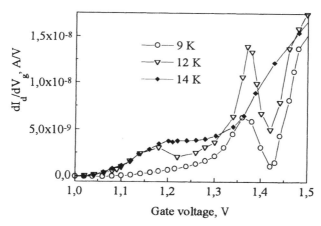

Figure 5. Transconductance measured at low temperatures when the shallow interface traps are filled.

It is noted that when the traps are filled meaningful data is only available at low temperatures, below which the TSCR peaks are observed. Therefore, in the figure 5, only the data recorded at temperatures less than 14 K are shown. It can be seen that peaks and valleys are present in the $g_m(V_g)$ curves at specific values of the gate voltage. Such a behaviour can be explained either by changes in the carrier concentration or channel mobility. Since these effects take place at relatively high currents, as was mentioned earlier, the carrier concentration appears not to be strongly affected by the filling of carrier traps. Instead they are related to the onset, under certain conditions, of an additional scattering mechanism leading to a reduction of the carrier mobility.

In reference [8] the effect during room-temperature operation of shallow traps located in the defect-rich non-stoichiometric layer at the Si-SiO_2 interface in small-area transistors was investigated. It has been shown that recharging of these centres lies at the origin of such phenomena as random telegraph signal (RTS) noise in sub-micrometer transistors and 1/f noise in larger transistors. As a basic reason of these phenomena the channel conductivity modulation, due to carrier scattering at the charged centre at the interface, was noticed. In our experiments we assume that the charge state of shallow centres remains unchanged, at least at temperatures below 12 K. The fact that a sharp reduction of transconductance occurs at a specific value of the gate voltage indicates that the interaction of the trapped electrons with free conduction electrons has a resonant character. The inelastic scattering of the electrons at the charged centre is known to be most efficient when its energy is close to the energy of the local state [9]. If the accumulation layer is formed at the front interface and if the drain voltage is small, the surface potential of the silicon film may be considered as constant and close to zero. Therefore, the whole gate voltage is developed across the gate insulator. Because the shallow traps are situated in the dielectric layer, at some distance from the silicon surface, their energy position in respect to the bottom of the silicon conduction band will vary with changes in the gate voltage. When the voltage, V_g, corresponds to the alignment of the traps containing the electrons with the bottom of the conduction band, the maximum scattering and, hence, the greatest depth in the valley in the $g_m(V_g)$ dependence will be observed.

4. Conclusions

It has been shown that the charge state of shallow electrically active oxide traps within accumulation mode n-channel SOI MOSFETs affects the low temperature operation of the transistors. For the first time the effect of resonant scattering of conduction electrons at charged centres resulting in mobility modulation in the accumulation layer has been observed.

5. Acknowledgments

The authors are grateful to T.E.Rudenko and A.N.Rudenko for fabricating the test transistors and to A.N.Nazarov for helpful discussions.

6. References

1. Alawneh, I., Simoen, E., Biesemans, S., De Meyer, K., and Claeys C. (1998) Comparison of the freeze-out effect in In and B doped n-MOSFETs in the range 4.2 – 300 K, *Journal de Physique IV* **8**, Pr3-3–Pr3-8.
2. Rotondaro, A.L.P., Magnusson, U.K., Claeys, C., and Colinge J.-P. (1993) Evidence of different conduction mechanisms in accumulation-mode p-channel SOI MOSFET's at room and liquid-helium temperatures, *IEEE Trans. on Electron Devices* **40**, 727–732.
3. Colinge, J.-P. (1990) Conduction mechanisms in thin-film accumulation-mode SOI p-channel MOSFET's, *IEEE Trans. on Electron Devices* **37**, 718–723.
4. Colinge, J.-P. (1991) *Silicon-on-Insulator Technology: Materials to VLSI*, Kluwer Academic Publishers, Dordrecht.
5. Lysenko, V.S., Sytenko, T.N., Zimenko, V.I., and Snitko O.V. (1982) Investigation of traps in the transition region of $Si\text{-}SiO_2$ structures at cryogenic temperatures, *Phys Stat. Solidi (a)* **71**, 619–625.
6. Gomeniuk, Y.V., Litovski, R.N., Lysenko, V.S., Osiyuk, I.N., and Tyagulski, I.P. (1992) Thermally stimulated field emission of charge from traps in the transition layer of $Si\text{-}SiO_2$ structures, *Applied Surface Sci.* **55**, 179–185.
7. Gomeniuk, Y.V., Litovski, R.N., Lysenko, V.S., Osiyuk, I.N., and Tyagulski, I.P. (1992) Current stochasticity of field emission of charge from traps in the transition layer of implanted MIS structures, *Applied Surface Sci.* **59**, 91–94.
8. Mueller, H.H., Schulz, M. (1995) Individual interface traps at the $Si\text{-}SiO_2$ interface, *Journ. of Materials Science: Materials in Electronics* **6**, 65–74.
9. Naito, M. and Beasley, M.R. (1987) Microscopic study of tunneling processes via localized states in amorphous-Si/SiO_x tunnel barriers, *Phys. Rev.* B **35**, 2548–2551.

NANOSCALE WAVE-ORDERED STRUCTURES ON SOI

V.K. SMIRNOV[1] AND A.B. DANILIN[2]

[1]*Institute of Microelectronics RAS,*
3 Krasnoborskaya Street, 150051 Yaroslavl, Russia

[2]*Centre for Analysis of Substances,*
1/4 Sretensky Blvd., 103045 Moscow, Russia

This work deals with the fabrication of wave-ordered structures (WOS) by bombarding SOI (SIMOX) with low-energy N_2^+ ions. The formation conditions of nanometer-scale silicon body in separate waves of WOS during ion bombardment and its modification during annealing have been studied. SIMS study of planar structures simulating the front and rear sides of waves in WOS annealed under various conditions permits us to identify annealing conditions under which a nanoscale silicon wire is formed which is rapidly depleted of implanted nitrogen.

1. Introduction

Relief formation during ion bombardment was observed earlier [1-5]. In our previous work [6] we studied the dependence of wave-ordered structure (WOS) formation during low-energy N_2^+ bombardment of silicon upon the main experimental parameters (ion energy, incidence angle and wafer temperature) and WOS formation dynamics. The WOS proved to be uniform and stable and their wavelengths were on the order of nanometers. The geometry and internal structure of individual waves were determined. We showed that WOS can be obtained of SOI near the silicon/insulator interface by N^+ ion implantation.

This work reports the effect of annealing on the internal structure of WOS, in particular, on the formation of the silicon body in separate waves of a WOS formed near the SOI-forming insulator. The difficulty of direct experimental determination of wave structure in WOS due to the extremely small test areas encouraged us to simulate the sides of separate waves using planar structures, by analogy with Bachurin et al. [7].

2. Experimental

Si(100) wafers and SOI (SIMOX) structures with the superficial silicon layer thickness $H_{si} = 193$ nm were bombarded with nitrogen ions in an ultrahigh vacuum (residual pressure about $2 \cdot 10^{-10}$ torr) on a SAM PHI-660 instrument. The N_2^+ ion beam with an energy of 9 keV was rastered to 100×100 and 500×500 µm² on the

P.L.F. Hemment et al. (eds.),
Perspectives, Science and Technologies for Novel Silicon on Insulator Devices, 315–320.

target surface. The incidence angle θ was 20 and 70 deg to the normal direction. The typical ion flux was 5×10^{14} cm^{-2}s^{-1}. The total dose for silicon samples was about 2×10^{17} cm^{-2}. The stabilisation of the surface composition was detected from the intensity of the LVV Si low-energy Auger peak. The nitrogen-to-silicon concentration ratio on the surface was determined *in situ* using Auger Electron Spectroscopy (AES) after the termination of the bombardment. For the SOI structures, the bombardment duration (and hence the total dose) was chosen so to preserve a silicon layer of a desired thickness, and, thus control the distance from the final sputtered surface to the Si/SiO$_2$ interface. The thickness of the unsputtered silicon layer was determined using a Talystep profilometer.

Some of the samples were annealed for 1 h in vacuum ($p = 1\times10^{-6}$ Pa) at 1000 $^\circ$C. Depth profiling of the implanted nitrogen was performed using SIMS on a Cameca IMS-4F instrument. The concentration measurement accuracy was about 25%.

3. Results and discussion

As was shown in [1], N_2^+ bombardment of silicon produces a nanometer-scale WOS on the sputtered surface at a certain sputtering depth which depends on the incidence angle, ion energy and target temperature. In the N_2^+/Si system, this WOS remains stable during further sputtering, i.e. it is reproduced with almost unchanged parameters (wavelength and amplitude) if a layer as thick as several tens of nanometers is sputtered after WOS formation. It was therefore suggested [1] that a SIMOX structure would be used with a silicon overlayer whose thickness H_{Si} is considerably greater than the WOS formation depth. Thus, by sputtering the silicon for an appropriate time, it will be possible to form the WOS on the SIMOX wafer where the wave-like structures are isolated from one another.

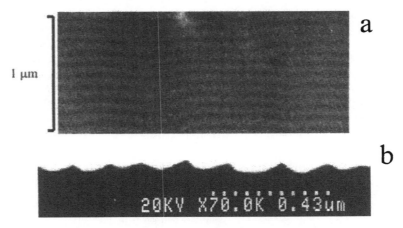

Figure 1. (a) Plan view of a fragment of a coherent WOS and (b) WOS cross-section.

Fig. 1 shows a SEM image taken from a fragment of a relatively coherent WOS formed in the N_2^+/Si system. The wave geometry was determined directly from a

SEM cross-section study. A cross-section profile for one of the samples is also shown in Fig. 1 (b). The inclination of the walls of the WOS to the plane of the wafer when viewed in cross-section is defined by the incidence angle θ. For $\theta = 45$ deg the wave cross-section is in the form of a symmetrical triangle with inclinations of about 25 deg [1]. For this wave geometry, the local incidence angles $\theta_{L1,2}$ are 20 and 70 deg for the front and rear wave sides relative to the beam, respectively.

Knowing the wave geometry we were able to investigate the internal composition of the structures. Because of the small wave dimensions their direct analysis is quite a difficult task, so, instead, the effect of nitrogen ions on wave sides was simulated by nitrogen bombardment of a silicon surface at incidence angles equal to the local ones for the wave (θ_{L1} and θ_{L2}). Elemental analysis of the surface was performed using AES, and the nitrogen depth distribution was analysed using SIMS.

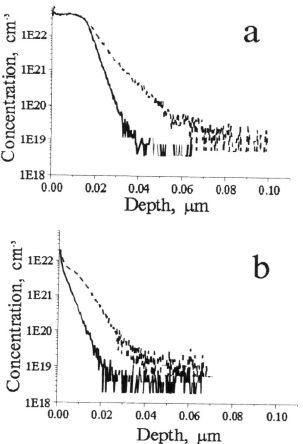

Figure 2. Nitrogen distributions in thin silicon nitride layers simulating (a) front and (b) rear wave sides. Dashed lines show as-implanted profiles and solid lines, after 1 h anneal at 1000 °C.

318

Figure 2a,b shows SIMS nitrogen profiles in silicon irradiated with N_2^+ at $E = 9$ keV and $\theta = 20$ (Fig. 2a) and 70 deg (Fig. 2b) as-implanted and annealed for one hour at 1000 °C. The time and temperature of the anneal were chosen on the basis of previous data [2]. Both for the former (simulation of nitrogen distribution for the front wave side) and the latter cases (the same for the rear wave side) it can be seen that the nitrogen profile is shrunk by the anneal and its edges become more abrupt. As shown earlier [2], a nitride film forms for $\theta = 20$ deg. The profile edges have exponential shapes for both cases. After the anneal, the typical exponential decay length is close to 3 nm. Such an abrupt drop of the SIMS signal intensity may only occur at very sharp interfaces and with δ-doped layers. Thus it is concluded that the silicon nitride / silicon interface is extremely thin and that the silicon has been rapidly depleted of nitrogen.

However, if a WOS is produced on SOI near the Si/SiO_2 interface and annealed, this latter interface may affect the nitrogen redistribution during annealing. To characterise the behaviour of nitrogen, we performed the following experiments. A SOI structure was bombarded with 9 keV nitrogen ions at an angle of 20 deg. The durations of sputtering were chosen such that the thickness of the remaining silicon layer (i.e. the distances from the final sputtered surface to the Si/SiO_2 interface) be different. The thickness of the unsputtered silicon layer was determined by depth profiling the SIMS craters using a Talystep profilometer.

The samples were annealed under the conditions described above, and the nitrogen distribution was determined using SIMS. The results are presented in Fig. 3. It can be seen that the presence of the Si/SiO_2 interface shows only a little effect on the thickness of the silicon nitride layer that forms near the surface. However, a nitrogen-rich layer forms at the Si/SiO_2 interface itself labelled 1, 2 and 3 in Figure 3, where nitrogen may act as a donor. At a first glance, comparison of the profiles suggests that the proximity of the superficial nitride layer and the nitrogen-rich layer at the Si/SiO_2 interface controls the release of nitrogen from the intermediate silicon layer. However, this may be only an apparent effect caused by the insufficient resolution of SIMS for such closely located layers. It nevertheless seems that these data on the release of nitrogen from the silicon body can be used for the restoration of the WOS composition on SOI.

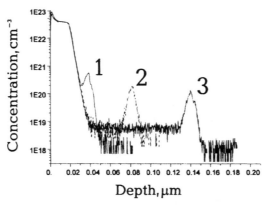

Figure 3. Nitrogen distributions in the front wave sides of WOS formed at (1) 40, (2) 83 and (3) 142 nm from the Si/SiO_2 interface in SIMOX.

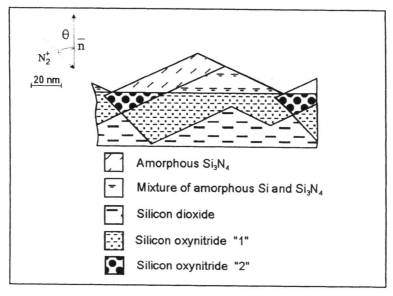

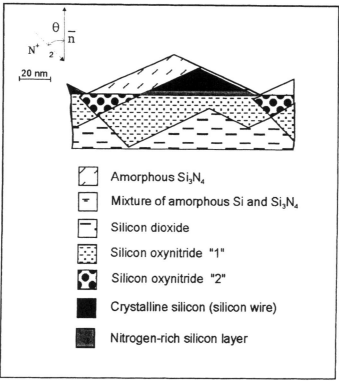

Figure 4. Schematic representation of the internal structure of WOS waves formed on SIMOX (a) before and (b) after annealing.

On the basis of the above results the internal structure of separate waves on SOI was restored for the as-implanted and the as-annealed states and is shown schematically in Fig. 4. It can be seen that the structure produced by annealing is an array of insulated silicon wires shaded black in Figure 4 (b) with nanometer-scale cross-sections.

4. Conclusions

1. The formation of nanoscale silicon wires in separate waves of wave-ordered structures (WOS) on silicon and SOI has been studied on planar structures by simulating the formation and annealing conditions of WOS, thereby reproducing the composition of wave side surfaces.
2. The redistribution of implanted nitrogen due to the formation of WOS has been studied for WOS formed near the Si/SiO$_2$ interface of an SOI structure. Experiments have shown that a layer slightly enriched with nitrogen forms near this interface.
3. It has been shown that the silicon wires of separate waves is rapidly depleted of implanted nitrogen during a vacuum anneal for 1 h at 1000 ºC.
4. Schematic of WOS on SOI cross-section has been provided and is shown in Figure 4.

References

1. Carter, G., Colligon, J.S. and Nobes, M.J. (1973) *J. Mater. Sci.* **8**, 1473.
2. Ducommun, J.P., Cantagrel, M. And Moulin, M. (1975) *J. Mater. Sci.* **10**, 52.
3. Carter, G., Nobes, M.J. and Arshak, K.I. (1979) *Wear* **53**, 245.
4. Smith, R. and Walls, J.M. (1979) *Surf. Sci.* **80**, 557.
5. Smith, R., Valkering, T.P. and Walls, J.M. (1981) *Phil. Mag.* **44**, 879.
6. Smirnov, V.K., Kibalov, D.S., Krivelevich, S.A., Lepshin, P.A., Potapov, E.V., Yankov, R.A., Skorupa, W., Makarov, V.V. and Danilin, A.B. (1998) *Nucl. Instr. Meth. B*, to be published.
7. Bachurin, V.I., Churilov, A.B., Potapov, E.V., Smirnov, V.K., Makarov, V.V. and Danilin, A.B. (1998) Formation of thin silicon nitride layers in Si by low energy N$_2^+$ ion bombardment, *Nucl. Instr. Meth. B*, to be published.

THIN PARTIAL SOI POWER DEVICES FOR HIGH VOLTAGE INTEGRATED CIRCUITS

F. UDREA, H.T. LIM, D. GARNER, A. POPESCU AND W. MILNE
University of Cambridge
Department of Engineering, Cambridge CB2 1PZ, England
Fax: 44 1223 332662   e.mail: fu@eng.cam.ac.uk

AND

P.L.F. HEMMENT
University of Surrey
School of Electronic Engineering, Information Technology &
Mathematics, Guildford, Surrey GU2 5XH, England

Abstract. The forward blocking characteristics of lateral SOI power devices on a very thin ($\leq 0.5 \ \mu m$) silicon layer are analysed. It is known that SOI power devices suffer from reduced breakdown voltage due to the less efficient RESURF effect. Partial SOI technology results in a higher breakdown capability. In addition, partial SOI devices do not require the drift region to be linearly graded. Moreover, the self-heating effect in partial SOI devices is drastically reduced since the patterned oxide layer is very thin and heat can dissipate via the silicon window to the substrate, which also helps to distribute the temperature more evenly in the drift region. It is concluded that thin partial SOI power devices could be strong candidates for high voltage ICs (HVICs).

1. Introduction

Over the past several years, high voltage device structures fabricated in thin ($\leq 0.5 \ \mu m$) SOI layers have received considerable attention. If the SOI thickness is sufficiently thin, the device isolation can be easily achieved by trench or LOCOS technology. The SOI power devices may however suffer a lower breakdown voltage compared to Junction Isolation (JI) power devices due to a less efficient RESURF (REduced SURface Field) effect [1].

P.L.F. Hemment et al. (eds.),
Perspectives, Science and Technologies for Novel Silicon on Insulator Devices, 321–327.
© 2000 *Kluwer Academic Publishers. Printed in the Netherlands.*

Although the oxide layer supports part of the off-state potential, the depletion region is constricted in the thin drift layer which leads to high electric fields and premature breakdown. To improve the breakdown voltage in thin SOI power devices several techniques have been proposed such as a linearly graded dopant profile in the drift region with lower doping at the source side and higher doping at the drain side [2] and SIPOS [3]. These techniques employ thick insulating layers which may however give rise to self-heating especially in harsh conditions such as inductive switching or short-circuit operation. For high power applications which involve high currents and high voltages operation, it is vital to overcome these problems.

We report here novel power device structures based on thin partial SOI technology. Partial SOI is a new technological concept for power integrated circuits (PICs) in which a patterned buried oxide is used to give reduced self-heating and increased breakdown in the power devices, while still providing the excellent isolation which is inherent in conventional SOI technology [4]. This work shows that this novel technology can be extended to ultra thin LDMOS power devices.

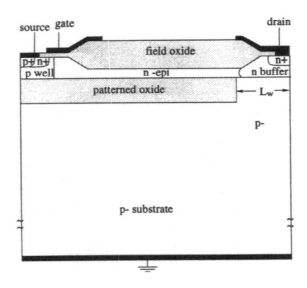

Figure 1. Schematic cross-section of a thin partial SOI LDMOS transistor. The buried oxide layer is patterned to form a small silicon window of length L_w at the drain side.

2. Device Structure and Its Off-State Operation

The cross-section of the thin partial SOI LDMOS is shown in figure 1. Its drift region is only 0.5 μm thick and has a uniform doping profile. A relatively thin buried oxide layer is above a lightly doped p^- type substrate. The key feature of this structure is that a small part of the buried oxide layer is etched away to open a window of 10 μm width at the drain side. For off-state operation, the source and gate are grounded while the drain is biased to a positive voltage. As the buried oxide is interrupted at the drain side, the substrate is not automatically connected to zero bias and is therefore floating. Consequently, an electrode must be added at the bottom of the substrate and is grounded.

The new device realizes an improved RESURF effect since, as in the case of Junction Isolation LDMOS case, the potential distribution at the surface is dictated by the fully-depleted vertical n+/n-/p- junction at the drain side. Therefore, unlike uniformly doped drift layer SOI power diodes where the potential lines are crowded towards the source and drain in the top SOI layer (figure 2(a)), in partial SOI structures, the potential lines penetrate into the lowly doped substrate through the "window" in the oxide layer, and thus part of the high voltage in the blocking mode is now supported by the substrate, releasing the pressure on the oxide and the thin silicon layer. In other words, the substrate plays an active role in uniformly distributing the potential at the surface as shown in figure 2(b). It is important to note that partial SOI devices do not require the drift region doping to be linearly graded.

Figure 3 shows a comparison of the breakdown I-V characteristics of both SOI and partial SOI devices. The partial SOI device with a combination of thin uniformly doped SOI layer and thin patterned oxide layer offers the highest breakdown voltage. It does, however, give a slightly higher leakage current than SOI devices.

3. Self-Heating Effect

Power devices built in a very thin SOI layer on thick buried oxide with a linearly graded drift region already have been shown to increase the self-heating problem under steady-state conditions [5]. The sharp increase of the temperature is due to the oxide acting as a heat barrier to the substrate. With a linearly graded drift region, the devices are found to have a much greater temperature rise near the source than near the drain. This substantial local heating near the source may aggravate device performance and cause reliability problems in the gate and source metallisation. Compared with the partial SOI devices, the temperature rise is lower and the temperature gradient is skewed more towards the drain side as shown in figure 4(a). The self-heating effect in partial SOI devices is drastically reduced because heat can now dissipate through the oxide window to the substrate. Besides, the uniform doping profile in the drift region also helps

324

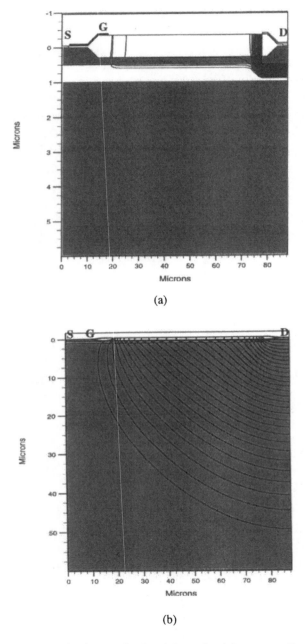

(a)

(b)

Figure 2. Potential distribution at the breakdown for (a) a conventional SOI LDMOS with a relatively thin buried oxide layer and (b) a thin partial SOI LDMOS with uniformly doped drift region on a thin oxide.

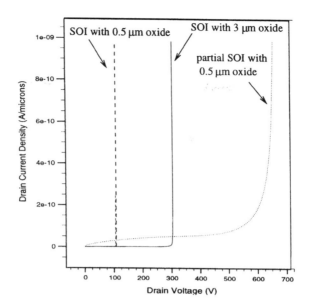

Figure 3. Breakdown characteristics. The partial SOI offers the highest breakdown voltage but a higher leakage current than SOI devices.

to distribute the temperature more evenly along the drift region. This will certainly reduce the problem of the local heating near the source region and improve the overall performance of the device.

For the transient case, especially when long power pulses are imposed, the SOI device temperature can increase significantly due to the reduced heat capacity of a thin SOI layer [6]. However, it is found that for partial SOI devices there is no significant temperature rise at the same power rating compared to their SOI counterparts as shown in figure 4(b). This again is because part of the heat is escaping via the silicon window, widening the heated area and putting less heating pressure on the thin SOI layer. Besides, as the temperature rise is very sensitive to the thickness of the buried oxide, the partial SOI devices which require only a relatively thin patterned oxide to achieve high breakdown are very promising for high temperature operations.

4. Power Integrated Circuits using Partial SOI

While the high power cells have the substrate open to realize a high breakdown voltage and reduce the temperature within the circuit, the low power cells are entirely encapsulated in oxide preventing interference with the high voltage cells. Therefore, individual isolation of low power devices in partial

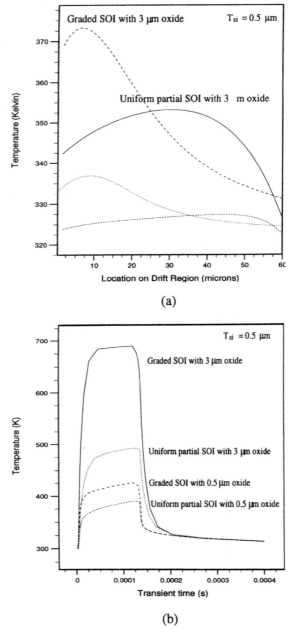

Figure 4. (a) Distribution of temperature in the drift region in on-state conditions for a power of 1mW/μm. (b) Temperature profiles for the SOI and partial SOI LDMOS devices following the application of 100 μs, 2.5 mW/μm power pulse.

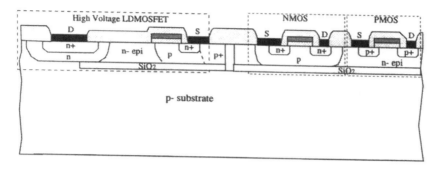

Figure 5. PIC cell using the partial SOI based technology with a thin SOI layer. The cell consists of a high voltage LDMOSFET and low voltage MOSFET devices integrated in the same chip.

SOI PICs is virtually the same as for conventional SOI PICs (figure 5).

5. Conclusions

A new LDMOS structure based on thin partial SOI technology is proposed and demonstrated through numerical simulations. It has been shown that a breakdown voltage as high as 650 V can be obtained with a 0.5 μm thick SOI and patterned oxide layer. The self-heating effect is drastically reduced under both steady-state and transient conditions. The results point to partial SOI technology as a viable solution for high-voltage ICs.

References

1. J. A. Appels et al. (1980), Thin Layer High-Voltage Devices (RESURF Devices), *Philips J. of Research*, **Vol. 35, No. 1**, pp. 1-13

2. S. Merchant et al. (1991), Realization of high breakdown (> 700 V) in thin SOI device, *Proc. 3rd ISPSD*, pp. 31-35

3. T. Matsudai et al. (1992), Simulation of a 700 V High-Voltage Device Structure on a Thin SOI, *Proc. 4rd ISPSD*, pp. 272-277

4. F. Udrea et al (1997), A new Lateral Insulated Gate Bipolar Transistor (LIGBT) structure based on partial isolation SOI technology, *IEE Dev. Lett.*, **Vol. 33, No. 10**, pp. 907-909

5. Y. K. Leung et al (1997), Spatial Temperature Profiles Due to Nonuniform Self-Heating in LDMOS's in Thin SOI, *IEEE Electron Dev. Lett.*, **Vol. 18, No. 1**, pp. 13-15

6. H. Neubrand et al (1994), Thermal Behaviour of Lateral Power Devices on SOI Substrates, *Proc. 6rd ISPSD*, pp. 123-127

KEYWORD INDEX

Topic	Paper Number
3D applications	4.4
a-Si(H)	2.1
Absorption coefficient	3.9
Accumulation	4.7
Activation energy	1.1
Alumothermic process	2.2
AMLCD	2.1
Amorphous carbon	2.3
Analytical model	3.10
Anodic etching	1.3
Anodic oxidation	1.3
Auger spectroscopy	3.9
Back-gate voltage	3.5
Bias-thermal stress	3.3
Blistering	1.1
Bond strength	4.4
Bonded wafer	1.4
Boron complex	1.4
Breakdown	3.6
Bulk substrate	3.10
Buried insulator	2.4, 3.7
Buried oxide	3.3, 3.4
Buried silicide	1.2
C-V characteristics	4.6
C-V measurements	1.4
Carbon (DLC)	2.3
Carrier depth profiles	1.4
Carrier lifetime	3.8
Carrier mobility	2.5
Cellular structure	1.3
Chemical processing	2.2
Chemical properties	3.9
CMOS	4.2
Complementary bipolar	1.2
Computer modelling	1.5
Computer simulation	3.2
Coplanar transmission line	3.10
Cryogenic devices	4.7
Cryogenic electronics	4.1
Cryogenic operation	3.5
Crystalline Si films	2.5
Defect engineering	3.2
Defects	1.4, 3.3, 3.4

Topic	Paper Number
Depletion capacitance	3.10
Deposition rate	2.5
Detrapping	1.1
Deuterium	3.4
Device characteristics	1.4
Diamond	2.4
Diamond like	2.3
Diffusion	1.1, 3.2, 3.4
Dilamination	1.1
Effective channel length	4.1
Effective medium theory	3.9
Electrochemical oxidation	3.6
Electrolytic solutions	3.6
Electronic applications	2.3
Electronics on glass	2.5
Ellipsometry	3.9
Energy saving	2.2
Epitaxy	3.6
Exothermic reaction	2.2
Field emission displays	2.3
Filtered cathodic vacuum arc	2.3
Fluorine	2.2
Fourier transform infrared (FTIR)	3.1
Fully depleted SOI	3.8
Generation lifetime	4.6
Glass substrate	2.1
High efficiency processing	1.5
High temperature	3.3, 4.5
High temperature electronics	4.2
Homoepitaxy	1.3
HTCVD	1.3
Hydrogen	3.4
Hydrogen related defects	1.4
Industrial process	1.1
Industrial waste	2.2
Instability	3.3
Internal stress	3.6
Ion-assisted deposition	2.5
Ion-beam modification	4.8
Ion beam synthesis	2.6
Large area electronics	2.1
Laser crystallisation	2.5
Laser processing	1.5
Laser recrystallisation	2.7
Lateral solidification	1.5

Topic	Paper Number
Layer transfer	1.1
LDD	3.5
Leakage current	3.8
Loss constant	3.10
Low cost devices	2.7
Low cost process	2.2
Low cost technology	2.1
Low-dimensional	4.3
Low dose SIMOX	3.2
Low temperature epitaxy	2.5
Low temperature operation	4.1
Low temperature processing	2.5
Low temperature technology	2.1
Low temperatures	4.3
LT-polySi	2.1
MBE	1.4
MEDICI simulations	3.5
MEMS	4.4
Mesoscopic devices	4.3
Microfluidic valves	4.4
Micromachining	4.4
MicroRaman probe	3.1
Minority carrier diffusion length	2.5
Mobility modulation	4.7
Monocrystalline Si films	2.5
Morphology	3.6
MOS devices	3.8
MOS transistors	2.4, 4.5
MOSFETs	4.3
Multilayer structures	1.1
Multilayers	3.9
Multiple internal reflection	1.1
n-MOSFET	4.7
Nanoelectronics	4.1
Nanoscale silicon wire	4.8
Nitride bonding	4.4
Off-state current	4.5
Optical analysis	3.1
Oxide traps	4.7
Oxygen redistribution	3.2
Oxygen transport	3.2
Partial SOI	4.9
Partially depleted MOSFET	4.6
Photovoltaic devices	2.2
Physical mechanisms	3.2

Topic	Paper Number
Piezoresistance	2.7
Point defects	3.2
Polish stop	1.2
Polycrystalline Si films	2.5
Polycrystalline silicon	2.2, 2.7
Polysilicon technology	2.1
Power devices	3.6
Power integrated circuits	4.9
Precipitation	3.2
Pressure sensors	4.4
Process compatibility	2.4
Proton implantation	1.1
Purification	2.2
Pyrolysis	2.2
Quantisation effects	4.1
Quasi-monocrystalline Si films	2.5
Radiation effects	3.7
Radio frequency applications	3.10
Raman scattering	3.1
Recombination	3.8
Recombination lifetime	4.6
Recycled waste	2.2
Resonant scattering	4.7
Retrograde doping profile	3.5
S-parameter measurements	3.10
SCSLS	1.5
Self-heating	2.4
Semiconductor on glass	1.1
Sensors	2.7
Sensors - large area	2.1
Series resistance	3.5, 4.1
Shallow traps	4.7
SiCOI	2.6
Silane synthesis	2.2
Silicide	1.2
Silicon film	4.1
Silicon on diamond	2.4
Silicon on glass	2.5
Silicon thin film	4.6
SIMOX	3.1, 3.3, 3.9, 4.2, 4.8
SIMS	4.8
Single electron transistor	4.1
Skin effect	3.10
SLS	1.5
SmartCut process	1.4

Topic	Paper Number
SmartCut®	1.1
SOI power devices	4.9
SOI substrate	3.10
SOI technology	4.9
Structural properties	3.9
Substrate conductivity	3.10
Subthreshold slope	4.5
Super thin silicon films	1.5
Synthesis	3.2
TFT	1.5
Tikhonov's technique	3.9
Total-dose response	3.7
Transport characteristics	4.1
Transport properties	4.3
Trapping	3.4
Ultra-thin SOI structures	4.3
Viscosity	2.2
Wafer bonding	1.1, 2.6, 4.4
Wafer scale SOI	1.3
Wave-ordered structures	4.8
Zone melting recrystallisation (ZMR)	3.3

AUTHOR INDEX

Author	Paper Number

M C Acero
Centre Nacional de Microelectronica CNM-CSIC,
Campus UAB, 08193 Bellaterra, Spain
2.6

I V Antonova
Institute of Semiconductor Physics, RAS 630090,
Lavrentieva 13, Novosibirsk, Russia
1.4

D Ballutaud
LPSB-CNRS, 1 Place Aristide Briand, 92195 Meudon
Cedex, France
3.4

I P Barchuk
Institute of Semiconductor Physics, Prospect Nauki 45,
252028 Kiev-28, Ukraine
3.3, 3.7

B Bekelov
Institute of Physics and Technology, Almaty 480082, Kazakstan
2.2

S Bengtsson
Solid State Electronics Laboratory, Department of
Microelectronics ED, Chalmers University of Technology,
S-41296 Goteborg, Sweden
2.4

M Bergh
Solid State Electronics Laboratory, Department of
Microelectronics ED, Chalmers University of Technology,
S-41296 Goteborg, Sweden
2.4

R B Bergmann
Institut fur Physikalische Elektronik, Universitat Stuttgart,
Pfaffenwaldring 47, D-70569 Stuttgart, Germany
2.5

A Boutry-Forveille
LPSB-CNRS, 1 Place Aristide Briand, 92195 Meudon
Cedex, France
3.4

V M Borisov
TRINITI, Troitsk, Moscow Region, Russia
1.5

E G Bortchagovsky
Institute of Semiconductor Physics, Nauka Av 45,
252650 Kiev, Ukraine
3.9

336

Author	Paper Number

J Boussey
Laboratoire de Physique des Composants a Semiconducteurs,
LEMO-LPCS, BP 257, 38016 Grenoble Cedex 1, France

3.10

M Bruel
CEA/LETI-Departement de Microtechnologies, 17 rue des Martyrs,
38054 Grenoble Cedex, France

1.1

C Claeys
IMEC, Leuven, Belgium
and
KU Leuven, ESAT-INSYS, Kard Mercierlaan 94,
B-3001 Leuven, Belgium

3.5, 4.1

C A Colinge
Department of Electrical and Electronic Engineering,
California State University, 6000 J Street, Sacramento,
CA 95819-6019, USA

4.4

J P Colinge
Department of Electrical and Computer Engineering,
University of California, Davis, CA 95616, USA

3.8, 4.2

C Collet
Thomson CSF, Central Research Laboratories, Domaine de
Corbeville, 91404 Orsay Cedex, France

2.1

S Cristoloveanu
LPCS, ENSERG, BP 257, 38016 Grenoble Cedex, France

3.8

A B Danilin
Centre for Analysis of Substances, 1-4 Sretensky Blvd,
103405 Moscow, Russia

4.8

R Dassow
Institut fur Physikalische Elektronik, Universitat Stuttgart,
Pfaffenwaldring 47, D-70569 Stuttgart, Germany

2.5

T Debski
Institute of Electron Technology, Al Lotnikow 32/46,
02-668 Warsaw, Poland

4.6

A I Demin
TRINITI, Troitsk, Moscow Region, Russia

1.5

Author	Paper Number

S C Djurenko
Institute of Semiconductor Physics, Prospect Nauki 45,
252028 Kyiv-28, Ukraine

3.7

A A Druzhinin
Lviv Polytechnic State University, Kotlarevsky Street 1,
Lviv 290013, Ukraine

2.7

A V Dvurechenskii
Institute of Semiconductor Physics, Pr Lavrent'ev 13,
630090 Novosibirsk, Russia
and
Novosibirsk State University, Pirogova 2,
630090 Novosibirsk, Russia

1.3, 3.6

A A Efremov
Institute of Semiconductor Physics, 45 Prospect Nauki,
252028 Kiev-28, Ukraine

3.2

A I El'Tsov
TRINITI, Troitsk, Moscow Region, Russia

1.5

T Ernst
LPCS, ENSERG, BP 257, 38016 Grenoble Cedex, France
and
ST Microelectronics, 38920 Crolles, France

3.8

J Esteve
Centre Nacional de Microelectronica CNM-CSIC,
Campus UAB, 08193 Bellaterra, Spain

2.6

G N Feofanov
Institute of Semiconductor Physics, RAS 630090,
Lavrentieva 13, Novosibirsk, Russia

1.4

L Fonseca
Centre Nacional de Microelectronica CNM-CSIC,
Campus UAB, 08193 Bellaterra, Spain

2.6

A A Franzusov
Institute of Semiconductor Physics, RAS 630090,
Lavrentieva 13, Novosibirsk, Russia

1.4

Author	Paper Number

H S Gamble — 1.2
School of Electrical and Electronic Engineering, The Queen's
University of Belfast, Belfast, BT7 1NN, UK

D Garner — 4.9
Department of Engineering, University of Cambridge, Cambridge,
CB2 1PZ, UK

J Gibki — 4.6
Institute of Microelectronics and Optoelectronics,
Warsaw Technical University, ul Koszykowa 75,
00-662 Warsaw, Poland

Y V Gomeniuk — 4.7
Institute of Semiconductor Physics, National Academy of
Sciences of Ukraine, Prospect Nauki 45, 252650 Kiev, Ukraine

R Grotzschel — 1.3, 3.6
Research Center Rossendorf Incorporated,
PO Box 510119, D-01314 Dresden, Germany

A Gutakovskii — 1.3, 1.4, 3.6
Institute of Semiconductor Physics, Pr Lavrent'ev 13,
630090 Novosibirsk, Russia

P L F Hemment — 4.9
School of Electronic Engineering, Information Technology
and Mathematics, University of Surrey, Guildford, Surrey,
GU2 5XH, UK

O Huet — 2.1
Thomson CSF, Central Research Laboratories, Domaine de
Corbeville, 91404 Orsay Cedex, France

A Jakubowski — 4.6
Institute of Microelectronics and Optoelectronics,
Warsaw Technical University, ul Koszykowa 75,
00-662 Warsaw, Poland

J Jomaah — 3.10
Laboratoire de Physique des Composants a Semiconducteurs,
LEMO-LPCS, BP 257, 38016 Grenoble Cedex 1, France

O B Khristoforov — 1.5
TRINITI, Troitsk, Moscow Region, Russia

Author	Paper Number
V I Kilchytska Institute of Semiconductor Physics, Prospect Nauki 45, 252028 Kyiv-28, Ukraine	3.3, 3.7, 4.5
V V Kirienko Institute of Semiconductor Physics, Pr Lavrent'ev 13, 630090 Novosibirsk, Russia	1.3, 3.6
Yu B Kirukhin TRINITI, Troitsk, Moscow Region, Russia	1.5
R Kogler Forschungzentrum Rossendorf, PF 510119, D-01314 Dresden, Germany	2.6
I T Kogut Lviv Polytechnic State University, Kotlarevsky Street 1, Lviv 290013, Ukraine	2.7
M Korwin-Pawlowski General Semiconductor Ireland, Macroom Co, Cork, Ireland	4.6
U Kreissig Research Center Rossendorf Incorporated, PO Box 510119, D-01314 Dresden, Germany	3.6
M A Lamin Institute of Semiconductor Physics, Pr Lavrent'ev 13, 630090 Novosibirsk, Russia	1.3
E M Lavitska Lviv Polytechnic State University, Kotlarevsky Street 1, Lviv 290013, Ukraine	2.7
P Legagneux Thomson CSF, Central Research Laboratories, Domaine de Corbeville, 91404 Orsay Cedex, France	2.1
J Lescot Laboratoire d'Electromagnetisme, MicroOndes et Optoelectronique, France	3.10
H T Lim Department of Engineering, University of Cambridge, Cambridge, CB2 1PZ, UK	4.9

Author	Paper Number
A B Limanov Institute of Crystallography, Russian Academy of Sciences, Leninski Prospect 59, 117333 Moscow, Russia	1.5
V G Litovchenko Institute of Semiconductor Physics, 45 Prospect Nauki, 252028 Kiev-28, Ukraine	3.2
V S Lysenko Institute of Semiconductor Physics, National Academy of Sciences of Ukraine, Prospect Nauki 45, 252650 Kiev, Ukraine	3.7, 4.5, 4.7
J A Martino Laboratorio de Sistemas Integraveis, Universidade de Sao Paulo, Brazil	3.5
I I Maryyamova Lviv Polytechnic State University, Kotlarevsky Street 1, Lviv 290013, Ukraine	2.7
V P Melnik Institute of Semiconductor Physics, Nauka Av 45, 252650 Kiev, Ukraine	3.9
W I Milne Department of Engineering, University of Cambridge, Trumpington Street, Cambridge, CB2 1PZ, UK	2.3, 4.9
L V Mironova Institute of Semiconductor Physics, RAS 630090, Lavrentieva 13, Novosibirsk, Russia	1.4
J R Morante EME, Department d'Electronica, Unitat Associdada CNM-CSIC, Universitat de Barcelona, Avda Diagonal 6450647, 08028 Barcelona, Spain	2.6, 3.1
B N Mukashev Institute of Physics and Technology, Almaty 480082, Kazakstan	2.2
A N Nazarov Institute of Semiconductor Physics, Prospect Nauki 45, 252028 Kiev-28, Ukraine	3.3, 3.4, 3.7

Author	Paper Number
F Ndagijimana Laboratoire d'Electromagnetisme, MicroOndes et Optoelectronique, France	3.10
E P Neustroev Institute of Semiconductor Physics, RAS 630090, Lavrentieva 13, Novosibirsk, Russia	1.4
A S Nicolett Faculdade de Tecnologia de Sao Paulo, Brazil	3.5
L Oberbeck Institut fur Physikalische Elektronik, Universitat Stuttgart, Pfaffenwaldring 47, D-70569 Stuttgart, Germany	2.5
V I Obodnikov Institute of Semiconductor Physics, Pr Lavrent'ev 13, 630090 Novosibirsk, Russia	3.6
Y Omura High-Technology Research Center and Faculty of Engineering, Kansai University, 3-3-35 Yamate-cho, Suita, Osaka 564-80, Japan	4.3
I N Osiyuk Institute of Semiconductor Physics, National Academy of Sciences of Ukraine, Prospect Nauki 45, 252650 Kiev, Ukraine	4.7
Y M Pankov Lviv Polytechnic State University, Kotlarevsky Street 1, Lviv 290013, Ukraine	2.7
A Perez-Rodriguez EME, Department d'Electronica, Unitat Associdada CNM-CSIC, Universitat de Barcelona, Avda Diagonal 6450647, 08028 Barcelona, Spain	2.6, 3.1
F Plais Thomson CSF, Central Research Laboratories, Domaine de Corbeville, 91404 Orsay Cedex, France	2.1
A Popescu Department of Engineering, University of Cambridge, Cambridge, CB2 1PZ, UK	4.9

Author	Paper Number
V P Popov Institute of Semiconductor Physics, RAS 630090, Lavrentieva 13, Novosibirsk, Russia	1.4
D Pribat Thomson CSF, Central Research Laboratories, Domaine de Corbeville, 91404 Orsay Cedex, France	2.1
C Reita Thomson CSF, Central Research Laboratories, Domaine de Corbeville, 91404 Orsay Cedex, France	2.1
T J Rinke Institut fur Physikalische Elektronik, Universitat Stuttgart, Pfaffenwaldring 47, D-70569 Stuttgart, Germany	2.5
A Romano-Rodriguez EME, Department d'Electronica, Unitat Associdada CNM-CSIC, Universitat de Barcelona, Avda Diagonal 6450647, 08028 Barcelona, Spain	2.6
S I Romanov Institute of Semiconductor Physics, Pr Lavrent'ev 13, 630090 Novosibirsk, Russia and Tomsk State University, Sq Revolution 1, 634050 Tomsk, Russia	1.3, 3.6
B N Romanyuk Institute of Semiconductor Physics, Nauka Av 45, 252650 Kiev, Ukraine	3.9
O Rozeau Laboratoire d'Electromagnetisme, MicroOndes et Optoelectronique, France and Laboratoire de Physique des Composants a Semiconducteurs, LEMO-LPCS, BP 257, 38016 Grenoble Cedex 1, France and ST Microelectronics, 38920 Crolles, France	3.10
A N Rudenko Institute of Semiconductor Physics, Prospect Nauki 45, 252028 Kyiv-28, Ukraine	3.7, 4.5

Author	Paper Number

T E Rudenko
Institute of Semiconductor Physics, Prospect Nauki 45,
252650 Kiev-28, Ukraine

3.7, 3.8, 4.5

C Serre
EME, Department d'Electronica, Unitat Associdada CNM-CSIC,
Universitat de Barcelona, Avda Diagonal 6450647,
08028 Barcelona, Spain

2.6, 3.1

E Simoen
IMEC, Leuven, Belgium

3.5, 4.1

W Skorupa
Forschungzentrum Rossendorf, PF 510119,
D-01314 Dresden, Germany

2.6

V K Smirnov
Institute of Microelectronics RAS, 3 Krasnoborskaya Street,
150051 Yaroslavl, Russia

4.8

L V Sokolov
Institute of Semiconductor Physics, Pr Lavrent'ev 13,
630090 Novosibirsk, Russia

1.3

V F Stas
Institute of Semiconductor Physics, RAS 630090,
Lavrentieva 13, Novosibirsk, Russia

1.4

V E Storizhko
Institute of Applied Physics, Petropavlovskaja St 58, 244030 Sumy

3.9

M F Tamendarov
Institute of Physics and Technology, Almaty 480082, Kazakstan

2.2

S Zh Tokmoldin
Institute of Physics and Technology, Almaty 480082, Kazakstan

2.2

D Tomaszewski
Institute of Electron Technology, Al Lotnikow 32/46,
02-668 Warsaw, Poland

4.6

I P Tyagulski
Institute of Semiconductor Physics, National Academy of
Sciences of Ukraine, Prospect Nauki 45, 252650 Kiev, Ukraine

4.7

Author	Paper Number

F Udrea
Department of Engineering, University of Cambridge, Cambridge,
CB2 1PZ, UK

4.9

A Vandooren
Department of Electrical Engineering, University of California,
Davis, CA 95616, USA

3.8

A Yu Vinokhodov
TRINITI, Troitsk, Moscow Region, Russia

1.5

Ya N Vovk
Institute of Semiconductor Physics, Prospect Nauki 45,
252028 Kyiv-28, Ukraine

3.7

C Walaine
Thomson CSF, Central Research Laboratories, Domaine de
Corbeville, 91404 Orsay Cedex, France

2.1

Yu I Yakovlev
Institute of Semiconductor Physics, Pr Lavrent'ev 13,
630090 Novosibirsk, Russia

3.6

I O Zabashta
Institute of Applied Physics, Petropavlovskaja St 58, 244030 Sumy

3.9

L A Zabashta
Institute of Applied Physics, Petropavlovskaja St 58, 244030 Sumy

3.9

Made in the USA
Columbia, SC
27 August 2021